Landscaping
Principles and Practices

6TH EDITION

Join us on the web at

Agriscience.delmar.com

Landscaping
Principles and Practices

6TH EDITION

Jack E. Ingels
State University of New York
College of Agriculture and Technology
Cobleskill, New York

THOMSON
*
DELMAR LEARNING

Australia Canada Mexico Singapore Spain United Kingdom United States

THOMSON
DELMAR LEARNING

Landscaping Principles & Practices, 6th Edition
Jack E. Ingels

Business Unit Executive Director:
Dawn Gerrain

Acquisitions Editor:
Zina M. Lawrence

Developmental Editor:
Andrea Edwards

Editorial Assistant:
Rebecca Switts

Executive Production Manager:
Wendy A. Troeger

Production Manager:
Carolyn Miller

Production Editor:
Kathryn B. Kucharek

Technology Project Manager:
Joseph Saba

Executive Marketing Manager:
Donna J. Lewis

Channel Manager:
Nigar Hale

Cover Image:
PhotoDisc

Cover Design:
Kristina Almquist

For permission to use material from this text or product, contact us by
Tel   (800) 730-2214
Fax   (800) 730-2215
http://www.thomsonrights.com

Library of Congress Cataloging-in-Publication Data
Ingels, Jack E.
   Landscaping principles and practices / [Jack E. Ingels].--
6th ed.
     p. cm.
   ISBN 1-4018-3410-8
     1. Landscape architecture 2. Landscape gardening
     3. Landscaping industry I. Title.

SB472.153 2003
712--dc21   2002041558

NOTICE TO THE READER

Contents

Preface

America's landscapes are rich, diverse, and dynamic. They include unspoiled natural regions, farmlands, urban and suburban population centers, recreational and industrial parks, historic settings, mountains, prairies, and seascapes. Our national landscape is as diverse as the American population, a mix that is never quite the same from one decade to the next. The individual who pursues a career in landscaping is a service professional who seeks to understand the needs of those who use and enjoy the land, while being sensitive to the large and small distinctions that different regions possess. Meeting those diverse needs and satisfying specific clients without harm to other people or the environment is the challenge of the modern landscape professional.

Like the previous editions, *Landscaping: Principles and Practices,* sixth edition, is intended as an overview of a complex and growing industry. The text acquaints the reader with the three major areas of practice within the business—design, installation, and management—by organizing the chapters into three main sections. The first section, "Landscape Designing," will acquaint the reader with the intellectual and physical processes that result in the creation of a new outdoor environment. It will offer training in the graphic skills needed to present ideas on paper so that others may understand them. It will also help the reader become familiar with ways of determining the needs of the users so that designing can begin. The second section, "Landscape Contracting," addresses the office and field skills needed to bring the landscape project to fruition. Contractual relationships between landscapers and others with whom they work in the construction trades are discussed. Calculative techniques needed to turn landscape drawings into quantitative measurements are described, and opportunities for the reader to practice while reading are presented. In addition, basic methods of planting and lawn installation are described, as well as consideration of irrigation and interior plantscapes. New to this edition, and included in this section, is a chapter that addresses an ongoing concern prevalent throughout the landscape industry, the human resource. Where are the men and women needed to staff the companies of the industry going to come from? The chapter looks at the topic from all sides, including both the training needed by someone desirous of making a career in the business, as well as how a company can effectively recruit and retain good employees. Section three covers "Landscape Maintenance," the key to modern landscape management. Plant maintenance, lawn care, winterization, pest control, and pricing are topics covered in this section. Also in this section, but relevant to all areas of practice, is a new chapter on Safety.

As an introductory text, *Landscaping: Principles and Practices,* sixth edition, is directed to traditional and nontraditional learners seeking to begin their training as landscape professionals. While much of the material covered is unquestionably horticultural, the text presents the profession first and foremost as a business. By following the text, readers progress from the basic principles of landscape design and graphics through methods of installation and maintenance to specific business practices such as recruiting, contract preparation and interpretation, estimating and pricing, and creating a safe workplace.

The text aids the reader's learning through the use of specific learning objectives, end-of-chapter reviews, extensive illustrations, and supplemental activities. Technical terms are defined as they are introduced within the text and again in the glossary at the back of the text. Appendices at the end of the text include examples of landscape plans prepared by professional designers, illustrations of common landscape tools, and an expanded listing of organizations and associations that have connections to the landscape profession.

Features of the Sixth Edition

- Chapters are organized in accordance with the three branches of the landscape industry, making it easier for the reader to go to the information that is of current interest or need.

- Additional end-of-chapter questions and suggested activities have been provided.

- Plant lists have been expanded to include more species, covering a wider geographic range of the country. In addition more specific information about plant hardiness zone ratings has been added.

- New information is provided to explain how to interview clients.

- Expanded coverage of hardscape materials includes information about water features and night lighting.

- New technologies that are changing the ways companies work are discussed.

- New material included in the discussion of site analysis includes Geographic Information

Systems, grade interpretations, and the preparation of base maps.

- Current research dealing with the installation and healthy maintenance of the full range of landscape plants has been included in this edition.

- A new chapter explains how to recognize and distinguish between the many causes of injuries to plants. It goes on to discuss the principles of controlling plant injuries.

- The safe use of chemicals is described in detail.

- Safe operation at all levels within a landscape company is stressed in a new chapter. The chapter covers the responsibilities of government, the company leadership, and the individual worker in assuring a safe workplace.

- The new chapter on Human Resources gives the reader advice on how to prepare both academically and personally for a successful career in the landscape business. It also looks at the issue from the other side, *i.e.,* how does a company find, train, and retain good employees.

About The Author

Jack E. Ingels holds a Board of Trustees' Distinguished Teaching Professorship at the State University of New York's College of Agriculture and Technology at Cobleskill. He is responsible for the College's Landscape Program, which prepares associate and baccalaureate degree students for career-track positions with landscape firms throughout the nation.

Professor Ingels completed his undergraduate degree at Purdue University and his graduate studies at Rutgers University. His postgraduate training was at Ball State University. His fields of specialization include ornamental horticulture, landscape design, landscape garden history, and plant pathology. He is an experienced university educator, named in 1995 as one of the Top Ten Landscape Educators in the United States. He is also the author of *Ornamental Horticulture: Science, Operations, and Management*, currently in its third edition, another successful text in Delmar's agricultural series.

Acknowledgments

The author wishes to express appreciation to the following for their help in the preparation of this text:

The Allen Organization

Matthew Bauwens, Landscape Architect

Steven Beattie, Landscape Architect

Robert C. Bigler Associates, Architects

Eric Blamphin, Designer

Michael Boice, Designer

Chapel Valley Landscape Co.

Cooperative Extension Service
Cornell University, Ithaca, New York

Vickie Davis

Vicki Harris

Russell Ireland, Landscape Contractor

Thomas Kenly, Designer

John Krieg, L. A.

James Lancaster, Designer

Lied's, Inc.

Mike Lin

Mark Magnone, L. A.

Lawrence Perillo, Designer

Edward J. Plaster

Robert Rodler, Landscape Contractor

Smallwood Design Group

Ricky Sowell

State University of New York
College of Agriculture and Technology
Cobleskill, NY

A.J. Tomasi Nurseries, Inc.

David Weston

Landscaping: Principles and Practices was classroom tested at the State University of New York College of Agriculture and Technology at Cobleskill, New York.

The following individuals devoted their time and considerable professional experience to reviewing the manuscript for this edition of the text:

David Henderson
Jefferson State Community College
Birmingham, Alabama

Mark Schusler
Tarrant County College
Fort Worth, Texas

Terry Bowlds
Franklin County High School
Frankfort, Kentucky

Introduction

An Overview of the Landscape Industry

When William Shakespeare posed the question, "What's in a name?", he could have been referring to the term *landscaper*. Few job titles conjure up such a variety of definitions and descriptions from well-meaning people, each certain that his or her definition is correct. Each would also probably believe that the term is not complex, because the industry itself is not complex. The same thing happens when people are asked to define the duties of a conservationist, a doctor, or a teacher. What seems so simple and straightforward at the outset grows in complexity upon realizing that there are endless things to conserve, dozens of ways to practice medicine, and students requiring educational services from nursery school through post-graduate programs.

Landscaping is a service industry. It serves people by fabricating environments where they can live, work, play, or just pass time. These environments are primarily in the outdoors or are in interior settings that seek to suggest the outdoors. Landscapers use the products of others to create habitat areas that can range from utilitarian to fantastic. While humans are the predominant clientele, others with whom we share the planet may also benefit from the landscapers' efforts.

Landscapers serve residential clients, civic clients, corporate clients, and others. They work as one-person, part-time operations and as international corporate giants, and in companies that fit somewhere in between those extremes. Landscape firms are listed in the yellow pages of small villages, cities, and world capitals. In short, there is no one way to be a landscaper or to practice landscaping. There are many different levels of practice and differing degrees of educational preparation required, depending upon how a person opts to work within the industry.

The popular image of the landscaper as a small, one- or two-person operation, with one truck, a lawnmower, and some hand tools with which to serve a nearby residential clientele, is a valid one. There will probably always be that type of landscaper because there will always be a demand for simple plant installation and maintenance services made by a clientele who would rather hire someone else to do the work than do it themselves. It isn't that those clients couldn't do the work, but rather that they do not want to do it. As long as there are people who can hire others to do work that they do not enjoy doing, there will be a market for those services. The small, one-truck landscape operation fills a need. The array and quality of their services vary, and the requirements for entry into the industry at that level are minimal. There is little or no requirement for investment in land or buildings. There is no industry accreditation necessitating formal schooling at this level of practice. Many of the needed skills are easily learned, often by apprenticing with another firm for a season or two. Companies at this level of practice have the greatest turnover rate of any in the industry. While statistics are sketchy, there is no reason to believe that their record of success is better than small businesses at large, where mini-companies with five or fewer employees fail more frequently than they succeed. Still, because it is an easy business to get into, with little financial risk even if it should fail, there is no reason to forecast the demise of this segment of the industry.

The larger segment of the industry is a multibillion dollar industry that serves a population that is increasingly discerning and demanding in its requirements for technical competency and service quality. Demographic trends in the America of the new century indicate that more people are living in nontraditional ways, such as singles, single parents, married without children, or married with fewer children. That, coupled with accelerating land and building costs, is resulting in homes being built on smaller lots, often clustered close together while sharing a common green acreage nearby. Landscapes are predicted to become increasingly smaller but more luxurious as the baby boom generation, which has seen the sights of Europe and elsewhere, begins to gray and settle down to a quieter lifestyle at home. Many have good incomes, and good pension programs, and it is expected that they will create a new luxury residential and recreational landscape market in the United States. The opportunities for employment within the industry will be good, but will require a higher level of professionalism than was typical of the industry in the past.

Concurrent with rising standards for residential landscaping are comparable expectations by commercial and civic clients. They use landscaping as mirrors to the public, hoping to reflect the quality of their products or services via the positive image conveyed through their corporate or institutional setting. While they may value weed-free lawns and properly pruned trees for entirely different reasons than a homeowner, their appreciation of companies and individuals capable of providing these features is as genuine.

As the demand for landscape services has grown in volume, client expectations, and financial commitment, traditional small companies have been unable to meet the needs of the expanding client pool. Larger companies have come into existence, most since the 1970s, knowing at the outset which client type they would serve, what specialty services they would offer, and how to provide those services in a qualitative and cost-effective manner. These companies are as attuned to business management techniques as they are to ornamental horticulture. They seek employees who know how to work with plants and people. The expectations for employee education are growing rapidly, and the competition between companies for trained potential employees is often fierce. At all levels within the organizational structure of larger companies, the Caucasian male, long the dominant component of the labor and management team, is being joined by men and women of other ethnic origins and from other countries. These nontraditional workers will soon represent the majority of workers in the industry. To facilitate and accelerate the training of the workers, many companies offer in-house educational sessions on topics that their workers need to understand. Using videos and/or instructor-led training sessions, frequently conducted in Spanish or other non-English languages, the companies encourage their employees to grow professionally and personally. Many companies offer their employees full access to corporate planning objectives, financial statements, marketing strategies, and records. Some companies will subsidize the costs of their employees taking courses at nearby colleges. All of this and more is done by companies to help their employees become better educated workers who understand their roles in the company. By feeling part of the organization, rather than estranged from it, employees are better able to understand the significance of their work in satisfying the needs of the clients and conveying a positive image of the company.

The landscape industry of today continues to offer the services that it has in the past, structured both traditionally and nontraditionally. The three branches of the industry are still the same as they have been for many years: landscape architecture, landscape contracting, and landscape management. **Landscape architects** are the professionals who conceptualize the outdoor spaces. They seek a perfect match between the desires of a client, the capabilities of a site, and the best interests of the environment. If perfection is unattainable, then they strive to come as close to it as practical. New landscapes have their origins in the mind's eye of the landscape architect or an allied professional, the **landscape designer**. In most states landscape architects are registered professionals. Landscape designers may be, too, depending upon individual state laws. Registration usually requires the accomplishment of formal education, a specified period of apprenticeship, and successful completion of a certification examination. Certification as a landscape architect is a more difficult achievement than designation as a landscape designer, although there is considerable variation among the states. Usually, the greatest apparent distinction between a landscape architect and a landscape designer is in the scope of their projects. Landscape designers usually work with residential clients, while landscape architects are more likely to be involved with institutional or corporate clients or very affluent residential clients seeking complex assistance that is beyond the capabilities of most designers.

Using the traditional tools of the draftsman and/or the new technology of computers, the landscape architect or designer transfers ideas and concepts from his or her mind onto paper where it can be seen and understood by clients and others. By using different views of the design, printed text, colored renderings, computer imaging, and old-fashioned persuasion to get the client excited about the proposal, the landscape architect or designer seeks to gain client acceptance of the ideas and approval to begin. Once approval is gained, the design professional must then prepare and present the design proposal in a way that will explain its intent to those who will eventually build it. Construction drawings, cost estimates, contract signing, and supervision of the construction on behalf of the client all follow as the designer's ideas come off the drawing board and go into the ground.

Landscape architects are the client's representative from beginning to end if the client so wishes. When the client is not one person, but instead is a corporate board or governmental agency, the landscape architect is especially valued as the one who provides a face and a voice to the others involved with completing the project.

Most states now require that a registered landscape architect receive formal training from a university program accredited by The American Society of Landscape Architects. There are about forty accredited programs across the nation, although the number changes from time to time.

The actual construction of landscapes requires skill and tools that are different from those of the landscape architect. It is the **landscape contractor** who accepts responsibility for transforming the landscape architect's or landscape designer's plans and ideas into reality. Today's landscape contractor wears several hats: horticulturist, construction craftsman, business manager, and service specialist. Landscape contracting companies are the most complex branch of the landscape industry, and the branch which has moved farthest and fastest from its traditional roots in recent years. Formerly a small landscape contracting firm would have attempted to provide all of the skills and services needed to accomplish the installation of a landscape project. Some companies still do. However, the economic realities of doing business for cost-conscious and service-savvy clients today has convinced some companies to only do what they can do better and more cost effectively than anyone else. While they may accept responsibility for complete construction of a landscape project, they will only personally do a part of it, such as plantings and pavements, and **subcontract** the remainder of the work to other specialty firms that can do the work better, faster, and cheaper. Pools or other water features, electrical work, stonework, or other requirements of the design may be subcontracted. It is often disappointing to young people who enter the industry with an enthusiasm for everything to discover that their companies' involvement with big projects has a more narrow focus than they were expecting.

Due to the diverse nature of landscape contracting, employees must possess a wide range of skills, with no one person likely to embody them all. The person most skilled in sales solicitation is unlikely also to be an expert at paver installation or fence construction. The educational requirements needed for employment as a landscape contractor are not as definitive as for a landscape architect. Few states have certification standards, although there is some movement in that direction currently being promoted by The Associated Landscape Contractors of America. They have established curriculum standards for colleges and universities that choose to offer an ALCA-approved major in landscape contracting. ALCA is also promoting the certification of landscape contractors at a state or regional level, which would meet national standards for competency established by ALCA. Certification would require periodic renewal to assure that landscape professionals stay up to date.

For employees working at the crew level in a landscape contracting company, the required horticultural skills can usually be learned on the job. Many people enter the industry by apprenticing. Should an opportunity or personal ambition cause the crew-level worker to move into a supervisory position, there may be a need for greater skills than can be learned on the job. As an employee moves through the ranks of a landscape contracting organization, the daily use of horticultural knowledge is often supplanted by the need for human resource and business skills. Therefore, someone seeking to maximize career opportunities in landscape contracting should be trained not only in ornamental horticulture, but in business and personnel skills as well. While it is still possible to have a career in landscape contracting without acquiring a college degree somewhere along the way, the opportunities are greater for men and women who bolster their personal credentials with solid academic training before or during the development of their careers.

Landscape management involves the extended care of existing landscapes. It begins with the expiration of the guarantees given by the landscape contractor, at which time the landscape architect has gone on to seek other clients and the landscape contractor has moved trucks, tools, and workers to break ground at another site. The landscape that they leave behind is far from complete. It is not yet grown, not yet weathered, not yet used . . . and abused. The new landscape must be sustained and managed, just as a building, farm, or office is managed. A budget must be established and a schedule put into place that will predict the number of mowings, snow plowings, prunings, litter sweeps, flower rotations, etc., that are needed to guide the design toward mature fruition at the standard of maintenance desired by the client. While landscape management may sound like the grown-up version of the neighbor's child who cut lawns for pocket money, it is the fastest growing area of landscaping today. It requires horticultural skills different from those needed by the landscape designer and contractor, and often necessitates equally different tools and equipment. The subcontracting of landscape management services is much less common than the subcontracting of landscape contracting services, but management companies may divide their workforce for specialization with either residential or commercial clients.

While there are numerous successful companies that offer only landscape contracting or landscape management services, even more companies offer both. They see a successful construction performance for a client as the way to acquire the long-term management contract. Over a period of several years, the latter may prove more financially lucrative than the former. Construction income may be a one-time profit, but the management contract represents ongoing income. As the economy fluctuates, companies that offer both construction and management can often alter their balance of work to meet the needs of the marketplace and remain profitable even as the more specialized firms begin to flounder. Aside from the differences in the tasks performed by the field crews, the organization and operation of landscape companies offering construction and management services are so similar that the inclusion of both under one corporate umbrella makes sense. The educational preparation required for career development within such companies is nearly identical as well.

Even landscape architects who have long regarded themselves as professionally distinctive from the rest of the industry are recognizing the benefits of closer collaboration with landscape contractors. Some landscape contractors now employ their own registered landscape architects, permitting them to offer a full range of services to a client and thereby relieving the client of the need and bother to deal with multiple companies in order to get a landscape project completed. One-stop shopping is desired by many private clients, and the large landscape firms are able to offer that convenience. Such companies also provide employment opportunities for landscape architects, many who value the complete control over a project that is provided when one company handles everything from initial contact with the client, through the design stage, and on to construction and long-term maintenance. Frequently referred to as **design-build firms**, such companies are becoming increasingly common nationwide.

In summary, the landscape industry in America is changing. Once composed entirely of small companies that served a local market and tried to provide all of the services required by their residential clientele, the industry is evolving into one that serves larger markets, often crossing state lines, and serving a more diverse clientele. It employs people from across the ethnic, gender, and educational spectrum. It actively strives to improve its professional image and price tag by hiring, training, and retaining talented people who are, in turn, seeking positions with companies that offer them good starting positions and the opportunity for advancement. It is an industry growing so quickly that there are few wise old men and women to learn from. As a result, company executives are teaching themselves and passing on what seems to work to their subordinates. Active national organizations such as the American Society of Landscape Architects, the Associated Landscape Contractors of America, the Professional Grounds Management Society, the Professional Lawn Care Association of America, and the National Landscape Association are contributing to the education of their members through newsletters, conferences, video productions, and book sales that bring new ideas and techniques to the forefront. As the industry upgrades its self image, it should follow that it will upgrade its public image. That is essential if the industry is to attract and retain the talent it needs from the college campuses of the nation. An upgraded public image is also essential to permit companies to charge fairly for their services. Customers who wouldn't flinch at bills for medical services or lawyers fees must be persuaded that the talent and skills applied to the development of their landscapes warrant a professional charge as well. That acceptance will only come from customers when it is warranted; thus the continuing need for professionalism at all levels within the landscape industry. It is an interesting point in time to observe the landscape industry. It still has a foot in the past, but it is both looking and moving toward a future that is promising, yet undefined.

SECTION 1

Landscape Designing

CHAPTER 1

Using Drawing Instruments

Objectives:

Upon completion of this chapter, you should be able to

- identify and use properly the traditional tools of the landscape designer
- measure and duplicate angles
- measure and interpret dimensions to scale

The drawing tools of the landscape architect and the landscape designer have traditionally been the same as those of the engineering draftsperson. With very little alteration over many years, traditional tools are still highly important to modern designers, and a knowledge of their proper use is essential to understanding how landscape design plans are created. However, in the 1980s landscape designers began supplementing traditional drawing methods with computer-aided methods. Computerized drawing technology has grown rapidly, necessitating a knowledge of both traditional and high-tech methods if the full spectrum of drawing options is to be available to the designer. This chapter will focus on the traditional tools. The use of computers in the landscape industry will be discussed in a later chapter.

The Purpose of the Drawing Tools

When an idea begins to form in the mind of a designer, it is seldom clearly defined. It is usually necessary for the designer to make some preliminary sketches for his or her own benefit in order to prepare something for eventual presentation to the client. As the design evolves from fuzzy idea to final presentation, it becomes important to represent the idea in a form that will enable the client to understand the proposal, envision how it will look, and later present it to the landscape contractor in sufficient detail to permit realistic cost estimation and actual construction. To be most beneficial to all users, a landscape design must be drawn to scale, representing actual dimensions and placements. Through its graphic styling it must create an image of how the new landscape will look and what it will be like to live within it. The plan must also convey an impression of professional competency on behalf of the landscape firm if the client is to be persuaded to entrust that designer and that company to steer the project to completion.

The Tools and How They Work

The **T-square** is a long straightedge that derives its name from its shape, Figure 1-1. When used with a smooth-surfaced **drawing board** or **drawing table** that has four 90-degree corners, the T-square can be used to draw a series of parallel horizontal or vertical lines, Figure 1-2. It is important that the T-square be kept flush with the edge of the drawing board at all times when parallel or perpendicular lines are being drawn. Common uses of the T-square include the representation of property lines, roads and drives, fences, and utility lines, as well as adding borders around drawings. The T-square is also used as a base of support for other tools such as triangles and lettering templates.

The **drawing pencil** is the most frequently used tool of the landscape designer, and probably the most important. It is the link between the designer's mind and the paper. It may be inexpensive and short-lived, as with wooden pencils, or more costly but longer-lasting, as with lead holders, Figure 1-3. Wood pencils have a graphite core (the lead), while lead holders have a chamber that holds replaceable leads. Both wood pencils and lead holders require some type of sharpening device to create the desired point. **Sandpaper pads** are often used, as are **specialized pencil sharpeners**, Figures 1-4, 1-5. Millimeter pencils hold thin leads of prescribed diameters, which create lines of comparable widths and never require sharpening. The consistent width can be a convenience as long as varied line widths are not required.

Drawing leads are given H or B ratings, which are measures of their comparative hardness or softness. The harder leads have H ratings, and softer leads have B ratings. The higher the H rating is, the harder the lead; while the higher a B rating is, the softer the lead. Leads may be rated as hard as 9H or as soft as 6B. While no rule exists, most landscape designers work with 2H or 3H pencils or leads for basic drawing. However, other leads are frequently employed for special needs, especially for interpretation techniques. The harder a lead is, the lighter will be the line and the longer the lead will retain its point. The softer a lead is, the darker

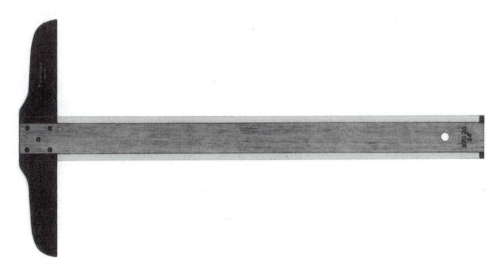

Figure 1-1 T-square (Courtesy of Staedtler, Inc.)

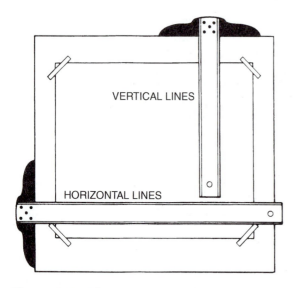

VERTICAL LINES

HORIZONTAL LINES

Figure 1-2 The T-square may be used to create parallel horizontal or vertical lines.

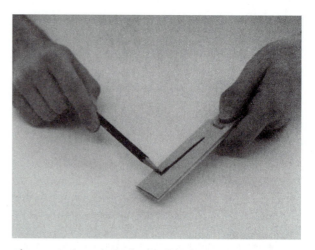

Figure 1-4 A wood pencil being sharpened using a sandpaper pad

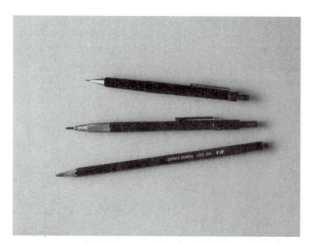

Figure 1-3 Types of drawing pencils. From bottom to top: Wood pencil, lead holder, and millimeter pencil

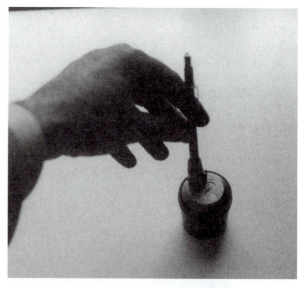

Figure 1-5 The lead in the lead holder is sharpened using a mechanical lead sharpener.

and more easily smudged will be the line, and the more often will the lead require sharpening.

Technical pens are used when ink is needed or preferred over pencil for the drawing. While there is sharper line definition and greater permanency with ink, it is more difficult to correct errors or

make changes to inked drawings. Pens are available in both disposable and refillable forms. Refillable pens have a reservoir that contains the ink. Like millimeter pencils, technical pens create lines of a prescribed width. That necessitates a set of pens, rather than one, if the drawing requires

Figure 1-6 Technical pens. Each one creates a different line width.

lines of varying widths, Figure 1-6. The tips of the pens are constructed of different metals. Which type of tip is appropriate depends on the type of drawing surface being used. Some papers and plastics are more abrasive to certain metals than to others, causing some pen tips to wear out sooner than others.

Lettering guides or templates are used to create a stenciled lettering style that may be appropriate for certain kinds of drawings. Whether using pencils or technical pens, the lettering template must be matched to the width of the lead or pen tip. The result is lettering that appears somewhat stiff and mechanical, Figure 1-7. Lettering templates are used in combination with the T-square to keep letters evenly aligned. When ink is used, the templates must have raised edges to allow them to slide back and forth along the T-square without smearing the letters.

The **Ames lettering guide**® is an inexpensive plastic device used in association with the T-square to produce guidelines for hand lettering. It can be set to create lines in varying numbers and of vary-

Figure 1-7 Lettering template

ing widths, which can then be replicated endlessly without requiring additional measurement, Figure 1-8. The Ames guide also has straightedge sides,

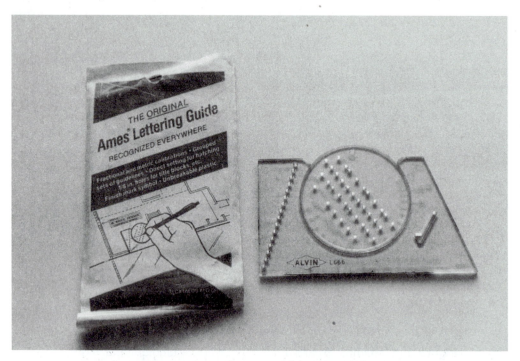

Figure 1-8 The Ames® lettering guide

Figure 1-9 The 45°/90° triangle (left) and the 30°/60°/90° triangle (right) (Courtesy of Staedtler, Inc.)

Figure 1-10 The triangle can be used with the T-square to create angles of 30, 45, 60, or 90 degrees.

which allow it to be used in actual creation of lettering as well.

Triangles serve many functions in landscape design and illustration. Commonly constructed of clear or colored plastic, the triangles will have either 30-60-90 or 45-45-90 angle combinations, Figure 1-9. The triangles may be used as straightedges by themselves or in combination with the T-square to create lines at consistent angles of 30, 45, 60, or 90 degrees, Figure 1-10.

A **compass** is used to create circles, Figure 1-11. Circles are the basis for the symbolization of many landscape features, including plants, patios, pools, and lawn areas. The metal pointed leg of the compass is placed at the center of the circle, while the leaded leg creates the arc. When drawing circles, remember that the distance between the two legs

of the compass should be half the desired diameter of the circle. For example, if a circle with a 4-inch diameter is desired, the compass legs should have a spacing of 2 inches, Figure 1-12.

The **protractor**, Figure 1-13, measures the relationship between two joined lines. The relationship is known as an **angle**; the unit of measurement is a **degree**. To measure an angle, the center of the protractor's base is placed at the point where the two lines of the angle join. The 0-degree mark of the protractor's baseline is aligned along the lower line of the angle. The angle is determined by reading **up from 0 degrees** to the point where the second line intersects the protractor. In Figure 1-14, the protractor is measuring a 25-degree angle. It is important to remember that the reading is always taken between two existing lines, starting at 0 degrees. In this way, confusion over the protractor's double scale is avoided.

Curves are hard plastic or flexible plastic supports for curvilinear lines, Figure 1-15. The hard plastic curves, known as French curves, are available in assorted shapes. They are moved freely to whatever

Figure 1-11 Compass

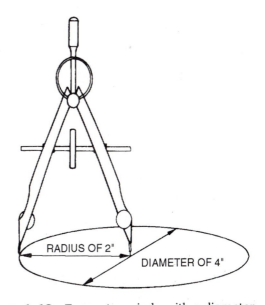

Figure 1-12 To create a circle with a diameter of 4 inches, the legs of the compass are spaced 2 inches apart. The radius of this circle measures 2 inches.

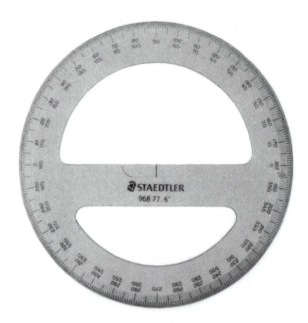

Figure 1-13 A 360° protractor. It is also available in a 180° model. (Courtesy of Staedtler, Inc.)

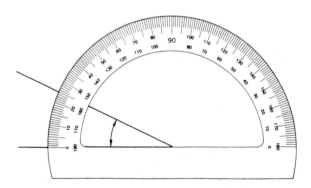

Figure 1-14 The protractor is placed along one line of the angle and aligned with zero (0 degree). The angle is read at the point of intersection of the second line.

placement best supports the pencil as it creates free-flowing lines for the design. The flexible curves allow nearly limitless support for the designer's pencil, since they can be twisted into practically any shape. Both French curves and flexible curves are available in different sizes and lengths.

The **scale** is used to represent actual dimensions in a reduced size. While the scale resembles an ordinary ruler, it has many more uses than merely measuring inches, Figure 1-16. The scale instrument has a triangular shape that provides six edges, thereby enabling each edge to create or measure drawings having a precise relationship between the drawn length and the actual length, such as 1" = 10' or or ¼" = 1'.

Two types of scales are commonly used by landscape designers: the engineer's scale and the architect's scale. The **engineer's scale** divides the inch into various multiples of 10. That permits dimensions to be read as whole numbers and decimal portions, such as 1.5 feet. The **architect's scale** divides the inch in a number of ways that permit dimensions to be represented as feet and inches. A drawing prepared with an architect's scale might use the side of the instrument that permits a scale relationship of 1/16" = 1' or 3/32" = 1'.

Depending upon the desired size of the drawing or the size of the property that must be represented on paper, a scale will be selected for the drawing to correspond with one of the six edges on either the architect's or engineer's scale instrument. For example, if a property line is 100 feet long, it would be represented on designers' drawings as lines of different lengths depending upon the scales of the drawing and the particular scale instrument used. The following table indicates the scale instrument most likely to be used for the plan scales selected and illustrates how the drawn line lengths will vary.

	Line Length to Represent 100 feet	
Scale of the Plan	**Architect's Scale**	**Engineer's Scale**
1" = 10'		10 inches
1" = 20'		5 inches
1" = 40'		2-1/2 inches
1/16" = 1'	6-1/4 inches	
1/8" = 1'	12-1/2 inches	
1/4" = 1'	25 inches	

In Figure 1-17, the same measurement of 60 feet is located on each of the six sides of an engineer's scale. Note that each unit represents 1 linear foot

Figure 1-15 Tools for guiding curvilinear lines. Left, the French curve; Right, the flexible curve

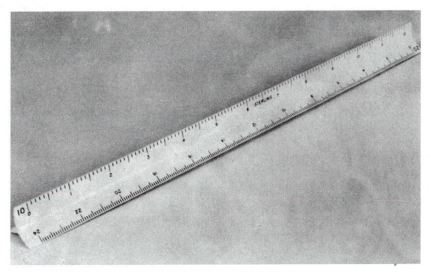

Figure 1-16 The scale. The triangular shape of this instrument creates six measuring sides.

regardless of the side of the instrument being used, as long as the scale of the drawing matches the side of the scale instrument. A similar comparison of measurements can be made using the architect's

scale; however, on that instrument, only the end units are divided into inches and fractions, so direct reading is a little more complex. The instrument must be repositioned to bring the end units into

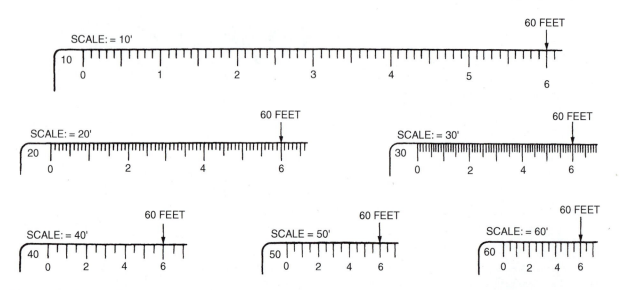

Figure 1-17   Reading the scale instruments. The six views show where one measurement (60 feet) appears on each side of the engineer's scale instrument. Individual foot units are not shown in the scales on the bottom line. (Drawing not to scale)

place for precise measurement when it is something other than a whole number.

Templates are available in many styles and sizes to assist the designer with the creation of a neat, attractive plan. Circle templates can be helpful for uses such as plant symbolization; irrigation templates for representing the components of irrigation systems; tree and shrub templates for use when original styling is unwarranted; geometric shape templates to represent assorted non-plant features in a design; burnishing guides to permit rapid representation of textured pavement surfaces; and assorted other templates for cars, people, arrows, etc. The variety is almost endless.

Erasers need no explanation of purpose; but it is worth noting that there are different kinds available, depending upon the designer's need. **Plastic erasers** have all but completely replaced the traditional rubber eraser as the workhorse in the drawing room. The plastic eraser wears away more slowly and is washable, permitting it to be cleaned of graphite and used again. It does not become hard and unusable over time. Should a sharp edge be required on the eraser, the plastic can also be

carved or whittled with a sharp knife or razor blade. Plastic erasers are produced by a number of companies and in a variety of forms, including plugs that fit into electric erasing machines, Figure 1-18. **Kneaded erasers** are soft and pliable and can be used to lift graphite from a drawing by placing the eraser on the line, pressing down, and then removing it. As a technique for lightening a line, it is very helpful. **Special purpose erasers** for the removal of inked lines or colored marker are available, but many do not perform as well as they claim. **Liquid erasers** are also available for removing old, dried ink lines or those created by copy machines.

An **erasure shield** can be helpful when erasing graphite lines, especially those made with soft B leads, which are prone to smearing, Figure 1-19. The thin metal shield is filled with punched out slits, which can be positioned over the line or word to be removed, thereby protecting the rest of the drawing from being smeared by the eraser.

Drafting powder also prevents smudging and keeps the drawing clean. The powder is available in bottles, shakers, and bags. A light dusting is

Figure 1-18 An electric eraser. It can be powered through a wall outlet or by batteries. Different eraser plugs are inserted for pencil, ink, or marker erasures.

Figure 1-19 An erasure shield reduces graphite smears.

sprinkled over the drawing as the designer works, absorbing skin oil and preventing the drafting tools from making direct contact with the graphite. As hands, arms, and tools move back and forth across the paper's surface, smudging is minimized.

Drafting tape is a low-adhesive paper tape produced in continuous rolls or as individual paper dots. The low adhesion is intentional and desirable, since its purpose is to hold the drawing on the board while the designer works, then release the drawing upon completion without tearing off the corners. Other paper tapes, such as masking tape, should not be used. Although they often look like drafting tape, they have much greater adhesion and can tear paper when removal is attempted. **Transparent mending tape** is also useful when drawings do get torn, and most designers keep a roll nearby. Sometimes the tape will leave a shadow on prints made from a repaired original.

The **rolling ruler** is a multi-use tool, and as such is favored over the more traditional ones by some designers. It is literally a straightedge on a roller. That permits the designer to make a series of parallel lines at any angle of choice without using another straightedge for base support,

Figure 1-20. The tool can also be used to make circles, Figure 1-21, although there is not the infinite choice of diameters offered by the compass. Skillful use of the rolling ruler requires some practice, because it is prone to slipping on the paper.

Rendering tools, such as colored pencils and color markers, are usually not required or used during the design stage of a landscape project. They are more commonly used when preparing the design for presentation to the client. Therefore the description and discussion of rendering tools and uses will be reserved for a later chapter.

Getting Started

Tools are manufactured in a wide range of sizes and qualities. Less expensive and lower quality ones are commonly available in neighborhood malls. High-quality, professional tools can be purchased in a drafting supply or art supply store or from a catalogue that carries professional instruments. As with most things, the value of the initial investment usually determines the long-term value of the purchase.

Figure 1-20   The rolling ruler being used to create parallel lines

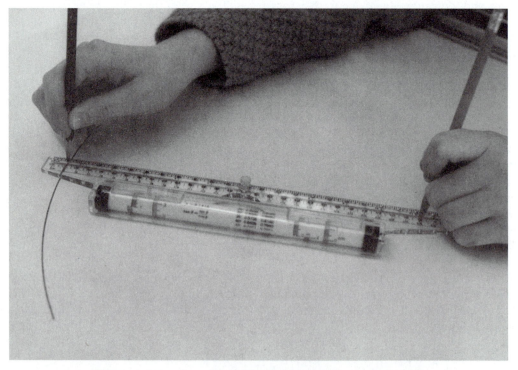

Figure 1-21   The rolling ruler being used to create circles

Practice Exercises

A. Tape a piece of drawing paper onto a drawing board. Practice drawing parallel horizontal lines and vertical lines using a T-square. Use a 2H pencil first and then a 3H or 4H pencil. Which pencil marks smear most easily? Which are easiest to erase?

B. With your protractor, measure the angles in the figure below. Duplicate the angles on a separate sheet of paper.

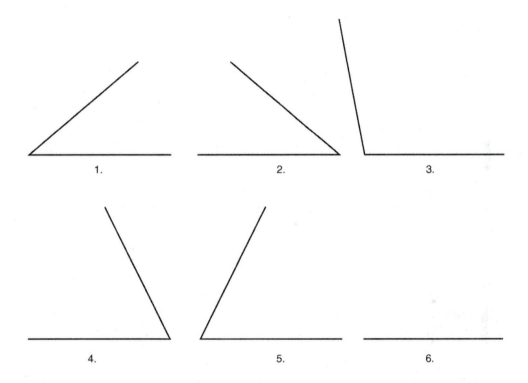

C. Using an engineer's scale instrument to measure and a triangle as a straightedge, draw a line 30 feet long to the scale of 1" = 10'. Draw another 30-foot line to the scale of 1" = 50'. Draw an 87-foot line to the scale of 1" = 40' and another to the scale of 1" = 30'.

D. Using an architect's scale instrument to measure and a triangle as a straightedge, draw a line 18 feet long to the scale of 3/8" = 1'. Draw a line 68 feet and 4 inches long to the scale of 1/8" = 1'.

E. Use a compass and a scale instrument to create a circle with a 20-foot diameter drawn to the scale of 1" = 10'. What is the radius setting for the compass? Draw another circle with a 20-foot diameter to the scale of 1" = 20'. Draw another circle with a 24-foot diameter to the scale of 1/4" = 1'.

Achievement Review

A. Explain the function of each drawing instrument listed.

1. T-square
2. triangle
3. compass
4. protractor
5. scale
6. curve
7. drafting powder
8. rolling ruler
9. technical pen
10. Ames lettering guide

B. From the choices offered, select the best answer to each question.

1. A compass set for a 2-inch radius forms a circle of what diameter?
   a. 2 inches
   b. 4 inches
   c. 1 inch
   d. 6 inches

2. If the scale of the drawing is $1" = 5'$, the completed circle described in question 1 would be how many feet wide?
   a. 5 feet
   b. 10 feet
   c. 15 feet
   d. 20 feet

3. Which of the following indicates the hardest pencil lead?
   a. 2H
   b. 3H
   c. 4H
   d. 5H

4. Which two instruments combine most easily to make a 90-degree angle?
   a. compass and scale
   b. T-square and triangle
   c. protractor and scale
   d. compass and T-square

5. Why is masking tape unsuitable for securing a drawing to the drawing table?
   a. Paper tape discolors the drawing paper.
   b. The adhesion is too strong.
   c. The adhesion is too weak.
   d. Cloth tape is better.

6. Why are plastic erasers preferable to rubber erasers?
   a. They are less expensive.
   b. They are disposable.
   c. They last longer.
   d. They eliminate the need for erasure shields.

7. If a driveway is 50 feet long in reality, how long will its symbol be on a drawing with a scale of $1" = 20'$?
   a. 2-1/2 inches
   b. 5 inches
   c. 7-1/4 inches
   d. 10 inches

8. If the scale of the drawing is changed to $1" = 5'$, how long will the symbol of the drive be?
   a. 2-1/2 inches
   b. 5 inches
   c. 7-1/4 inches
   d. 10 inches

9. If the scale of the drawing is changed to $1" = 10'$, what will be the actual length of the drive?
   a. 2-1/2 inches
   b. 5 inches
   c. 10 inches
   d. 50 feet

10. Which is the softest lead?
    a. 5B
    b. 4B
    c. 3B
    d. 2B

CHAPTER 2

Lettering

Objectives:

Upon completion of this chapter, you should be able to

- explain the importance of designer lettering
- describe different methods of lettering designs
- create a variety of lettering styles

The Importance of Lettering

Today's sophisticated graphic technology has made everyone consciously or subconsciously aware of the importance of lettering as a support element in the overall visual presentation of a product or service. Many corporations obtain trademark protection for not only the names of their products, but for the lettering styles as well. Seemingly endless variations of style can transform the 26 characters of the alphabet into words that lend interest and excitement to graphic displays. Landscape plans and drawings also require lettering that matches the sophistication and importance of the work they are intended to explain. While never overshadowing the plan or drawing, neither should the size and style of lettering used by a designer diminish the worth of the

project being proposed. As illogical and unfair as it seems, many clients will associate the graphic quality of a presentation with the quality of the design. Therefore anyone who would be a good landscape designer should also strive to be a good graphic artist. The landscape designer must master enough graphic art skills to permit all aspects of a landscape proposal to be conveyed fully to a client. Foremost among those graphic skills is the development of a good hand lettering style, as well as knowledge of other ways that letters can be applied to drawings.

Methods of Lettering

Several factors affect the selection of a lettering method by a designer, not the least of which are the choices available. Not every firm possesses a full state-of-the-art range of selections. That aside, the choice of method is most likely to be affected by the importance of the immediate drawing and the time schedule for its production. Preliminary drawings that will stay in-house and will not be seen by the client require lettering that is fast and functional. Presentation drawings for the client and the public require lettering that is carefully styled and sized to fit the presentation graphics. In nearly every case, however, a landscape designer will seek and select a lettering method that is easy to create. There is little justification for spending a long time on lettering. Time is money, even for artists and designers.

The methods of applying letters to landscape drawings include

- lettering templates.
- waxed press-on letters.
- lettering machines.
- transfer film.
- hand lettering.

Each method has certain characteristics that make it suitable for certain circumstances or types of drawings.

Lettering Templates

Lettering templates are guides that reproduce the same letter size and style over and over. They may be used with a drawing pencil, lead holder, millimeter pencil, or technical pen. The drawing instrument must be selected to match the width of its tip with the width of the lines punched into the template; otherwise the lines of the letters will be wobbly. If technical pens and ink are used, lettering templates must have raised edges so they can be used without smearing the ink, Figure 2-1.

Lettering templates in various sizes and styles are available on the commercial market. A single landscape drawing may require letters of various sizes. Thus, a selection of lettering templates may be needed to complete a drawing. Using mixed styles of letters within the same drawing is usually avoided because it creates visual complexity.

The use of lettering templates by landscape architects and designers is limited. Templates are time-consuming to use, and they create stiff, mechanical-looking letters, lacking the visual

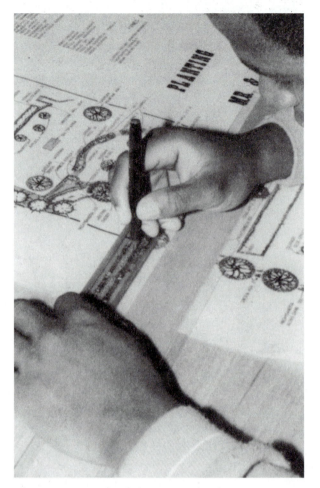

Figure 2-1 Using a lettering template

interest that hand lettering creates or the style that more sophisticated technology can now produce.

Waxed Press-on Letters

When a hand-drawn landscape drawing requires stylized lettering to accompany it, one of the methods available is the use of **waxed press-on letters**. Quick and easy to use, the letters are commercially manufactured in an assortment of styles and sizes, Figure 2-2. The letters are mounted and sold on sheets of plastic or waxed paper, Figure 2-3. They are transferred to the drawing surface by

Folio Extra Bold	**Gill Extra Bold**	News Gothic Bold	*Berling Italic*
Folio Bold Condensed	Grotesque 7	News Gothic Condensed	**Berling Bold**
Franklin Gothic	**GROTESQUE 9**	**Pump Medium**	Beton Medium
Franklin Gothic Italic	***Grotesque 9 Italic***	**Pump**	**Beton Bold**
Franklin Gothic Cond.	Grotesque 215	**Simplex Bold**	**Beton Extra Bold**
FRANKLIN GOTHIC EX COND	**Grotesque 216**	**Standard Medium**	Bookman Bold
FRANKLIN GOTHIC COND	Helvetica Ex Light	**Standard Extra Bold Cond.**	*Bookman Bold It.*
Futura Light	Helvetica Light	Univers 45	**Carousel**
Futura Medium	**Helvetica Medium**	Univers 53	Caslon 540
Futura Medium Italic	**Helvetica Bold**	Univers 55	**Caslon Black**
Futura Bold	*Helvetica Light Italic*	Univers 57	**Century Schoolbook Bold**
Futura Bold Italic	***Helvetica Med. Italic***	Univers 59	Cheltenham Old Style
Futura Demi Bold	***Helvetica Bold It.***	Univers 65	Cheltenham Med

Figure 2-2 Styles and sizes of press-on letters

the heat of friction created when a pencil is rubbed over the surface of the sheet, Figure 2-4. To improve their adhesion, the letters are then rubbed again with a pencil through a sheet of waxed paper that usually accompanies each sheet of letters, Figure 2-5.

Advantages of Waxed Press-on Letters

- easy to apply
- create a professional graphic image
- easily removed with cellophane tape, permitting correction of errors

Disadvantages of Waxed Press-on Letters

- expensive
- crack with age
- wasteful; vowels commonly depleted before the consonants

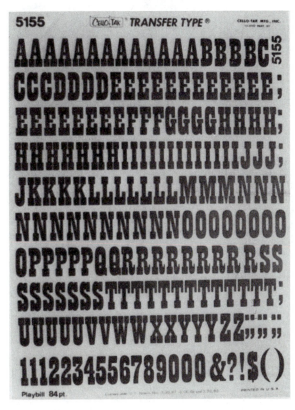

Figure 2-3 A sheet of one style of press-on letters

Figure 2-4   Press-on letters are transferred onto the drawing by the heat of friction.

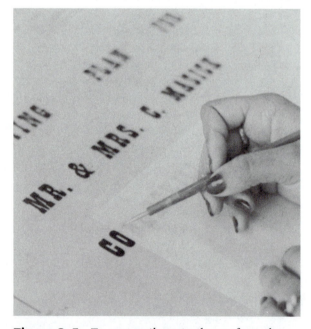

Figure 2-5   To secure them to the surface, the letters are rubbed again through a sheet of waxed paper.

Lettering Machines

Designers who prepare large numbers of presentation drawings are apt to favor a **lettering machine** over waxed press-on letters. While the initial expense is greater, over time the machine is more cost effective. The lettering machine is simple to operate, since it functions much like a typewriter. A font style and size are selected, and the desired letters or words are entered using a keyboard. They print onto a tape that is contained within a cartridge. When the desired words are completed, the tape is clipped off and applied to the drawing. Because the tape is transparent and thin, it is nearly invisible after application, Figures 2-6 and 2-7. Unlike waxed press-on letters, the lettering machine involves no waste. Prices vary greatly between manufacturers and models.

Transfer Film

Transfer film is paper-thin, transparent, plastic sheeting with a clear adhesive backing and a sheet of protective waxed paper over it. It enjoys a number of uses in graphic art since anything that is drawn on it can then be placed and affixed anywhere desired by the designer. Commonly referred to as "sticky back" by landscape architects and designers, transfer film is used for plant lists, construction notations, client names and addresses, corporate logos, and dozens of other design components. The initial lettering can be prepared on a typewriter or word processor and printed on a standard sheet of paper. It is then replicated onto the transfer film by passing it through a photocopy machine, using the film in place of conventional copy paper. Later the protective backing is pulled away and the film positioned where desired on the design, Figure 2-8. The use of transfer film by landscape designers is common. It allows lettering that is used repeatedly, such as the company name and logo, scale indicators, slogans, page markings, and standard notations, to be added quickly whenever and wherever they are needed by simply pulling the master sheet from a file and passing it through a copy machine. Transfer film is also an easy way to merge the modern technology of computer-generated lettering with traditional forms of

Figure 2-6 The lettering machine permits many styles and sizes of letters to be produced. (Courtesy of Kroy, Inc.)

Figure 2-7 The lettered tape is peeled from its backing and applied to the tracing surface.

designing and drawing. Selecting from the many font styles of modern word processors and adjusting the size of the letters to match the drawing being labeled, designers can use transfer film in the same way they used waxed press-on letters in the past.

Hand Lettering

One of the hallmarks of professionally trained designers and graphic artists is a distinctive style of **hand lettering**. While it does not make the design better nor does it cause the proposed project to

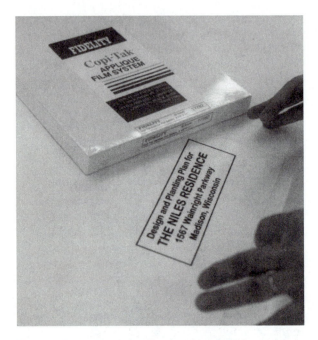

Figure 2-8 Transfer film being applied

cost a penny more or less, good hand lettering offers visual evidence of designer competency to clients. Perhaps it is because most people do not have good lettering skills that they view favorably those who do. Whatever the reason, the development of a lettering style that is interesting, lively, and rapidly executed should be a goal of anyone who aspires to practice professionally as a landscape architect or landscape designer.

Hand-drawn letters are created with single strokes of the pen, pencil, or marker. To attempt more complex styles would be counterproductive because the lettering would take too long and the letters would probably be inconsistently styled. Instead, designers use styles that are simple variations of the basic block style that they learned as children in school, Figure 2-9. In that style, letters are constructed with single stroke lines, and their heights and centers are controlled by the location of three **guidelines**. The guidelines may be pre-printed on the paper or may need to be drawn on lightly before lettering begins. If that early school training allowed any variation of the basic block

form, it was to slant letters to the right at approximately 60 degrees to suggest a more natural, humanistic interpretation of the style, Figure 2-10.

To create letters that spell out words while simultaneously contributing to the overall appearance of the design presentation requires that the basic block style be altered. Alterations take one of two forms:

- changing the location of the center guideline.
- changing the number of guidelines from three to five.

Changing the Center Guideline

Maintaining three guidelines, but raising or lowering the center one, causes the weight of the letter

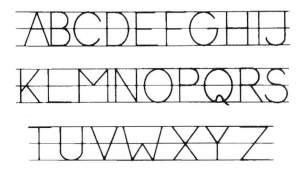

Figure 2-9 The alphabet in basic block style

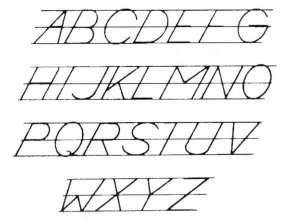

Figure 2-10 Slanted block alphabet

to shift down or up accordingly. Figures 2-11 and 2-12 illustrate the standard block style of lettering with the center lines changed. Figure 2-13 illus-

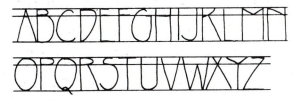

Figure 2-11 Distorted block alphabet with midline raised

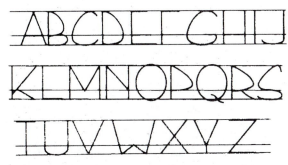

Figure 2-12 Distorted block alphabet with midline lowered

Figure 2-13 Slanted, distorted block alphabet with guidelines. Half of the alphabet has the midline above center, and half has the midline below center.

trates the changed appearance of slanted block letters when the center guideline is either raised or lowered. These styles require some practice, but are easily learned.

Changing the Number of Guidelines

Since the purpose of guidelines is to aid the designer in proper shaping of the letters, it follows that changing the number of guidelines from three to five will significantly alter the appearance of the letters. In Figure 2-14, two guidelines are added to the three used in the standard block style. Note that line 2 is very close to line 1 and that line 4 is very close to line 5. The main purpose of the two additional lines is to introduce a slight upward slant to the horizontal strokes of the letters and numerals. In the three-guideline style, all horizontal strokes are aligned directly on the guidelines. The five-guideline style creates an intentional horizontal angle while keeping the degree of angling consistent throughout, Figure 2-14. The five-guideline style has a looser appearance due to the slight horizontal angle of the lines and the inclusion of several deliberately distinctive letters, such as the *K* and *R*. It is only distinctive when done as upper case (capital) letters. It is not adaptive to lower case (non-capital) letters.

Hand Lettering Techniques and Tips

Good hand lettering is a graphic skill that too many beginning designers fail to master, largely because of their impatience and their desire to get on to bigger and better challenges. With practice and adherence to a simple but effective regime, everyone can develop a rapid and attractive lettering style that will enhance the total graphic presentation.

  a. *Always use guidelines.* Whether drawn with a straightedge or with a tool such as the Ames lettering guide, all lettering should be constructed using guidelines. They will keep the lettering consistently sized and styled throughout the plan. They will also keep the lettering straight as long as the drawing is secured to the drawing board and properly aligned.

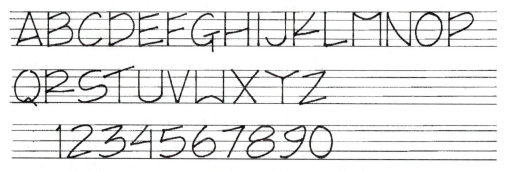

Figure 2-14 Lettering and numerals styled using five guidelines

b. *Keep the guidelines light.* It is usually unnecessary to erase lettering guidelines if they are drawn lightly. They need only to be dark enough to be visible to the person doing the lettering. A pencil with very hard lead is suitable for drawing guidelines. If the work is to be copied, guidelines can be made with a light blue pencil or pen, since photocopy and ozalid process machines do not reproduce blue well.

c. *Use straightedge support for all vertical lines.* All letters look best when their vertical alignment is parallel. A predictable mistake made by beginners and impatient designers is the belief that they do not need to use a straightedge for lettering once the guidelines are drawn. With the exception of very small letters, all letters need the sharpness and uniformity that straightedges provide. Figure 2-15 illustrates the use of a T-square and triangle to assure consistent vertical alignment during lettering.

d. *Continually monitor line quality.* Different types of drawing tools create different types of lines, and the designer must know what constitutes quality for each type. Maintaining an even line tone can be difficult when pencil is the drawing instrument. Lines drawn confidently with the support of a straightedge may be darker than curving lines drawn more cautiously. Curving lines often tempt the beginning designer to sketch the line rather than drawing it confidently with a single stroke. Pencil leads also widen if not sharpened frequently.

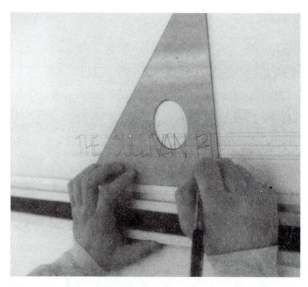

Figure 2-15 The triangle, resting on the T-square, supports the vertical lines of the letters.

That causes varying line widths within words. Unless deliberate line-width variation is being introduced, single-stroke lettering should also be single line-width lettering. Also, keep line intersections free of overlaps to avoid creating an inconsistent, shaggy appearance.

- Maintain even line tone.

- Do not sketch lines.

- Sharpen pencils to maintain consistent line width.

- Do not overlap line intersections.

e. *Space letters and lines to permit comfortable reading.* Letters that are crowded together or spaced too far apart create words that are difficult to read, Figure 2-16. The spacing between letters should be consistent throughout the word. The spacing between words should be greater than that between letters within a word. There must also be space between lines of words. The spacing common to printed lettering, such as this text, applies to hand lettering. If it is to communicate, the lettering must be easy to read.

f. *Maintain case consistency.* When upper and lower case lettering is used within a single presentation, the cases should not be mixed within words and sentences unless it is the first word of a sentence or is a word that warrants capitalization, Figure 2-17.

Introducing Line-Width Variation

An additional variation on the hand lettering styles described above can be incorporated with the deliberate addition of **line-width variation**. The

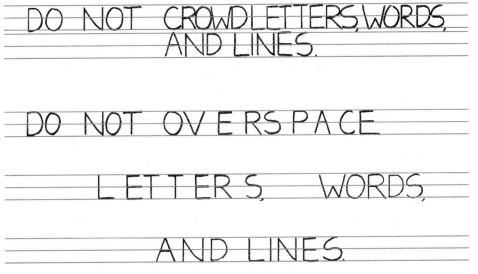

Figure 2-16 Incorrectly spaced letters, words, and lines

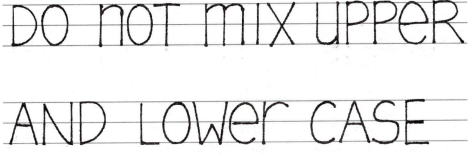

Figure 2-17 Incorrectly mixed upper and lower case letters

word *deliberate* is significant because the designer is intentionally creating a style, not just making a mistake in violation of the earlier concern for line quality.

To introduce deliberate line-width variation, the designer must either select or create an instrument with a chiseled tip. If pencil is the lettering instrument, the point needs to be widened and flattened with a sandpaper pad, Figure 2-18. If markers are used, those with broad nibs are the appropriate choice. Technical pens and millimeter pencils cannot be used for lettering when line-width variation is the style of choice, because their tips cannot be modified.

Using the chiseled tip, lines that are either thin or wide can be made by the same instrument, depending upon how it is held and the direction of the stroke, Figure 2-19. The designer must make a

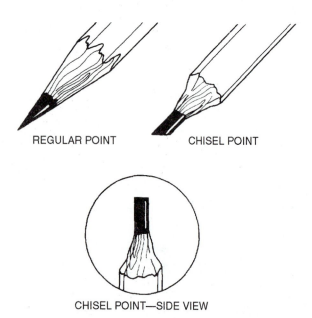

REGULAR POINT   CHISEL POINT

CHISEL POINT—SIDE VIEW

Figure 2-18 Chisel-point drawing pencil

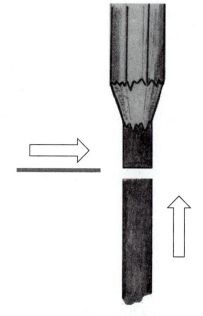

Figure 2-19 With a chisel-tipped lead, the pencil can make both broad and thin lines.

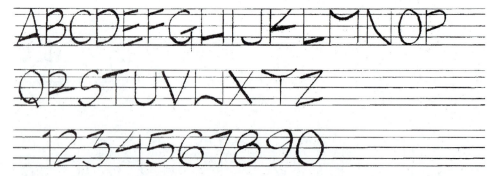

Figure 2-20 Line-width variation in lettering can be accomplished using a chisel-tipped pencil.

decision about whether to make the vertical strokes thin and the horizontal strokes wide, or vice-versa. Then, as long as there is no alteration of that decision, a distinctive and consistent lettering style will result. Figure 2-20 illustrates an alphabet and set of numerals created with thin vertical lines and wide horizontal strokes, using straightedge support for all vertical lines.

Practice Exercises

A. On a sheet of drawing paper, lay out three horizontal guidelines, each 1/4-inch apart. Following the examples shown in this chapter, create a full upper case alphabet, 1/2-inch height. Use a hard lead for the guidelines and a softer lead for the lettering. (Drafting powder can be used to prevent smearing.) Use a T-square to support your triangle for straightedge support of the vertical lines. Letter the alphabet several times and repeat as often as necessary those letters that you find difficult.

B. Repeat the exercise above using a 60-degree triangle to support the vertical lines, thus creating a slanted block alphabet. Repeat as necessary to improve your skill.

C. Repeat exercises A and B using hard lead pencil for the three guidelines and a felt-tipped pen to create the letters. Repeat as necessary to improve your skill.

D. Lay out five guidelines for lettering a complete alphabet and set of numerals. The letters and numerals should be 1/2-inch in height. Make lines 1 and 2 and lines 4 and 5 close together, with line 3 at the midpoint. Refer to the example in the chapter for guidance. Using the T-square to support the triangle for straightedge guidance of the vertical lines, create the letters and numerals as shown in Figure 2-14.

E. Repeat exercise D using a chisel-tipped soft lead pencil. Remember to chisel the tip frequently to retain the consistency of the style.

F. Prepare a sheet of drawing paper for the lettering of a paragraph with words that will be 1/4-inch in height. Select one of the styles of lettering that you have practiced in the above exercises. Copy the first paragraph under the heading *Hand Lettering* in this chapter. Remember to space properly within words and between words. Also remember to leave space between lines of words.

Achievement Review

A. List five methods of lettering commonly used by landscape architects and designers.

B. Select the best answer to the following questions.

   1. Assuming that all lettering choices were available to a landscape designer, what would most likely determine the method selected?
      a. the cost of the lettering method
      b. the level of graphic sophistication needed
      c. the preferences of the client
      d. the type of project being designed

   2. What characteristic must all methods of lettering possess if used in landscape designing?
      a. colorful   c. rapid
      b. fancy   d. large

   3. Which lettering method is most wasteful of material?
      a. lettering machine
      b. waxed press-on letters
      c. templates
      d. transfer film

4. What is the purpose of guidelines in hand lettering?
   a. to control the size and style of the letters
   b. to monitor line quality
   c. to match the lettering with the presentation style
   d. to assure correct placement of all labels and wording

5. What happens when the middle guideline of a three-line style is raised or lowered from the center?
   a. Nothing happens.
   b. The letters become wider.
   c. More space is needed between the letters.
   d. The weight of the letter shifts.

6. What does the deliberate introduction of line-width variation do to the letters created?
   a. It makes the letters wider.
   b. It destroys line quality.
   c. It creates a desirable difference in the lettering style.
   d. It eliminates the need for straightedge support.

C. Indicate if the following are characteristic of
   a. hand lettering
   b. lettering templates
   c. waxed press-on letters
   d. lettering machines
   e. transfer film
   f. all of the above
   g. both a and b
   h. types c, d, and e
   i. none of the above

1. Create stiff, mechanical-looking letters
2. Commercially manufactured in varied styles and sizes. Sold on sheets.
3. Transferred by heat of friction
4. The most personalized and distinctive style
5. Often referred to as *sticky back*
6. The letters must be removed from a backing to apply
7. The fastest way to duplicate logos or trademarked lettering
8. They have raised edges to prevent smearing
9. May be created with either pencil or pen
10. Must be easy to create in order to have practicality for use

CHAPTER 3

Graphic Materials and Techniques

Objectives:

Upon completion of this chapter, you should be able to

- distinguish between plan views, elevations, perspective views, and axonometric views of landscape proposals
- create plan view symbols for major landscape features and label them correctly
- lay out a complete landscape design plan
- compare different levels of graphic presentations and the proper uses of each
- describe the characteristics and uses of graphite pencils, colored pencils, felt pens, markers, pastels, pressure graphics, foam board, and the papers and films used for both original and copied work
- arrange the components of a graphic presentation

Same Design, Different Views

The interpretation of a designer's proposal for a project advances through numerous levels of refinement as it moves from abstract idea to acceptance by the client and eventual implementation by the contractor. Each person or group that has input into the evolution of the project, from vague concept to finished product, possesses a different level of design expertise and/or understanding. Each person also is involved with the project in a different way. For some the involvement is emotional; that is usually the client. For others the involvement is technical; that is most likely the contractor. The client wants to love the project. The contractor wants to know how to build it. Along the way, the designer also is continually questioning his or her own thinking. Will this idea work? Will this feature fit the space? What will be the result if I put these two features together? The landscape designer must be able to prepare drawings that will speak to each constituent.

Plan Views

The most common drawing prepared by landscape designers is the **plan view**. It is drawn to scale and is therefore measurable. In a typical plan view, objects are seen as though the viewer is above the drawing, viewing it at a perpendicular angle to the ground. It is not unlike that experienced when viewing the ground from an airplane.

The plan view drawing is an excellent way for a designer to show the proposed placement of plants and other elements of the design in relation to each other and to the existing features such as the buildings, roadways, utility poles, and similar items. Horizontal measurements are readily made. The exact width and length of objects or the distance between objects can be determined by anyone who understands how to use a scale instrument, Figure 3-1. Landscape architects, landscape designers, and landscape contractors all understand how to read plan views. For them to use the plan correctly, it need not be attractive, only accurate. Clients, however, frequently do not understand how to read plan views, since they have only horizontal dimensions. The lack of vertical measurements means that the third dimension is missing. People untrained in reading plans or who lack knowledge of plant heights are unable to envision how the design proposal will look when completed. For such clients, the designer may need to embellish the plan view with graphic tricks such as symbols that suggest how a plant or other feature will actually look. Other techniques such as the use of colors and shadows may aid client understanding. In many instances it may be necessary to prepare a different type of drawing that is better suited to the client's understanding.

Elevation Views

An **elevation view** is also a scaled drawing, thereby making its dimensions measurable. It has two dimensions, horizontal and vertical, Figure 3-2. Although lacking one of the horizontal dimensions possessed by plan views, elevations are generally more helpful to clients for visualizing a design proposal. They are well suited for displaying size relationships between plants, buildings, and people. They are also helpful in showing vertical relationships between various features of the landscape, such as raised decks and recessed patios, or hills and valleys. For contractors, the combination of plan views and elevations of the same project provides them with most of the physical measurements and explanations they need to get the design built. As with the plan view, the elevation view may be done with simple, minimal graphic effort or with elaborate presentation techniques. The designer makes the decision of how much time to spend on the graphics, depending on the importance of the project, its current stage of development, and the viewer for whom it is intended.

Perspective Views

Other than an actual model of the proposed landscape project, nothing allows a better visualization of the designer's intentions than a **perspective** drawing. A perspective allows multiple sides of a proposed project to be seen in one drawing, Figure 3-3. While it can be scaled, the dimensions are not measurable because most of the lines of the drawing are seen to recede away from the surface of the paper. Receding lines in a perspective are termed *foreshortened*, meaning that they are not drawn to their correct length. To do so would cause the dimensions to appear unrealistic in the drawing. Also in a perspective drawing, lines that are actually parallel nevertheless appear to converge if extended far enough. Controlling the angle of line convergence requires the designer to have sufficient knowledge of drawing technique to create a realistic interpretation. As a drawing to aid construction, the perspective has only limited value. However, as a means of helping both the designer and the client visualize the proposal's finished appearance, the perspective has great value. Ranging in graphic sophistication from a quick sketch to a highly detailed interpretation, the perspective drawing plays an important role in landscape design graphics. Frequently, however, it is the one least used because the designer fails to recognize that the client is not seeing the proposal as clearly as he or she sees it. Other times, the project's value does not justify the time required for construction of a perspective view.

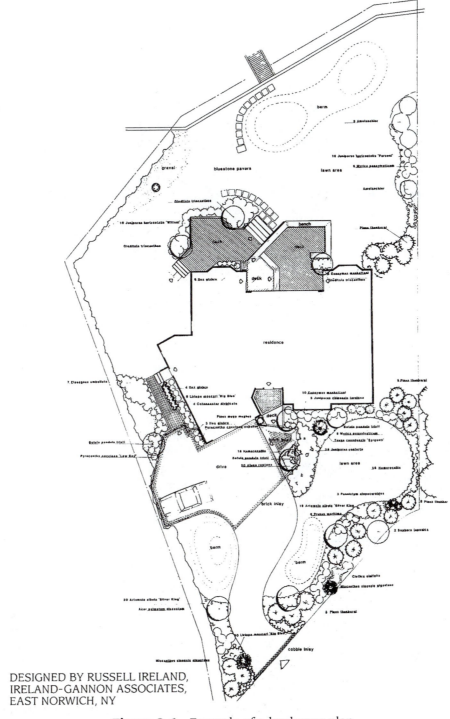

DESIGNED BY RUSSELL IRELAND,
IRELAND-GANNON ASSOCIATES,
EAST NORWICH, NY

Figure 3-1 Example of a landscape plan

Figure 3-2 A front elevation of a residence and plantings (Courtesy of Lawrence Perillo, Naples, Florida)

Figure 3-3 A perspective view permits easier visualization of how the proposed design will appear.

In some cases the lack of a perspective view may be due to the designer's inability to construct the drawing. With the continual advances in computer graphic technology, however, lack of hand drawing skills may not serve as an excuse much longer.

Axonometric Views

While not as realistic to the observer as perspective drawings, **axonometric views** offer a look at multiple sides of objects while also providing measurable dimensions to scale. Lines of the design that recede are not foreshortened to create the realism of a perspective, nor do parallel lines appear to converge. Object lines that are parallel in reality are left parallel in the drawing. With no perspective tricks, axonometric drawings are less helpful to the client who wants to see how the proposal will appear when built but the drawings are of much greater assistance to contractors, who can take measurements directly from them, Figure 3-4. Since all lines are drawn as actual scaled dimensions, and since the technique for constructing the drawing is much easier than with perspectives, many landscape architects and designers favor axonometrics as a rapid alternative to perspectives. Although axonometrics often look distorted, they still offer a semblance of realism and in many ways are much more usable by a greater number of people than are perspectives.

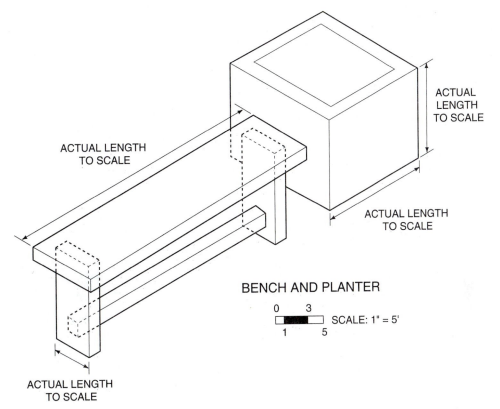

ACTUAL LENGTH TO SCALE

ACTUAL LENGTH TO SCALE

ACTUAL LENGTH TO SCALE

ACTUAL LENGTH TO SCALE

BENCH AND PLANTER

0   3

1   5

SCALE: 1" = 5'

Figure 3-4 Axonometric drawing. All dimensions are measurable to scale.

Landscape Symbols

When preparing landscape designs in plan view, designers use symbols that offer a suggestion of how the proposed or existing features would appear if seen from above. To be functional, the symbols must be drawn to their true size. The size is actual in the case of existing objects such as buildings and streets, and eventual with plants, which are usually drawn to their mature size rather than their size at the time of installation. For preliminary drawings that will not be seen by clients, but help the designers work their way through their own thought processes, or for construction plans that provide installation information for landscape contractors, the symbols need only be of correct size and placement. However, for presentation plans that will be used to help the client understand the design proposal and aid in selling the proposal, more elaborate and realistic styling of the symbols is needed.

To make the client's understanding of the design as easy as possible, the symbols need to be suggestive of the features they represent, yet not so graphically complex that they distract the client from full appreciation of the design. Whether drawn by hand or selected from a computer-generated menu, the symbols become the principal way that the designer explains a proposal to the client.

Needled Evergreens

When recalling the appearance of a pine or spruce tree, it is easy to understand the symbols commonly used to represent needled evergreens, Figure 3-5. The symbols suggest the spiny leaves and rigid growth habit of these plants, which are green throughout the year. These symbols can be used to represent both trees and shrubs as long as they are needled and evergreen.

Broadleaved Evergreens

Another group of plants are green throughout the year, but have wider and usually thicker, fleshy leaves. Plants such as the hollies, rhododendrons, and camellias exemplify the category of broadleaved evergreens. The symbols used to represent them suggest the larger leaf size and semirigid growth habit of the plants, Figure 3-6. As with the needled evergreens, these symbols can be used for both broadleaved evergreen trees and shrubs.

Deciduous Shrubs

Deciduous shrubs drop their leaves in the autumn. As a category of plants, they are much more numerous and diverse than either of the evergreen categories; therefore the symbols used to represent them tend to be more generic or general. The edge of the symbol is loose and irregular, suggestive of the less rigid growth habit of most deciduous shrubs, Figure 3-7. With this symbol, as with all plant symbols, a prominent dot in the center marks the spot on the plan where the plant is to be set into the ground.

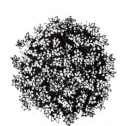  

Figure 3-5 Symbol styles for needled evergreens

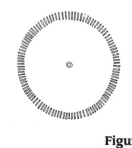 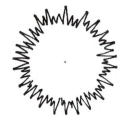 

Figure 3-6 Symbol styles for broadleaved evergreens

 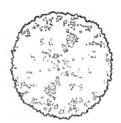 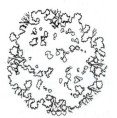 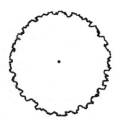

Figure 3-7 Symbol styles for deciduous shrubs

Deciduous Trees

Like deciduous shrubs, deciduous trees also lose their leaves during the winter season. Unlike most shrubs, trees usually have only a central trunk rather than multiple stems. Also, trees are generally taller than shrubs. All of these differences are suggested in the symbols selected by designers to represent deciduous trees, Figure 3-8. The symbols for trees are usually wider since trees are commonly larger than shrubs. Their line width may be greater, too, reflecting their greater height and prominence within the design. Depending upon how much detail must be shown beneath them, the tree symbols may be simple and plain, or intricately detailed to show branches and/or leaves.

Vines

Trees and shrubs tend to grow radially out from their centers. That is why the compass or circle template is so useful in forming their symbols. However, vines grow in a linear manner, and they do not hold to a predictable shape. Their symbols are shaped to suggest that rambling linearity, Figure 3-9.

Trailing Groundcovers

Groundcovers are those plants, usually 18 inches or less in height, that fill the planting bed beneath the trees and shrubs. Like vines, many groundcovers are shapeless, linear plants that would have little impact on the design if used alone, but when grouped in masses and given time to fill in, they become an important textural component of the design. The symbols used to represent trailing groundcovers are therefore more textural than structural, Figure 3-10. The texture is applied to all areas in the design where groundcover plants will be used.

Hardscape Materials

Design materials that are not living plant materials are often referred to as **hardscape**. They include such things as pavings, fencing and wall materials, furnishings, lighting, and water features. Like plant

Figure 3-8 Symbol styles for deciduous trees

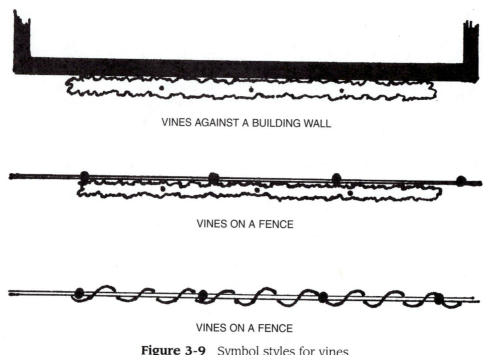

VINES AGAINST A BUILDING WALL

VINES ON A FENCE

VINES ON A FENCE

Figure 3-9 Symbol styles for vines

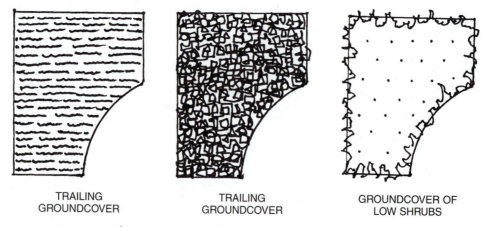

TRAILING
GROUNDCOVER

TRAILING
GROUNDCOVER

GROUNDCOVER OF
LOW SHRUBS

Figure 3-10 Symbol styles for groundcovers

symbols, hardscape symbols attempt to suggest how the materials or objects will actually appear in the landscape. They need to be scaled appropriately to allow the graphic to look realistic. When the scale of the plan is so small that it becomes impossible to draw each brick, stone, or similar feature to its exact size, then a textural interpretation may be used to suggest the hardscape rather than interpret it literally. Figure 3-11 illustrates typical hardscape features as represented in landscape plan views.

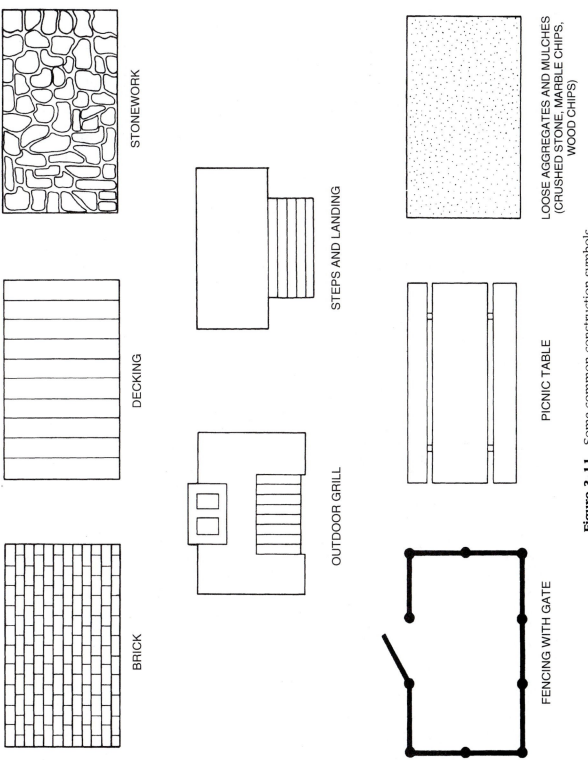

STONEWORK

DECKING

BRICK

STEPS AND LANDING

OUTDOOR GRILL

LOOSE AGGREGATES AND MULCHES
(CRUSHED STONE, MARBLE CHIPS,
WOOD CHIPS)

PICNIC TABLE

FENCING WITH GATE

Figure 3-11 Some common construction symbols

Explaining an Idea With Symbols

Once designers have learned to style symbols that others can interpret, they can begin using that skill to explain design concepts to others. While there are few hard-and-fast rules in graphics, over time certain techniques have been proven effective in assuring that the viewer is not misled or confused when interpreting the plan. Some of these techniques are more easily accomplished by hand drawing than by computer drawing, Figure 3-12.

- When one landscape object passes beneath another, the upper symbol is drawn with a wider, darker, and/or solid line, while the symbol for the object hidden beneath it is drawn with a thinner, lighter, and/or broken line. This is termed *line-weight variation*.

- Plants that are exactly alike and are massed will overlap slightly, and the area of overlap is not drawn. That permits a single label to identify all of the plants in that massed grouping.

- Plants that are not exactly alike, but are massed, are not overlapped. Each is drawn completely and separately, since separate labeling will be required.

Labeling

To make the landscape plan as useful and understandable as possible, it is usually necessary to add labeling and other notations to the plan. The goal is

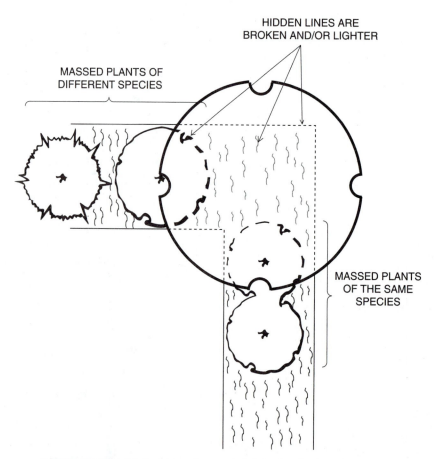

Figure 3-12 Techniques for explaining the designer's intent

to use only as much labeling as necessary to convey the required information, while assuring that the plan reads easily and quickly. Several types of lettered information should appear on the finished plan. Included are the symbol labels, the plant list, a directional arrow, the scale, the client's name and address, and the designer's name and firm.

Symbol Labeling

All of the graphic symbols used on the landscape plan must be labeled, and the labels must be on or as close to the symbols as possible. The closer the labels are to the symbols, the easier the plan will be to read. When the scale of the drawing results in very small symbols, it may be necessary to code the labels. That means that a number is used on or near the symbol that corresponds to a lettered label placed elsewhere on the plan, Figure 3-13. When more than one plant is being labeled, as in a grouping, the plant label or code number is followed by the number of plants in the grouping, Figure 3-14. By using one label for several plants or groupings of the same species, the number of labels can be minimized and clutter avoided. An excess of labeling lines (leader lines) can be confusing, often mistaken for object lines. They should be eliminated whenever possible.

Hardscape materials are labeled similarly to plant materials, but are seldom coded. There is no need to indicate specific numbers of bricks, stones, or boards. Instead, the feature to be constructed should be briefly described by such factors as height, color, or special treatment. For example, a label might read "Vertical louvered fence–5' tall/grey stain." Another might be "Concrete with exposed aggregate surface."

Plant Lists

From the plan's symbols and labels, a tally can be made of how many of each species of plant are required in the total plan. The tally permits the development of a **plant list**, Figure 3-15. An important component of the layout sheet, the plant list always includes the first two items (in three columns) and frequently includes the last two:

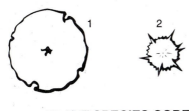

PLANT SPECIES CODE

| #1 | COMMON LILAC | SYRINGA VULGARIS |
| #2 | CREEPING JUNIPER | JUNIPERUS HORIZONTALS |

Figure 3-13 Coding the labeling of plant symbols is less desirable than direct labeling, but is sometimes necessary. Try to avoid using direct labeling and coding on the same plan whenever possible.

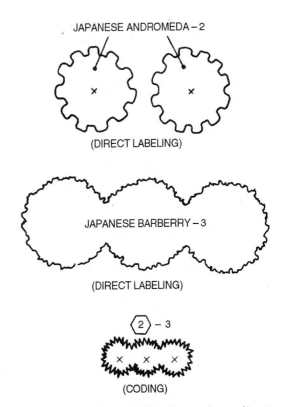

Figure 3-14 When the label or code applies to more than one plant of that species, the number should be indicated.

- Two listings of the plant species; one column listing them in alphabetical order by botanical names and an accompanying column listing the common names

- A column noting the total numbers of each plant species used

- A column noting the sizes of the plants at the time of installation

- A column containing special notations about certain species that the designer does not want overlooked

PLANT LIST

BOTANICAL NAME	COMMON NAME	QTY	SIZE	COMMENT
Acer ginnala	Amur Maple	8	6'	B&B
Betula lenta	Sweet Birch	2	8'	B&B/Spring Inst.
Carpinus betulus	Eur. Hornbeam	14	6'	B&B/Matched
Deutzia gracilis	Slender Deutzia	25	18"	Containers
Euonymus alatus	Winged Euonymus	25	3'	B&B
Ilex glabra	Inkberry	19	3'	BR
Magnolia stellata	Star Magnolia	3	4'6"	B&B/Tagged
Pinus nigra	Austrian Pine	8	8'	B&B

Figure 3-15 A typical plant list. It is one of the components of the final layout.

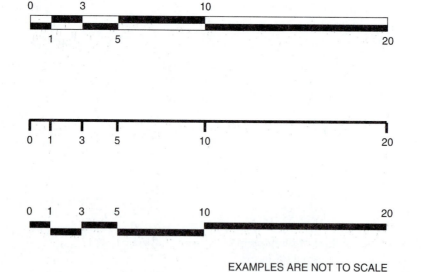

EXAMPLES ARE NOT TO SCALE

Figure 3-16 Examples of scale indicators

Scale Indicator and Directional Arrow

To permit the landscape plan to be measured and eventually installed, the scale to which it was drawn must be known and noted on the layout sheet. Figure 3-16 illustrates typical scale indicators.

Also important to full comprehension of the design is an awareness of its directional orientation. That is provided by the directional arrow, Figure 3-17. While it only needs to be accurate in order to be functional, it is traditional for designers to make the directional arrow somewhat ornate. Some firms copyright their arrows as one way of lending distinction to their graphic presentations. The arrows should be neither too ornate nor too large, but they can be fun to design.

Client and Designer Names

The inclusion of the names of the clients and the designer is important. Not just words, these labels should be regarded as important components of the plan layout. The client's name usually includes the address of the project and may include addi-tional descriptive text. It should be done using the largest and/or most ornate lettering on the plan. It is important to the customization of the plan and should be prominently featured.

The designer's name may include additional information, such as the firm name, address, and telephone number. It should not be formed with letters as large as those used for the client's name, and it should be located away from the client's name on the layout sheet.

Design Sheet Layout

There is no single correct way to arrange the design and various lettering components on the sheet. Certainly the design is the most important component and needs to be featured. The other components will vary greatly in size and prominence from layout to layout.

Traditionally, landscape designers do their actual designing on opaque paper. The paper may be gridded or plain. Then a sheet of thin, transparent tracing paper called **vellum** is placed over it and the entire work is traced, Figure 3-18. At the same time, the plant list and other labeling components

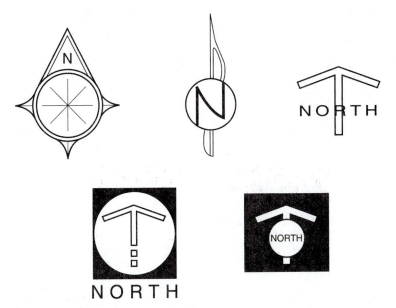

Figure 3-17 Examples of directional arrows

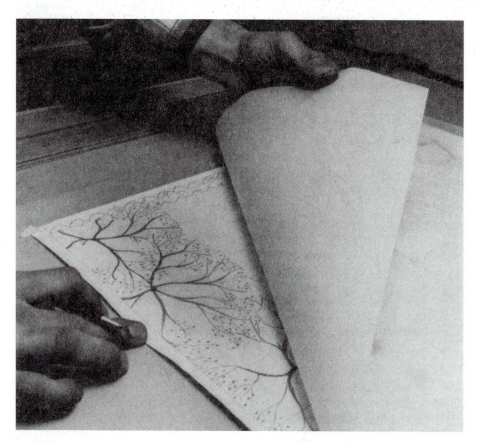

Figure 3-18 Vellum is placed over the original design for tracing.

are added, either by tracing or by the use of waxed press-on letters or transfer film. At this stage in the development of the layout the designer must decide what to place where on the sheet. If a computer creates the layout, the same decisions must be made. While use of the computer negates the need to trace the plan onto vellum, the need still remains for the designer to decide where to place the various components and what prominence to assign to each. Figures 3-19, 3-20, and 3-21 illustrate different sheet layouts. While different, there are some similarities:

- The design is placed on the left side or at the top of the sheet.

- The plant list is sized to help counterbalance the design on the sheet.

- The client's name and address block, with the largest lettering, is placed near the bottom to give stability to the layout.

- The scale indicator and directional arrow are placed near the design and located to balance the white space on the sheet.

- The designer's name and/or firm name are placed where they can help balance the white space.

NOTE: Some firms will group all of the labels together in a box, which is repeated on each sheet of a multiple-sheet design proposal.

- An imaginary line drawn down the center of the sheet would find the graphic weight equal on each side of center.

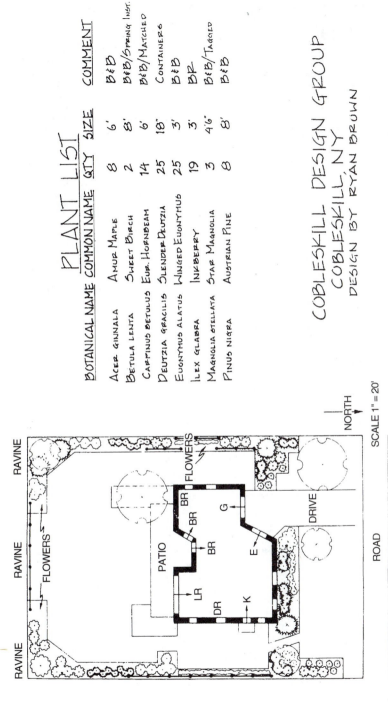

PLANT LIST

BOTANICAL NAME	COMMON NAME	QTY	SIZE	COMMENT
Acer ginnala	Amur Maple	8	6'	B&B
Betula lenta	Sweet Birch	2	8'	B&B/Spring Inst.
Carpinus betulus	Eur. Hornbeam	14	6'	B&B/Matched
Deutzia gracilis	Slender Deutzia	25	18"	Containers
Euonymus alatus	Winged Euonymus	25	3'	B&B
Ilex glabra	Inkberry	19	3'	BR
Magnolia stellata	Star Magnolia	3	4'6"	B&B/Tagged
Pinus nigra	Austrian Pine	8	8'	B&B

COBLESKILL DESIGN GROUP
COBLESKILL, NY
DESIGN BY RYAN BROWN

DESIGN AND PLANTING PLAN FOR
THE WHITESTONE RESIDENCE
DUANESBURG, NEW YORK

SCALE 1" = 20'

Figure 3-19 A typical plan view sheet layout

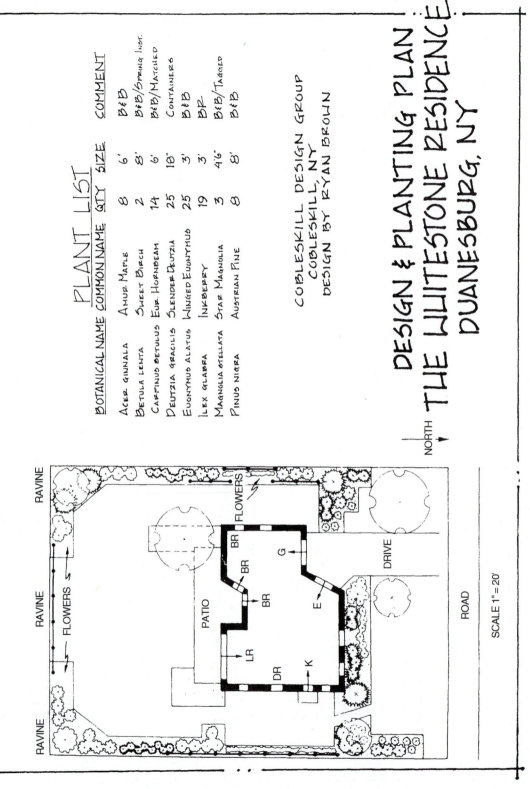

PLANT LIST

BOTANICAL NAME	COMMON NAME	QTY	SIZE	COMMENT
Acer Ginnala	Amur Maple	8	6'	B&B
Betula lenta	Sweet Birch	2	8'	B&B/Spring Inst.
Carpinus betulus	Eur. Hornbeam	14	6'	B&B/Matched
Deutzia gracilis	Slender Deutzia	25	18"	Containers
Euonymus Alatus	Winged Euonymus	25	3'	B&B
Ilex glabra	Inkberry	19	3'	BR
Magnolia stellata	Star Magnolia	3	4'6"	B&B/Tagged
Pinus nigra	Austrian Pine	8	8'	B&B

COBLESKILL DESIGN GROUP
COBLESKILL, NY
DESIGN BY RYAN BROWN

NORTH

DESIGN & PLANTING PLAN
THE WHITESTONE RESIDENCE
DUANESBURG, NY

RAVINE
RAVINE
RAVINE
FLOWERS
FLOWERS
BR
BR
BR
G
E
PATIO
LR
DR
K
DRIVE
ROAD
RAVINE

SCALE 1" = 20'

Figure 3-20 A variation on the horizontal sheet layout

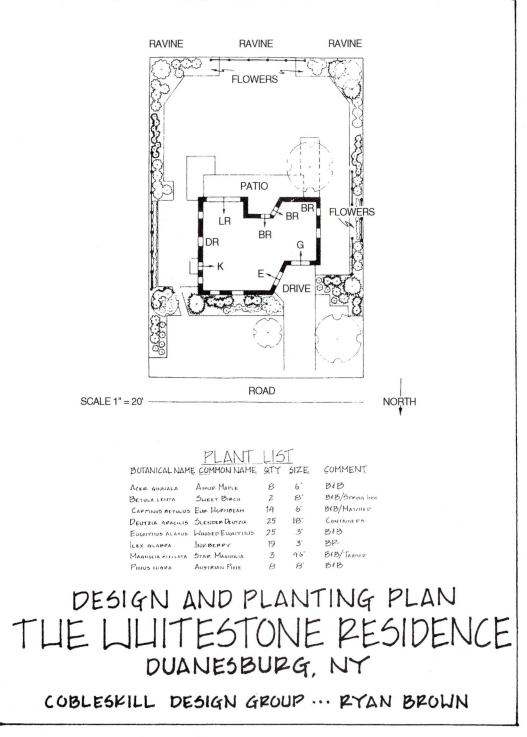

RAVINE · RAVINE · RAVINE

FLOWERS

PATIO

LR · BR · BR

FLOWERS

DR · BR

K · G

E

DRIVE

ROAD

SCALE 1" = 20' · NORTH

PLANT LIST

BOTANICAL NAME	COMMON NAME	QTY	SIZE	COMMENT
Acer ginnala	Amur Maple	8	6'	B&B
Betula lenta	Sweet Birch	2	8'	B&B/Strong limb
Carpinus betulus	Eur. Hornbeam	14	6'	B&B/Matched
Deutzia gracilis	Slender Deutzia	25	18"	Containers
Euonymus alatus	Winged Euonymus	25	3'	B&B
Ilex glabra	Inkberry	19	3'	BR
Magnolia stellata	Star Magnolia	3	4-6'	B&B/Tagged
Pinus nigra	Austrian Pine	8	8'	B&B

DESIGN AND PLANTING PLAN
THE WHITESTONE RESIDENCE
DUANESBURG, NY

COBLESKILL DESIGN GROUP ··· RYAN BROWN

Figure 3-21 A vertical layout of the same design. This is less common because it is sometimes awkward to unroll and read in the field.

■ A border line is often drawn around the sheet to tie together all of the components on the sheet.

Different Drawings for Different Requirements

Graphic presentations range from quick sketches to elaborate, detailed drawings. Some emphasize design concepts, while others explain construction methods. Some are seen by no one but the designer, while others appear in full-color brochures distributed to the public. It is foolish to spend more time on a graphic presentation than is warranted by the importance or value of the project. Figures 3-22, 3-23, and 3-24 illustrate a

complete sheet layout, a perspective, and a construction detail drawing, respectively. All have required precision by the designer, and all are intended for viewing and/or use by someone other than the designer. They justify the time required for their preparation. Figure 3-25 illustrates two common elements of landscape graphics: a plan view plant symbol and a human figure. Each is shown as it might be interpreted in a quick sketch, a simple presentation plan, and a refined presentation for a large project. The graphic style used should be consistent throughout the piece of work. If the building and plants in a drawing are being done with elaborate detail, it would be inappropriate to insert a human figure done as a loose sketch. The use of color is also an arbitrary decision that the designer must make. The first level of the decision deals

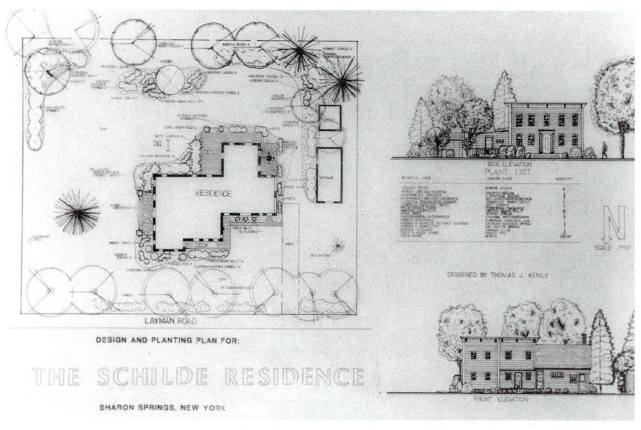

Figure 3-22 A complete presentation drawing

Figure 3-23 A perspective view can be time-consuming to prepare if done by hand and in this detail, but it provides a client with the best understanding of what the designer is proposing.

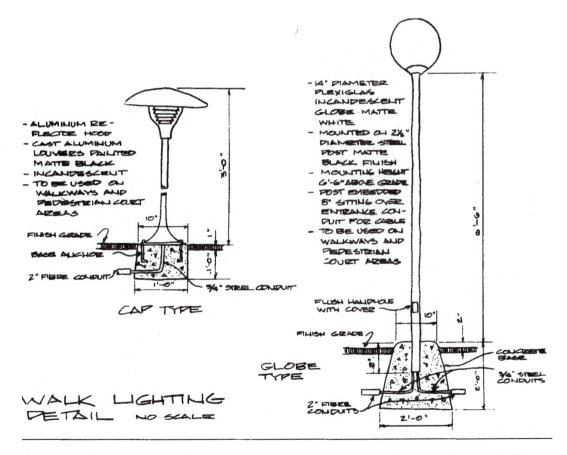

Figure 3-24 A detail drawing for walk lights (Courtesy James Glavin, Landscape Architect, Cranberry Lake, NY)

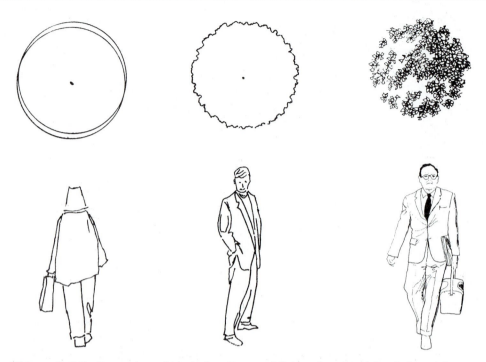

Figure 3-25 Examples of symbol styling and figure drawings for (left) quick sketches, (center) simple presentations, and (right) refined presentations

with whether to use color at all. The second level of the decision concerns what medium to use, since some media are faster or easier to apply than others. Familiarity with the media of the graphic artist is important for landscape architects and many landscape designers.

Graphic Media

The media of graphic art fall into two broad categories: those used to draw and delineate and those used to render and interpret. There is no sharp line of distinction between these two categories, since some media can be used either way.

Graphite Pencil

Graphite pencils were described as drawing instruments previously. Their center cores of graphite (lustrous carbon) are manufactured with varying degrees of hardness. Firmer leads have high H numbers and are more suitable for drawing than rendering. Softer leads, with high B numbers, are better suited for sketching and creating tonal variations needed to render drawings realistically.

Softer leads (such as 4B to 6B) can create a thin, fine line if kept sharpened, but sharpening is needed frequently. If chiseled to a wedge, soft leads will create wider, softer-appearing lines. Flat sketch pencils cannot be sharpened to a point, but are excellent for making wide strokes when tones of grey are needed, Figure 3-26. The apparent softness of graphite tones is also influenced by the drawing paper. Hard-surfaced papers will make the graphite tone appear even and uniform. Softer, textured papers will introduce tonal variation. Thin papers that permit light to pass through (translucent) will cause graphite tones to appear weaker than they would if used on heavier paper that light cannot penetrate (opaque).

Pencil has long found favor with landscape designers because it is the most versatile medium, usable for both drawing and rendering. It is also the

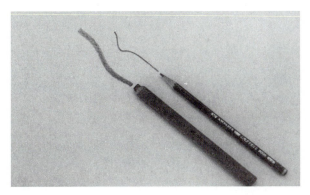

Figure 3-26 Soft leads with chiseled tips and flat sketch pencils can make wide graphite lines.

least expensive, most commonly available, and least intimidating of the art media. Used alone or in combination with other media, the graphite pencil is capable of tonal subtleties that are difficult to represent with other media.

In switching from writing use to sketching or rendering use, the pencil needs to be held differ-ently by the designer. As held for writing, the pencil is controlled mainly by the fingers. For graphic use, the pencil needs to lie lower in the hand, thereby permitting greater wrist action with less dependency on the fingers, Figure 3-27. As the angle of the pencil changes, the width and the crispness of the lines change, becoming wider and more fuzzy as more of the lead's edge makes contact with the paper. Also, by allowing a soft graphite lead to rotate slightly and/or by varying the pressure applied to the pencil, the designer is able to introduce additional interest into a pencil line. The resulting width and tonal variations within a single line suggest that the line is turning, advancing, or receding from the surface. That is in contrast to the static appearance of a straight-edge-supported line. Both types of lines have their uses, and the graphite lead is capable of producing both. A pencil sketch commonly uses both types of lines plus a range of tones, all resulting from skillful use of the inexpensive graphite drawing pencil, Figure 3-28.

Figure 3-27 The drawing pencil in two common positions: Left, for rendering, and Right, for lettering

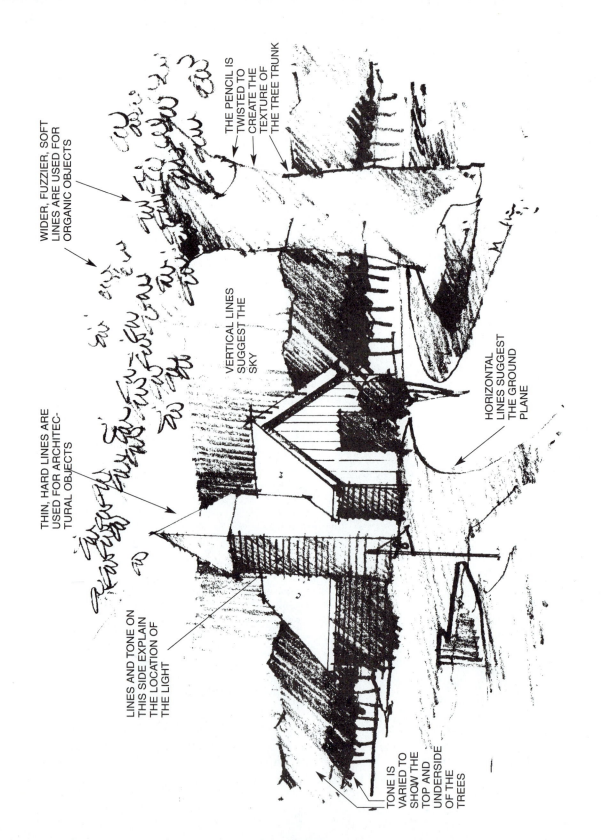

WIDER, FUZZIER, SOFT LINES ARE USED FOR ORGANIC OBJECTS

THE PENCIL IS TWISTED TO CREATE THE TEXTURE OF THE TREE TRUNK

VERTICAL LINES SUGGEST THE SKY

HORIZONTAL LINES SUGGEST THE GROUND PLANE

THIN, HARD LINES ARE USED FOR ARCHITECTURAL OBJECTS

LINES AND TONE ON THIS SIDE EXPLAIN THE LOCATION OF THE LIGHT

TONE IS VARIED TO SHOW THE TOP AND UNDERSIDE OF THE TREES

Figure 3-28 Pencil sketches use the full range of graphite capabilities. (Courtesy of Mike Lin, The Mike Lin Graphic Workshop, Manhattan, Kansas)

Colored Pencil

Several brands of colored pencils are available from art supply stores. Prismacolors (reg. trademark) are perhaps the best known and enjoy the widest use. Colored pencils are wax based and sold in sets that can include as many as 120 different colors. They are available in traditional pencil form, with round color cores enclosed in wood, also with thinner color cores enclosed in wood, and as long, square solid color sticks with no wood. The pencils are best sharpened with a pocketknife or single-edged razor blade and pointed with a sandpaper pad or knife, since their soft wax core is quickly chewed up by conventional graphite pencil sharpeners.

With little exception, colored pencils are used for the rendering of design plans, elevations, and perspective illustrations, rather than actual creation of designs. While designers may find them helpful for quick design studies, their principal value is to add color excitement to works that are to be used to sell the design to clients. Both beginners and experienced designers appreciate colored pencils because they allow the designer maximum control in the application of the color. Unlike markers that create brilliant color the instant they touch the paper, colored pencils can be applied slowly and lightly, allowing the designer to build color intensity to the desired level.

Colored pencils can be blended by overlaying different colors. As long as the designer is careful to apply each layer lightly so that light can penetrate it, the colors will mix, resulting in different tones and frequently in new colors. Wise designers carefully note and record which pencils were used to create desirable mixes so that they can duplicate the effect at a later time, perhaps on a different piece of work.

To gain the smoothest application of colored pencils, they should be applied uniformly in a single direction, much like a paint roller is used when painting a wall. Additional colors or layers should be applied in the same way, but at 90 degrees to the previous direction of application.

When smoothness is not the objective, colored pencils can create textures in numerous ways. With the point sharpened, textures can be created using

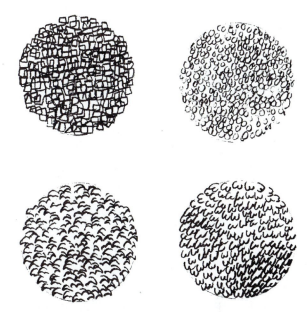

Figure 3-29 Colored pencil textures. Top left: The line creates endlessly repeating box shapes. Top right: The texture is formed from tiny circles. Lower left: The line is like a flock of birds in flight. Lower right: The line creates repeating W's.

short, repetitive strokes. While the lack of color in this text makes it impossible to illustrate the color mixing features of colored pencils, Figure 3-29 does illustrate various textures created with colored pencils. These textures can be used to render plant materials, hardscape surfaces, clothing, and many other features in landscape illustrations. Colored pencil works effectively to add color and textures when used alone or when used in combination with other media. It is wise to apply a spray fixative over the completed rendering to prevent it from smearing.

Felt Marker Pens

These instruments are fine-point markers that deliver lines similar to those created by technical pens, yet marker pens are disposable and much less expensive. Line widths vary from medium to extra fine. The inks used may be water soluble, water resistant, or waterproof, depending upon the manufacturer. All dry instantly upon contact with

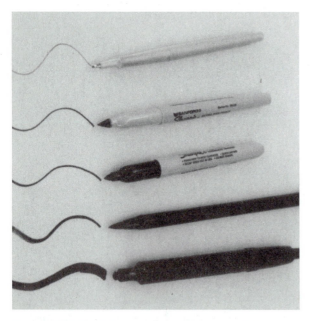

Figure 3-30 Felt pens can create varied line widths.

Figure 3-31 Different types of marker nibs. Each creates lines of different widths.

the surface, and most are applicable to a variety of drawing surfaces, Figure 3-30. Felt pens are gradually replacing technical pens as the instrument of choice for creating plans and other drawings that require ink. They are good for lettering, bordering, outlining, and symbol styling. They can also be used for sketching and texturing, although the required skill level of the artist is greater since erasing is nearly impossible.

Color Markers

Technically, the tool being described here is just another felt marker. However, color markers are used differently from the point-tipped instruments described as marker pens.

Color markers contain a reservoir of pigment that is liquified in assorted solvents that vary between manufacturers. They are available in a wide range of colors, tips, and qualities. It is important for the beginning landscape designer to realize that the inexpensive color markers used by children are unsatisfactory for use in professional ren-

dering. Here is one case where the higher price does translate into better quality. Design, Berol, Faber Castell, and Pantone are among the most popular markers used for landscape rendering today, and there are many other equally good brands. The markers are available with a variety of tips (nibs) that include broad chisel tips, pointed and ultra-fine pointed tips, and flexible foam tips suggestive of a paint brush, Figure 3-31. Some markers are available with two different nibs, one at each end of the same marker. The color selections are extensive, but there are great inconsistencies between manufacturers. Though called by the same name, a marker color from two different manufacturers can be two different shades. That sometimes makes it difficult to mix marker brands. Also, differences in the pigment solvents can make it difficult to use different brands of markers for a single rendering.

The first-time use of color markers can be a bit intimidating. Unlike colored pencils, which can be applied lightly and, if necessary, erased, markers soak in, dry instantly, and defy erasure. Color markers are a wet medium, with similarities to water colors, leaving novice users with the same feeling of reduced control over the outcome of the rendering. They require some practice and adhering to some basic rules of application in order to apply them effectively, but it is worth the effort. Used incorrectly, marker renderings are streaked, splotchy, gaudy, and amateurish. Used correctly, markers bring brilliance and realism to renderings that no other media can create as quickly.

Color markers can be used on all drawing surfaces common to landscape illustration. However the resulting color and the level of brilliance are directly related to the color of the paper, its degree of translucency or opaqueness, and the type of light being cast upon the illustration, Figure 3-32. In general, this could be said of all color media, not just markers.

Most landscape renderings rely upon the broad-nib color marker as the workhorse, with the finer pointed markers used as supplements. It is important to understand how to use broad-nib markers in order to maximize their use and longevity. The markers should be used with only an edge touching the paper, not the full stub end. The chisel tip offers the option of making thin or broad lines, depending upon which way it is drawn across the paper, Figure 3-33. Like colored pencils, markers should be applied evenly in a single direction across the surface to reduce streak-

ing. A straightedge may be used for support, but it is better to use something disposable, such as heavy cardboard or chipboard, and change it frequently. Plastic triangles are not satisfactory because they eventually begin to leave streaks of past colors on the new drawings. Also, the solvents in some markers will dissolve the plastic edges and make the triangles unusable later for line work with pencils and pens.

Color markers should be used with the lightest colors applied first, followed gradually by darker colors and darker tones. Like colored pencils, color markers can be layered to create darker tones of the same color or new colors from mixing different colors. A colorless blender, which is a marker containing only solvent, can dilute a marker color applied over it while both are still wet. The result will be a lighter tone of the color than would otherwise be produced. Also, colored pencils can be applied over markers to change colors and to lighten or darken the marker colors. Many mistakes made with markers can be hidden by applying colored pencils over the marker. Pencils also work well to add textures to marker surfaces. In landscape graphics, the combined use of colored pencils and color markers is more common than the exclusive use of either.

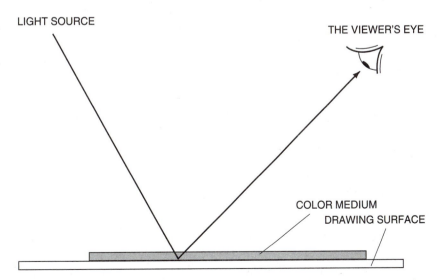

Figure 3-32 The three factors in determining how the eye perceives color are the quality of light, the choice of medium, and the paper selected.

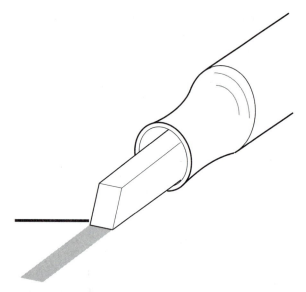

Figure 3-33 The broad-nib marker can make wide or thin lines depending upon the edge used and the direction of movement.

Figure 3-34 Pastel squares can create both wide and thin lines.

Pastels

Pastels are chalky crayons made of powdered pigment. They are available in a variety of colors, in round or square sticks. Some have an oil additive that reduces some of the chalky quality. For landscape renderings, the best pastels are square sticks without an oil base. The messiness of chalky pastels is exactly what makes them useful for landscape graphics. They allow the designer to add color quickly to large areas such as lawns, skies, and bodies of water, where the shape of the areas is indeterminate and a subtle, blended color effect is desired.

Application techniques for pastels vary. Using a square stick rather than a round one provides the designer with some control when a line is desired rather than a tone. By breaking the pastel stick into two uneven pieces and applying the color with the edge of the stick, not the end, varied line widths can be drawn, Figure 3-34. Another application technique takes advantage of the tendency of pastels to smear by scraping off particles of pigment from the stick, then using fingers or paper blending stumps to create a smooth texture in the area desired, Figure 3-35. Once applied, pastels require

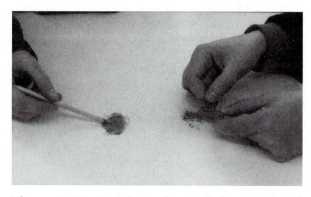

Figure 3-35 Pastels can be applied as powdered pigment by scraping with a coin and then blending with a paper stump or finger.

the application of a spray fixative to prevent unwanted smearing.

Pastels are comparatively inexpensive and require some practice to use effectively. Minimal application is always better than heavy-handed application. Pastels can be applied over dried marker colors and over colored pencil, but pencils and especially markers do not work well when applied over pastels.

Pressure Graphics

A vast assortment of material is available to the landscape graphic artist who wishes to speed up the development of a presentation while maintaining a professional appearance. Some of the materials have been described in earlier chapters, such as waxed press-on letters, lettering machines, and transfer film. To that list can be added printed border tapes, dry transfer sheets of plant symbols, cars, and people, as well as sheets of patterns and tones that can be used to create textures or shades on the surfaces of landscape features. Rubbed off from sheets with a pencil or stylus much like the press-on letters, which are also a dry transfer product, pressure graphics are a fast, though somewhat expensive, means of instantly enhancing a landscape presentation. Extensive use of pressure graphics is usually reserved for presentation designs.

Foam Board

For the presentation of designs to clients and in any other situation where it becomes necessary to mount drawings temporarily or permanently, foam board is frequently the mounting board of choice. It is a lightweight but sturdy board composed of two sheets of thin cardboard permanently bonded to a central sheet of foam board. Drawings can be mounted using glue, dry wax, pins, or other devices. Unless the mounting is permanent, the foam board sheets can be stored and used again for other presentations. Foam board is also used in landscape model construction. A sharp knife, metal straightedge, and protected cutting surface are needed to cut foam board properly.

Drawing Papers and Films

Many different drawing and copy surfaces are available for use by landscape designers. The choice of a surface depends on one or more of the following factors:

a. Whether the drawing is temporary or permanent

b. What drawing instrument is being used

Quadrille Paper

- Soft-surface ruled paper divided into 4, 5, 8, or 10 squares per inch.
- Inexpensive. It is available in pads or large sheets.
- Used for plan view designs, usually in first-draft form. Simplifies drawing to scale when selected to match with the drawing's scale.
- Most often used with drawing pencils.
- Opaque.

Newsprint Paper

- Rough surface. Absorbent.
- Inexpensive. It is available in pads or large sheets.
- Useful for sketching and preliminary design work.
- Accepts soft pencil and pastel. Markers bleed.
- Opaque.

Sketch Paper

- Smooth, hard surface. Less absorbent.
- Top brands such as Strathmore can be expensive.
- Available in pads or sheets.
- Useful for original drawings.
- Accepts most drawing and rendering media with very little bleeding. Markers will soak through.
- Opaque.

Marker Paper

- Smooth, hard surface. Little to no absorbency.
- Comparatively expensive. Available in pads.
- Used for marker renderings. Does not bleed or soak through.
- Also accepts other drawing and rendering media.
- Semi-translucent. May be used with laser printers and copiers.

Vellum

- Smooth, hard surface.
- Available in weights from light to heavy. Lightweight vellum is commonly used for sketching. Heavier weights are used for tracing and for making copies. Price varies by weight and size of roll.
- Accepts drawing pencils and pens.
- Transparent to translucent, depending upon weight.

Mylar Film

- Polyester drafting film in .003 and .005 gauge.
- Comparatively expensive. Available in sheets and rolls.
- Used for pencil and pen drawings where durability of the drawing is important or copies are required.
- Matte finished on one or both sides.
- Translucent.

c. Whether the drawing will be copied and, if so, with what process

d. Whether the drawing will be rendered and, if so, with what rendering media

Some of the papers used most commonly for landscape illustration are described below. Many other specialized art papers may be selected, but these are the ones most frequently used.

Print Papers

Frequently the paper that the design is initially drawn on is not the paper used for the finished work. While it may be best to draw on a paper or film that accepts pencil or pen, it may be preferable to transfer the drawing to different paper for rendering. For some projects, it may be necessary to reproduce the original for use by other parties. Sometimes dozens of copies are required as a com-

plex project moves through the design and approval process.

Two duplication processes are commonly used to reproduce landscape drawings in large numbers: the **diazo process** and the **photocopy process**. Both processes make direct positive copies, unlike the old blueprint process that made negative copies. The diazo process requires a specially treated paper, ultraviolet light, and an ammonia vapor developer to make copies, Figure 3-36. The photocopier operates much like any office copier. It works best with high-quality bond paper and uses black toner to reproduce the image, Figure 3-37. Each process has some advantages and some limitations. The diazo process is less expensive and less prone to problems common to photocopiers, such as streaking, jamming, or depleted toner. However, the diazo process necessitates periodic changing of the ammonia to maintain its strength,

Figure 3-36 The Diazo Ozalid copy process requires specially treated paper and ammonia (contained in the bottle) to make a copy. The original drawing must be on translucent paper.

Figure 3-37 The photocopier works much like an ordinary office copier. It requires toner to reproduce black line copies. The original drawing need not be on translucent paper.

and it is unusable if the special paper is out of stock or has been stored too long or exposed to light (it is light sensitive). In general the photocopy process requires a greater initial financial investment, ongoing maintenance contracts, and periodic service calls. Diazo copiers usually give longer periods of uninterrupted service. Photocopies are somewhat sharper copies than diazo copies. Some marker solvents will dissolve the toner lines of the photocopies, so caution is necessary if doing marker renderings on photocopies.

The Layout of Graphic Presentations

Much of what was described earlier in this chapter about the layout of a design presentation applies to the layout of any graphic presentation. The various drawings, lists, enlarged views, name blocks, and informational items are the components of the layout. Some are more important than others and should be featured prominently. All have shapes,

usually some type of rectangular shape, although components such as directional arrows or even property lines may not be rectilinear. In deciding where components should be placed within the total presentation, an attempt should be made to arrange the components so that they complement the rectilinearity of the other components and collectively create a larger rectangle. Figure 3-38 illustrates a presentation composed of several different components. Notice how the long side of each component parallels the long side of the presentation layout, and the short sides of the components and the layout sheet are also parallel. The outside edges of the components are kept in alignment so that the viewer's eye connects them, and the rectangle is again repeated.

Parallel lines and shapes are an important part of the presentation layout, and so is the ratio between the length of the layout sheet and its width. Layout sheets that are long and thin or perfectly square do not work well. The preferred shape for a layout presentation is a rectangle that is two-thirds as wide as it is long. For example, a sheet

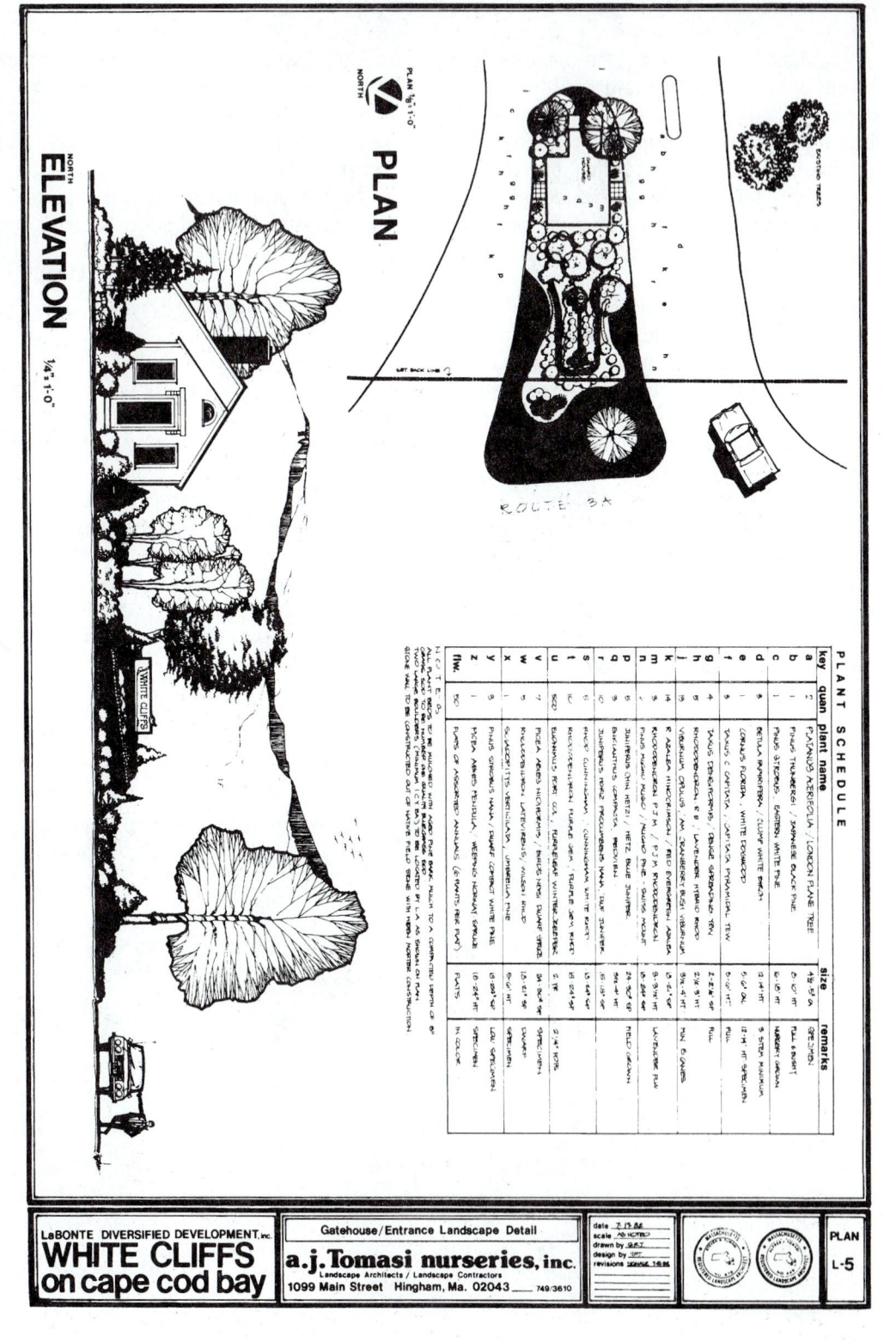

Figure 3-38 Careful balance of the white space around the components is important to proper sheet layout. (Courtesy of A.J. Tomasi Nurseries, Inc., Pembroke, MA)

that is 18 inches wide should be 27 inches long in order to create the desired ratio. Another example would be a sheet that is 24 inches × 36 inches. Some designers favor a slightly different ratio that gives a little more length to the layout. Their preferred width to length ratio is 3 to 5, such as 18 inches × 30 inches or 36 inches × 60 inches.

The third element of a successful layout presentation is the balance of white space between the components. Equally as important as the components to the success of the layout, white space must be evenly distributed through the presentation. It need not occupy as much area as the components, but it should be evenly distributed throughout the layout. Figure 3-38 illustrates an even distribution of white space.

Practice Exercises

A. On a sheet of drawing paper, use a compass or circle template, a scale, a straightedge, and a graphite pencil to create the symbols described below. Copy the examples illustrated in this chapter.

1. A needled evergreen tree that is 20 feet wide, drawn to the scale of 1" = 10'.
2. A needled evergreen shrub that is 5 feet wide, drawn to the scale of 1" = 5'.
3. A broadleaved evergreen shrub that is 8 feet wide, drawn to the scale of 1"= 4'.
4. A deciduous shrub that is 10 feet wide, drawn to the scale of 1"= 20'.
5. A deciduous tree that is 30 feet wide, drawn to the scale of 1"= 10'.
6. Three needled evergreens of the same species, each 10 feet wide, massed and drawn to the scale of 1"= 10'.
7. Three deciduous shrubs, each 10 feet wide and each a different species, massed and drawn to the scale of 1"= 10'.
8. A span of fencing 15 feet long with posts every 5 feet, drawn to the scale of 1"= 5'. Growing along it is a deciduous vine that spreads 10 feet wide. Draw the vine to the same scale as the fence.
9. A deciduous tree that is 25 feet wide, drawn to the scale of 1"= 10'. Beneath the tree are three 10-feet-wide broadleaved evergreen shrubs, drawn to the same scale. Position them so that part of each shrub is hidden by the tree.
10. Go back and label each plant or group of plants in the previous nine exercises. Make the lettering 1/8-inch tall. Use guidelines, straightedges, and the lettering style of your choice. Remember that when there is more than one plant to which the label applies, the number of plants becomes part of the label. The plant names to be used are as follows:

   In #1: Colorado Spruce
   In #2: Savin's Juniper
   In #3: Carolina Rhododendron
   In #4: Showy Border Forsythia
   In #5: Silver Maple
   In #6: Globe Arborvitae
   In #7: Vanhoutte Spirea, Beautybush, and Ninebark
   In #8: Shadow fencing, 5', and Boston Ivy
   In #9: Flowering Dogwood and Japanese Andromeda

B. Draw scale indicators for the following: 1"= 5', 1"= 10', 1"= 20'.

C. Draw four different directional arrows to indicate where north is located on a plan view. Make each 2 inches long. Use pencil for two of them and felt pen for two of them.

D. Prepare a client name and address component. Use a five-guideline lettering style and straightedges for line support. Make light guidelines with a 3H or 4H pencil. Do the letters in 2H or H pencil. Use the following wording and letter sizes.

DESIGN AND PLANTING PLAN FOR	[1/2" letters]
MR. AND MRS. JOHN DOE	[1" letters]
1234 MAIN STREET	[1/2" letters]
HEARTLAND, INDIANA	[1/2" letters]

E. Practice applying different rendering media to develop skill in both smooth and textural application. Apply large swatches of graphite and color tones to sketch paper or white drawing paper and practice until the following techniques are mastered:

1. Smooth, consistent, streakless tones
2. Smooth application with gradual value change from very light to very dark
3. Lighten a dark Prismacolor tone with an overlay of white pencil
4. Lighten a dark color marker tone with an overlay of white pencil
5. Create a grainy texture with colored pencil applied over a wash of smooth colored pencil
6. Create an architectural texture with colored pencil applied over a wash of color marker

Achievement Review

A. Indicate if the following characteristics describe

    a. Plan views
    b. Elevations
    c. Perspectives
    d. Axonometrics

1. Measurable drawing that allows comparison of building heights with heights of nearby plants and other features.
2. Measurable drawing showing the spacing between all features used in the landscape design.
3. Measurable drawing displaying multiple sides.
4. The vantage point of the viewer is directly above the drawing.
5. Receding parallel lines are not made to converge.
6. Receding parallel lines appear to be converging.
7. The most realistic view.
8. The dimensions are not measurable.
9. Two measurable dimensions: horizontal and vertical.
10. All measurable dimensions are horizontal.

B. Select the best answer to the following questions.

1. Which of the following is frequently used as both a drawing tool and a rendering tool?
    a. graphite pencil
    b. colored pencil
    c. color marker
    d. pastels

2. Which of the following is applied wet?
    a. colored pencil
    b. pastels
    c. color marker
    d. waxed press-on letters

3. Which of the following is the softest and messiest to apply?
    a. graphite pencil
    b. colored pencil
    c. color marker
    d. pastels

4. Which of the following is most costly per unit?
    a. graphite pencil
    b. colored pencil

    c. color marker

    d. pastels

5. Which of the following is used for finished presentation drawings?
   a. mylar film
   b. bond print paper
   c. newsprint
   d. quadrille paper

6. Which of the following is used for tracings?
   a. vellum
   b. diazo paper
   c. quadrille paper
   d. sketch paper

7. Which of the following is used to mount and display presentation drawings?
   a. diazo paper
   b. marker paper
   c. bond print paper
   d. foam board

8. Which of the following is non-absorbent?
   a. newsprint
   b. marker paper
   c. sketch paper
   d. quadrille paper

9. Which is the most durable drawing surface?
   a. marker paper
   b. vellum
   c. mylar film
   d. diazo paper

10. Which of the following has a chemically treated surface that must be developed in order to see the drawing?
    a. marker paper
    b. foam core board
    c. mylar film
    d. diazo paper

CHAPTER 4

The Site

Objectives:

Upon completion of this chapter, you should be able to

- define the word *site* and explain its significance in the development of a landscape
- list the typical features that must be evaluated on most sites
- describe sources of site information
- explain geographic information systems
- define the following terms: setback, zoning regulations, property lines, right-of-way, easement, and zero lot line
- describe the limitations that the terrain imposes upon human activities
- understand the basic concepts of land grading
- describe how to prepare a base map

The word **site** refers to a piece of land that has the potential for development. Sites come in all sizes and shapes and serve a myriad of purposes from residential to commercial, institutional, recreational, and beyond. While there may be a striking similarity between sites, one thing is definite: no two sites are identical. The dissimilarity may be macroscopic or microscopic; it may be cultural, physical, or natural; it may be above ground or below ground. Regardless of what the differences are, they are real. Each site has its own personality that must be identified by a designer or anyone whose work will impact the site. The site is the single most important determinant of what the landscape can and cannot do. It often holds secrets that must be discovered before development can begin. It is the stage upon which the landscape is set, so it must be inventoried, analyzed, plotted, and fully understood if its use is to be totally optimized.

Site Features and Characteristics

The first step in becoming acquainted with a site is to take an inventory of what is there. While the property owners can be an important source of information about the site, it is best for the designer to make the first visit to the site without the owner or client tagging along. It is too easy for the designer to begin seeing the site through the eyes of the owner and thereby assimilating his or her opinions and biases about the site. The designer needs to remain apart from the preconceptions of the site's potential and instead have an organized method to assess the many and varied nuances of each site.

Separating the features and characteristics of a site into categories is a logical place to begin. Some of the site's characteristics are *natural features*, while others are *man-made*. Still others are *cultural* (resulting from or associated with human society). Some features are totally *physical*, while others are most significant as *visual* features. Some features are unmistakably positive factors, and others are definitely negative in their impact. Many have a neutral quality until they are judged in the context of the proposed design.

A partial list of the features and characteristics that are inventoried and later evaluated during the site analysis includes the following.

Natural Features

a. terrain (rise and fall of the land)
b. topography (the record of an area's terrain)
c. slopes (their steepness as measured at different locales within the site)
d. erosion (both present and potential areas on the site)
e. directions of surface water drainage
f. areas of puddling or drought
g. soil qualities (pH, nutrient level, stoniness, depth of the topsoil, texture)
h. existing plant materials (quantity, quality, species names, sizes, locations)
i. microclimates (protected or exposed area, where plant growth may be affected)
j. prevailing winds
k. annual rainfall and snowfall on the site
l. depth of the frost line
m. off-site views

Man-made Features

a. existing buildings (size, architectural style, color, materials)
b. utilities (above and below ground)
c. paved areas, such as drives, existing patios, basketball or tennis courts
d. existing landscape features, such as walls, pergolas, fences, pools
e. building details, such as the location of doors, windows, utility meters, air conditioners, downspouts, dryer vents, exterior mounted lights
f. current storage spaces for trash containers, garden tools, recreational vehicles
g. adjacent property development

Cultural Features

a. property lines (legal lines that define the parameters of a lot)
b. setback (the minimum distance that structures, including walls, fences, pools, and outbuildings, can be located from a property line)
c. zoning regulations (legal restrictions on specified uses of land in an area)
d. deed restrictions (limitations on the use of certain materials in an area, or the requirement for a certain material or quality standard, usually imposed by an association of local owners)
e. right-of-ways (municipally owned strips of land that are usually between the road and the front property line of a private property—there are usually restrictions on what the property owner can place within the right-of-way)
f. easements (strips of land within the site that are legally accessible by utility companies—property owners are usually restricted in what they are able to place within an easement in case the utility companies want to use it to place or access their lines)
g. zero lot lines (the absence of any side-yard setback requirements on adjacent sites, thereby permitting two buildings to be built

right next to each other—such absence is common to multiple housing developments)

h. off-site noises or odors

i. historical significance of the site or nearby area

Sources of Site Information

As the above list illustrates, the features of a site are varied. Some are easy to measure and evaluate, but others are more complex and may even be beyond the capability of the designer or landscape firm's staff to assess without assistance. It is not uncommon for the input for a site inventory to come from multiple sources. Typically, things that require on-site counting or measuring or sample collecting can be done by the landscape company. Features that require expertise not common to the company may be described and recorded by other professionals, and that information may then be used by the designer. For example, the topography of a site can be read from a topographic map that has been charted and recorded nationally by the U.S. Geological Survey (USGS), Figure 4-1. The USGS maps are produced in several scales (1 inch = 2000 feet, 1 inch = 24,000 feet, and smaller). They are most useful for describing the terrain of large areas and they benefit designers working with large, non-residential sites. Other sources of topographic maps may be local state, county, or city administrative offices. For smaller sites, the government maps are often not as helpful as a topographic survey map prepared specifically for the site by a licensed land surveyor. The surveyor may work within the landscape company or may be hired just for that job. These maps are drawn to a larger scale and provide much more usable information about the site. On other projects, a surveyed record of the entire site may be unnecessary, but the designer may need to know the steepness of a single slope in order to develop the plans for a retaining wall or a set of steps. On another site, existing grades may need to be modified in order to allow the development of a drive or a patio, or existing buildings may need to be measured and located on the base map. In such situations, it is important that the designer know how to measure the key features on the site and take simple surveys of the land.

Soil factors may also be evaluated using data obtained by the landscape company entirely or with the aid of other agencies and services. On a residential site, the most important informational needs may be soil pH, texture, drainage, and depth of the topsoil. That information can be gathered easily by the company. If soil nutrition is a concern, the landscaper may take a soil sample and test it herself or send it to a private laboratory for analysis. On large-scale projects, more in-depth information may be needed. Some of that data may be obtained from soil survey maps available from the U.S. Department of Agriculture's Natural Resources Conservation Service. These maps can provide information like the depth of the bedrock on a site, the location of the water table, the permeability of the soil to water, and the general soil type(s) on and near the site.

Weather data cannot be gleaned from a visit to the site. If the landscaper does not have a long-term familiarity with the area and its year-round weather, it can be helpful to consult the local weather records. Those records can provide information about the average annual rainfall, snowfall, and days of sunshine. Also available will be information about predictable temperature ranges, average humidity, and wind records.

Locating the actual property lines of a site can be difficult at times. Often what appears to be the property line, or even what the owner believes is the property line, is not. Hedgerows, fieldstone walls, fences, pavement edges, and similar landmarks that separate properties may be the accepted separation between lots but not be the legal separation. The county clerk's office is usually the place of filing for property records. The legal boundaries of a site can be found there.

Other cultural characteristics of a site can be legal, architectural, historic, or the simple trappings of our civilization. How the land was used in earlier years may have a bearing on how it can be used in the present. Was it a landfill? a cemetery? a disposal site for chemicals? a battlefield? Are there tax advantages or other incentives for using the site for certain purposes? Are the buildings or other structural features on the site illustrative of a distinctive style of architecture or period of time, such as the Victorian era or

Figure 4-1 U.S. Geological Survey topographic map (Courtesy of U.S. Geological Survey)

the Art Deco period? Are there underground wires or pipes or telephone lines? Are there nearby sounds that are welcome or offensive, such as church bells, factory whistles, or heavy traffic? The same concerns can apply to smells. It could be a bakery or a slaughterhouse, but the impact of periodic aromas would be part of the character of a site being inventoried. Many of the cultural features of a site can be ascertained by a designer simply visiting the site. Other features may need to be researched. Local historic societies, owners' cooperatives, local architects, utility companies, longtime neighborhood residents, and various municipal agencies can provide useful information about the site.

Geographic Information Systems

In recent years, the analysis of the suitability of certain sites for large-scale development has been aided by a powerful technology known as **geographic information systems**. These systems have gained widespread acceptance and use by landscape architects during the past two decades. While the data provided by the systems can be applied to the analysis of small sites, their most frequent use is with larger nonresidential projects.

Geographic information systems are computerized systems that combine both hardware and software to provide mapping data about geographic regions of the country. Unlike aerial photography that merely locates the physical features of a site, geographic information systems provide data on a site's topography and physical resources, wetlands, aquifers, surface water features, physical infrastructure, administrative boundaries, historic landmarks, and regulated areas such as airports, government properties, and underground storage areas. Landscape architects have embraced the technology of the systems because they can access an immense amount of data in a short time and gain a faster understanding of the site than through any other means. Again, it is a technology that is not commonly used in residential design, but it has a broad spectrum of use in most other areas of design and development because it offers rapid and documented evidence of a site's suitability or lack of suitability for various uses.

Reading the Terrain

As noted previously, the rise and fall of the land describes its terrain and is recorded as its topography. Topographic maps, whether prepared by the USGS or a private surveyor, will represent vertical changes in the terrain as broken lines termed contour lines. The lines represent a vertical rise or fall over the horizontal distance measured from the map's scale. Each contour line connects all of the points of equal elevation on that map, and each is labeled to indicate its elevation. The vertical distance between contour lines, the contour interval, is always stated on the map. Steep slopes are identified by closely spaced contour lines. Gradual slopes are denoted by more widely spaced contour lines. See Figure 4-2.

For large sites, the USGS maps may provide satisfactory data for a designer to use in a landscape. The maps are available for most areas of the United States and may be purchased at a nominal cost from

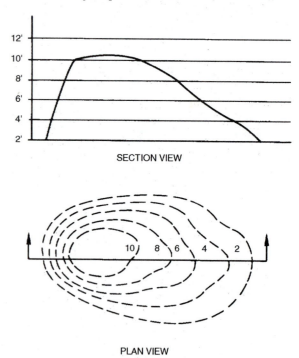

SECTION VIEW

PLAN VIEW

Figure 4-2 The contour interval is 2 feet. Closely spaced contour lines represent a steep slope. Widely spaced lines show a gradual slope.

regional offices of the Survey. For smaller sites, the scale of the map may need to be smaller and the contour interval as precise as 1 foot between the lines. Those types of topographic maps typically require preparation by a professional surveyor.

Figure 4-3 illustrates some of the land forms recognizable from a topographic map. In order to interpret the map fully, students should know the following points regarding contours and contour lines.

- Existing contours are always shown as broken lines.

- Proposed contours are always shown as solid lines.

- Contours are labeled either on the high side of the contour or in the middle of the line.

- Spot elevations are used to mark important points.

- Contour lines neither split nor overlap (except in overhangs).

- Contour lines always close on themselves. The site map may not be large enough to show the closing, but it does occur on the land.

- Runoff water always flows downhill along a route that is perpendicular to the contour lines.

Once the contours of a site are known and plotted, slopes can be measured and analyzed. **Slopes** are measurements that compare the horizontal length (measured from the map's scale) to vertical rise or fall (as determined by the contour lines and contour interval). Slopes may be expressed as ratios or gradients and percents. As a **ratio** or **gradient**, the horizontal space required for each foot of vertical change in elevation is compared as $V{:}H = R$ where V is the vertical distance, H is the horizontal distance, and R is the ratio or gradient. Ratios are commonly expressed as 1:3, 1:4 and so forth, Figure 4-4. As a **percent**, the vertical distance is divided by the horizontal distance and the answer is expressed as 33 percent, 25 percent, and so forth. Another way to visualize the percent of slope is to picture the slope extending along a horizontal distance of 100 feet. The vertical distance then becomes comparable to the percent of slope, Figure 4-5.

The Need for Terrain Information

The ease or difficulty of development depends upon whether the land is level or rolling, rocky or sandy, forested or open. A study of the terrain also supplies answers to basic questions: Where does the surface water flow? Will water collect in puddles anywhere? What types of human activities can take place? Will grass grow on that slope? Can a car be parked safely on that slope?

Most human activities require that the land be level or nearly level. Land that is 5 percent or less in slope is perceived by users as flat. Flat land is the easiest terrain to develop, but it may be difficult to move off surface water. A slope of at least 1 percent is usually necessary to drain away surface water on turf and other planted landscape surfaces.

Human activities can usually take place on a slope of 5 percent to 10 percent, but users will sense the uneven footing. Land that slopes more than 10 percent may require alteration (grading) to make it more usable. Figure 4-6 lists acceptable slopes for various landscape components.

Grading the Land

When the land is not suitable for the activities planned for the site, it may be necessary to reshape it. The form of the land is changed by a process called **grading**. Grading can be as simple as one worker leveling and smoothing a small area of earth with a spade and a rake and hauling away the leftover soil in a cart. It also can be so extensive that massive bulldozers and dump trucks are required to chew up and haul away entire mountains. Regardless of the extent of the project, grading is usually done for one of four reasons:

- To create level spaces for the construction of buildings

- To create the level spaces required for activities and facilities such as parking lots, driveways, swimming pools, and playing fields

- To introduce special effects or improved conditions into the landscape, such as better drainage, earth berms, tree wells, and ponds

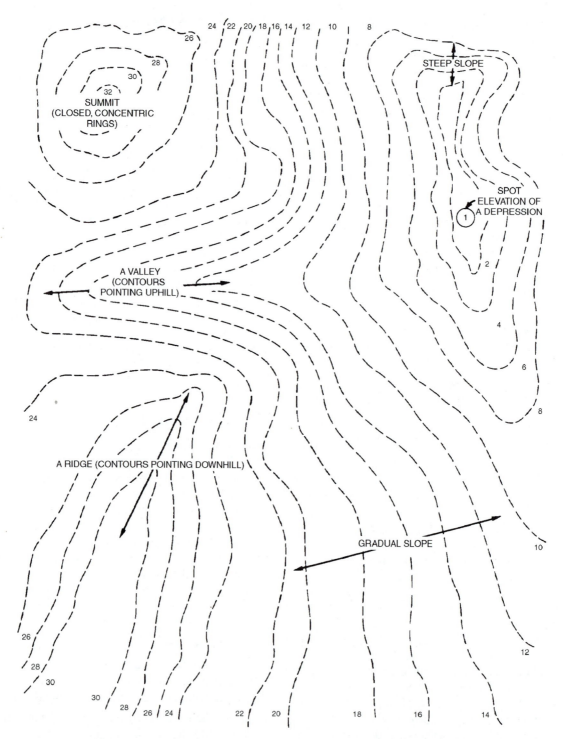

Figure 4-3 A topographic map labeled to show different land forms

- To improve the rate and pattern of circulation by means of better roads, ramps, tracks, or paths, Figure 4-7

When earth is removed from a slope, the grading practice is called **cutting**. When earth is added to a slope, the practice is called **filling**. On a contour map, a cut is shown as (1) a solid line diverting

RATIO: COMPARING THE HORIZONTAL AND VERTICAL MEASUREMENTS OF A SLOPE

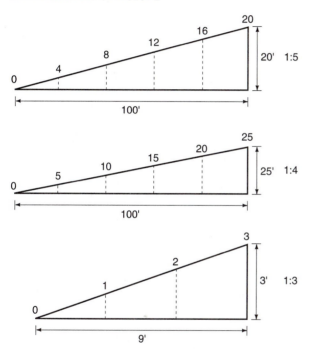

Figure 4-4 Ratio of slope

$$\text{PERCENT OF SLOPE} = \frac{\text{VD} = \text{(VERTICAL DISTANCE)}}{\text{HD} = \text{(HORIZONTAL DISTANCE)}}$$

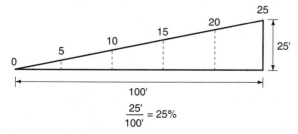

$$\frac{25'}{100'} = 25\%$$

Figure 4-5 Percent of slope

from and then returning to an existing contour line, and (2) moving in the direction of a higher contour. A fill is shown as (1) a solid line diverting from and then returning to an existing contour line, but (2) moving in the direction of a lower contour line. See Figure 4-8. A typical graded slope is illustrated in Figure 4-9.

Because the grading process can involve the movement of tons of soil and rocks, designers should approach such specifications cautiously. In cut and fill operations, the soil that is removed from the cut should be used to create the fill whenever possible. This practice minimizes the need for costly hauling and disposal. If possible, the topsoil layer should be stripped away and stockpiled before grading begins. The topsoil can then be spread over the finished grade before the site is replanted.

When land is graded, not only is the topsoil disturbed, but the surface water drainage and vegetation are disturbed as well. Water must drain away from the buildings, not toward them. Freshly graded slopes must be stabilized to guard against erosion. The surface roots of valuable trees must be protected from the destruction of cutting and the suffocation of filling. Figure 4-10 shows a typical slope before and after grading. Figures 4-11 and 4-12 show common techniques for dealing with trees that exist before the grading of a site.

Preparing a Base Map

A **base map** is a graphic depiction of the site features that were collected, measured, and inventoried as described previously. It enables the designer to take a look at the big picture after the individual snapshots of the site are pieced together. It is a plan view drawing that locates the existing buildings, their windows, doors, and other significant features; the existing hardscape such as drives, parking areas, walls, patios, etc.; all existing plant materials; and other physical items noted during the site inventory. The base map can also plot out the location of the setbacks, the easements, the underground utility lines, and overhead wires. It is also an opportunity to make notations about the

ILLUSTRATED EXAMPLE		2%	3%	4%	10%	30%	33%	50%
PERCENT OF SLOPE	IDEAL	1/2% TO 2%	2% TO 3%	1% TO 4%	1% TO 11%	16% TO 33%	20% TO 33%	33% TO 50%
	ALLOWABLE	1/2% TO 3%	1% TO 5%	1/2% TO 8%	UP TO 11%	UP TO 33%	UP TO 50%	UP TO 65%
LANDSCAPE COMPONENT		SITTING AREA, PATIOS, TERRACES AND DECKS	LAWNS	WALKS	DRIVEWAYS AND RAMPS	BANKS PLANTED WITH GRASS	BANKS PLANTED WITH GROUND-COVERS AND SHRUBS	STEPS

Figure 4-6 Recommended slopes for common landscape components

Figure 4-7 Rough grading is the first step in adapting the site to human use.

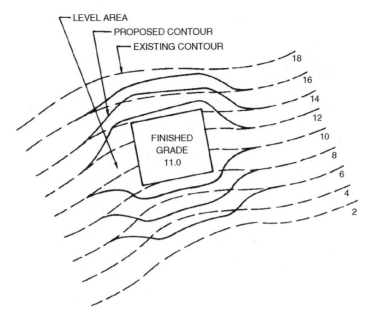

Figure 4-8 Cut and fill as shown on a topographic map

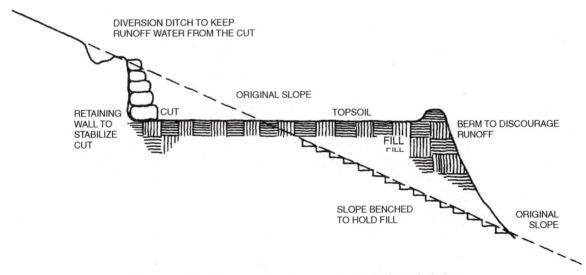

Figure 4-9 Cross-sectional view of a typical graded slope

direction of prevailing winds, surface water patterns, and attractive or offensive off-site features that impact the site. Only after this physical arrangement of the inventoried site features is prepared can the designer begin the next stage of the design process: the analysis of the site and the determination of how suitable or unsuitable it is to accommodate the client's program requests.

Most designers will follow the preparation of the base map with a drawing developed as an overlay

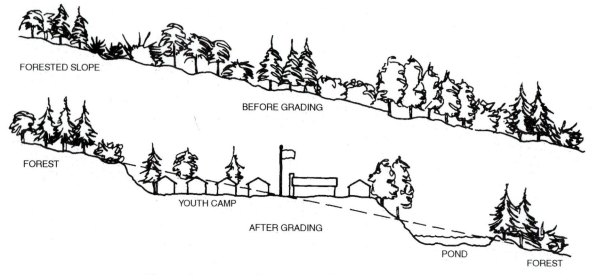

Figure 4-10 A typical slope before and after grading

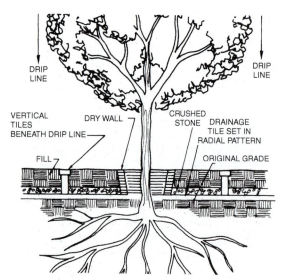

Figure 4-11 Raising the grade around an existing tree

Figure 4-12 Although the level of the lawn has been lowered, the tree's roots remain at the original level because of the retaining wall.

that eliminates all of the existing features that they plan to eliminate from the new landscape. The terminology for the revised drawing is not uniform. Some designers continue to refer to it as the base map, while others give it a new name, the *base sheet*. Regardless, both versions are of great value to

the designer and must be carefully prepared and drawn to scale. Neither is likely to be seen by the clients.

Throughout the site data inventorying process, it will be helpful if the designer takes photographs of the site. These photos will have varied and repeated uses throughout the design process, not the least among them being the benefits they will afford as the base map and the later overlay are prepared.

Achievement Review

A. Indicate if the site characteristics listed are (N) natural, (M) man-made, or (C) cultural.

1. condition of the turf
2. nearness of public transportation
3. rock outcroppings
4. swimming pool
5. terrain features
6. off-site views
7. buildings on the site
8. style of the architecture
9. existing shade
10. prevailing breezes
11. soil conditions
12. historical features
13. provisions for parking
14. nearness of neighbors
15. presence of wildlife
16. traffic sounds
17. zoning regulations
18. presence of large, old trees
19. existing lighting
20. surface water patterns

B. Define the following terms.

1. terrain
2. topography
3. contour line
4. contour interval

5. slope
6. property line
7. setback
8. zero lot line
9. easement
10. zoning regulations

C. Complete the following sentences that describe the characteristics of contours and contour lines.

1. Existing contours are always shown as _____ lines on topographic maps.
2. Proposed contours are always shown as _____ lines on topographic maps.
3. Contours are always labeled on the _____ side of the contour or in the _____ of the line.
4. Important points on topographic maps are marked by _____.
5. A 1:3 or 1:4 comparison is the _____ of a slope.

D. Label the parts of a typical cut and fill in the diagram below.

E. Identify the land forms A through E on the contour map on the following page.

F. List five examples of landscape projects that might benefit from the data obtainable through the use of geographic information systems.

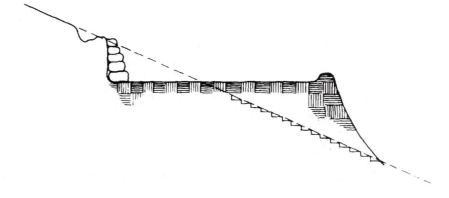

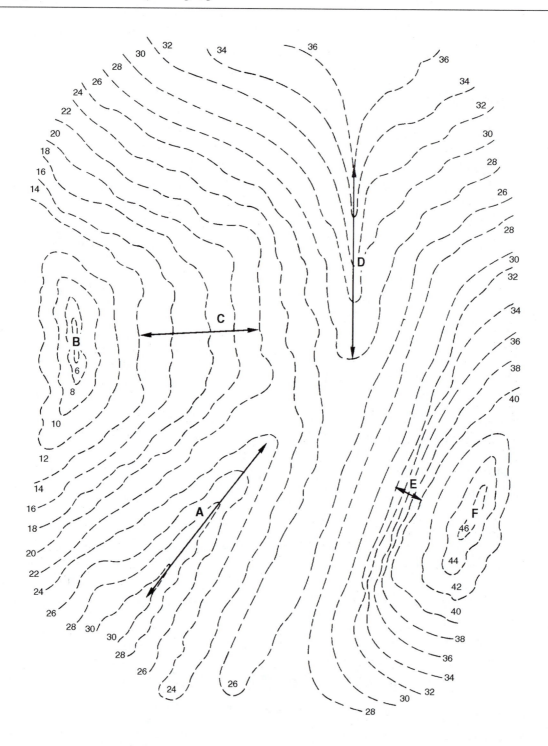

Suggested Activities

1. Fill several greenhouse flats with loamy soil. Plant several with a local turf grass. Obtain another flat (perhaps from a local nursery) that is filled solidly with English Ivy, Periwinkle, or similar vining ground-cover. Plant or obtain another that has an established growth of a typical roadside embankment stabilizer, such as crown vetch. Tip the flats to create various slopes up to 50 percent. Water all flats with the same amount of water applied from a sprinkling can to simulate rainfall. Record the erosion noted over the weeks at each degree of slope. Record also the quality of the turf grass growth at increasingly steeper angles.

2. Build a small contour model from the information on a topographic map. Use cardboard of a thickness comparable to the contour interval on the map.

3. Order topographic maps of the local area from the U.S. Geological Survey. Find recognizable local features on the map to help students visualize how the contour lines describe terrain variations.

4. Obtain soil survey maps of the local area and let students match the soil types described on the maps with actual samples that they take.

5. Invite a land surveyor to visit the class to demonstrate the use of surveying instruments and explain how land surveys are done.

6. Have students take measurements of a nearby property and then prepare a base map of the site.

The Landscape Process

Objectives:

Upon completion of this chapter, you should be able to

- explain the cyclic nature of the landscape process
- list the component steps of each phase of the cycle
- conduct a client interview
- distinguish between functional diagrams, preliminary designs, and final plans

Landscaping as a Process

A **process** is a sequence of steps, in the form of decisions or activities, that result in the accomplishment of a goal. A process may also be a sequence of stages during which something is created, modified, or transformed. Both interpretations of the term apply to the development of a landscape. The sequencing of decisions is what happens as a landscape architect or designer works to sort out the client's needs and desires as well as the assorted characteristics of a specific site and somehow fit one to the other. The sequencing of stages is what happens as a landscape project, beginning as the need or desire of a client, moves to the design stage, then to the construction stage, and finally into continuing maintenance and management.

Unlike a manufacturing process, landscaping is frequently a cyclic process. When an automobile rolls off the assembly line, the process that led to its construction is completed. However, the completion of a landscape project at one level of client satisfaction often sows the seeds of desire for a continuation or upgrading of the same project to a higher level of satisfaction, Figure 5-1. For example, consider two scenarios, one public and one private. A community that has no public parks desires to create a park. Open green space is the first priority of the community. Their sequential process would include such stages as finding a central location where the greatest number of residents could be served, acquiring rights to the land, and clearing and planting the area. In like manner, a homeowner who has no patio may desire one. Longing for a place to sit and cool off on a warm day, the client has little else in mind except a place where a few lawn chairs can be set out to enjoy

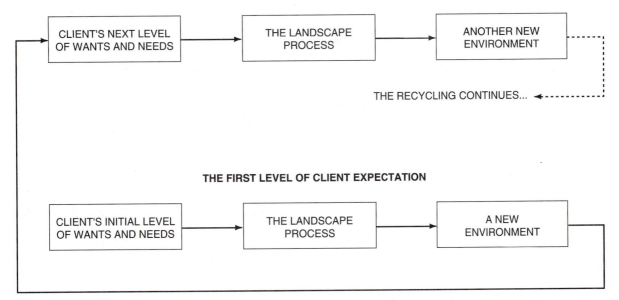

Figure 5-1 The landscape cycle

conversation with friends. The sequence of stages would be short. A size would be determined, a paving material selected, and the construction completed. In both cases the story could end here, but it is highly probable that it would not. The community, which formerly had no park but now does, may begin to compare its park to those of other communities. Seeing that the parks of other communities have features that their park does not have, such as a swimming pool or tennis courts, will stir new desires. Once a new wish list is generated, the process begins again, but with a higher level of satisfaction as the goal. The homeowner who had no patio but now does may be very contented sitting in his lawnchair conversing with friends, until one of the friends describes the fishpond or cooking center that is part of her home patio. Suddenly the homeowner begins to compare his patio to hers, and his satisfaction level drops. Next week he may call back the landscape company to begin the development of a bigger and better patio. Landscaping cycles and recycles.

Project Development as a Process

The development of a landscape project has many stages. A typical sequence is as follows.

1. The client formulates a need or series of objectives to be accomplished by a particular time.

2. A designer is contacted and the design is developed through a sequence of steps known as "the design process."

3. The client accepts the design, and the project is assigned to a landscape contractor through competitive bidding or noncompetitive negotiation.

4. The landscape contractor arranges for suppliers and subcontractors to provide the materials and/or work needed to accomplish the project.

5. The project is built, often with the client and/or the designer monitoring the contractor to assure that it is being built as anticipated.

6. Acceptance is given by the client. Final billing and payment follow.

7. A separate maintenance contract may be negotiated with a landscape management firm.

Project Maintenance as a Process

In similar fashion, the desire for continuing maintenance of a property by a landscape management firm also sets in motion a necessary sequence of actions and activities.

1. The client desires a level of maintenance quality and/or an array of services that he or she is unable or unwilling to accomplish personally.

2. Through competitive or noncompetitive means a landscape management company is selected.

3. The company assesses the needs of the client and presents a proposal, which may be accepted or modified to suit more closely the desires and budget of the client.

4. The work to be done, the crews, equipment, and materials required, and the frequency of performance are all scheduled by the company. Work is carried out on schedule, and the client is billed at regular intervals.

5. Regular meetings with the client allow for changes and upgrading of the contract as desired by the client.

Design as a Process

Discovering that creativity is a process rather than a sudden flash of inspiration may disappoint some who prefer to regard it all somewhat mystically. However, it should reassure others that good designing can come from nearly anyone who takes the time to maximize the contribution of each stage of the design process. Accepting that design is a sequential accumulation of data, followed by sequential analysis of the data, and the equally sequential organization of ideas and solutions, per-

mits the designer to fit client needs with site capabilities in a logical manner.

The design process occurs in two stages, which are ongoing simultaneously, but which are necessarily independent of each other for awhile. The two phases, **site analysis** and **program analysis**, are diagrammed in Figure 5-2. During the site analysis, the designer seeks a compilation of the natural, man-made, cultural, physical, and visual characteristics of the site. The compilation should be a thorough inventory of the site's attributes made without regard to whether they are positive or negative features. They will later be analyzed to determine what options and opportunities they offer for development. At the same time, but separate from the site analysis, the program analysis begins with a detailed compilation of the client's needs and desires. At this stage, the designer should make no attempt to judge the propriety, possibility, or logic of the client's needs. It is first necessary to make the inventory. Once that is done, then the analysis of the needs can be undertaken and a program written. The program organizes the client's disparate wish lists into related groupings, which will serve to guide the development of the design.

Up to this point, it is important that the designer not let one analysis influence the other. Allowing the client's program requirements to enter into the designer's thought process too soon can cause the designer to overlook potentially important uses of the site that might not have occurred to the client and that in turn may never occur to the designer if he or she sees the site through the client's eyes.

Interviewing the Client

The techniques and components of the site inventory and analysis are discussed in Chapter 4. So this is an appropriate place to acquaint the student with some of the techniques used to gain information about the client's program desires. While public and nonresidential projects may use surveys, group meetings, conference calls, and other impersonal means to ascertain the uses of the site by groups of users, the residential landscape lends itself nicely

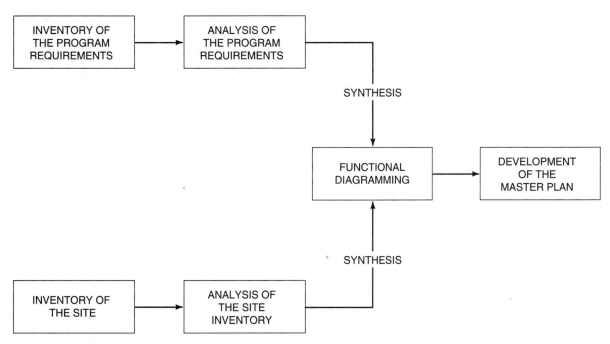

Figure 5-2 The two-track design process: program analysis and site analysis

to the one-on-one interview process. It warrants restating that the initial interview is an opportunity for data collection, not opinion formulation. The clients will have information about the site that the designer needs and that must be obtained during the interview. The client will also have opinions, dreams, expectations, preconceptions, preferences, and attitudes that need to be noted and measured during the interview. Later in the landscape process, these can all be factored into the designer's eventual proposals for the property. Still other clients may bring a different persona to the interview—that of having little or no idea of what they want. That type of client will rely heavily on the designer to take the lead in developing the program for the site. For the designer, this is an opportunity to educate the clients about the many possibilities for outdoor living that their site may offer. It is also a sales opportunity for the designer. By explaining to the clients the benefits of providing for shade, or wind diversion, or privacy control, or other features of outdoor room development, the designer serves

the clients by helping them to formulate better their program requirements. In turn, having more focused program requirements may enable the designer to expand the size of the project. As long as the clients are not pressured to approve a project that is beyond their financial means, the designer as educator and salesperson is a helpful and ethical player in the landscape process.

Basic Information Gathering

The designer needs to be prepared for the interview. It is not enough to merely sit down with the client and say, "Okay, tell me what you want." Some of the needed information is basic. The checklist below summarizes the points to be covered in nearly all first interviews.

- composition of the family, including numbers, ages, and genders

- pets, including breeds, and whether kept indoors or out, and the clients' attitudes about pet confinement

- attitudes about outdoor living, including past uses of the outdoors as well as intended future uses

- active or passive uses of the landscape, e.g., lawn games, dining, reading, child play

- attitudes about privacy

- hobbies or special interests, e.g., greenhouse, rebuilding old cars, lap swimming

- service area needs, e.g., clothes line, vegetable garden, cut flower production, pet yard, compost pile, tool and trash storage

- frequency, type, and size of outdoor entertainments

- attitudes toward landscape maintenance

- preferences, e.g., colors, materials, styles, brands, themes

- important views into the landscape from major rooms within the house

- the specifics: What do they want in their landscape? Why did they call you?

- budget

Interpreting the Clients

Just as each site has a personality that distinguishes it from all others, so too are all clients different. The checklist above, although applicable to most residential client interviews, cannot fully accomplish the program data gathering. The designer must make other observations that will influence the design while meeting and working with the clients. Not all of the observations will occur during the initial interview, so the program data may be acquired over a period of time.

The designer should look for subtle clues about the preferences or lifestyle of the clients that the clients may not recognize as being distinctive to them. Their behavior and/or decor indoors can be indicative of how they will use and enjoy their life in the outdoors. For example, is the home very formal or ultra-casual in its furnishings and arrangements? Are there floral arrangements around the rooms, or original art? Is there a distinctive theme to the interior design? What magazines are on the coffee table . . . ones that feature architecture, gardening, travel and leisure pursuits, or personal health? Are the children's toys picked up or strewn around all major rooms? They may give clues to the interests of the clients that can be incorporated into the landscape plan. If the clients offer coffee or tea, is it served in a mug or a china cup and saucer? Are beverages placed on coasters or not? All of these client idiosyncrasies are clues to what clients personally value. Some clients are more formal or more casual than others. Some have ethnic preferences that are manifested in their fondness for certain colors or styles. A wise designer will pick up on these differences during the program stage of the landscape process.

What Aren't They Telling You?

The interview with the clients often has several hidden layers that must be cleared away if the designer is to produce a plan that will truly satisfy them. What makes it difficult is that the clients are unlikely to offer any assistance in this regard. At issue is the reality that many clients are uncomfortable or unwilling to reveal to the designer their full or true purpose for doing the project. Some designers believe that there are as many as three levels of client revelations about their motivation for a project. The first level is the one the clients are comfortable talking about. They want a nice setting for their home and a pleasant patio and pool area where they can enjoy entertaining their friends and family members. The second level of motivation may never surface unless the designer is sensitive enough to recognize it. The clients are doing this landscape as self-affirmation that their lives and careers have meant something. Because of their education, their hard work, their loyalty, their surmounting the sacrifices and challenges of life, they are able to afford this landscape project and provide something really nice for their family. The clients may never speak this "We made it!" reason aloud, but it is genuine and they will want a landscape proposal that justifies it. The third layer of client motivation is the one most deeply hidden. Most people will not admit to it for fear of looking egotistical or braggish, so the designer must phrase his or her questions carefully in order to probe for this real reason for doing the project. The clients want to impress their friends and neighbors. In fact, they don't want to just impress them, they want to knock them over. They want their guests to turn green with envy when they come to visit. A designer who only heard, "We want a nice setting for the house and a pleasant patio and pool area where we can entertain a few friends" and went no further in probing for their real motivation would probably not satisfy them. The program data would be incomplete.

The Budget

Only in classroom projects are there landscapes to be designed where "money is no object." Money is almost always an object, and it should be addressed early in the information-gathering portion of the process. Experienced clients who have had landscaping done before and perhaps worked with a certain company several times will be more aware of costs and better prepared to set a reasonable expenditure range for the designer to work within. Inexperienced clients, with a preconceived cost in mind, are likely to want more than their budget allows. Others who have specific requests at the outset of a project may be unprepared for the cost estimate when it arrives. It is an important part of the designer's role as educator to keep the clients aware of the costs of their needs and desires as the landscape program requirements are outlined. There is no benefit to selling the clients more than they can or should afford. Important questions to ask during the interview include "Do you have a budget range in mind?" "Do you want to complete this project all at once or over a period of time?" "Of the numerous features you have expressed a desire for, which one(s) are most important at the outset of the project?"

Bringing the Site and Program Inventories Together

Eventually the client's program and the site's potential for development must be meshed, if possible. That is shown in Figure 5-2 as **synthesis**. At

this stage reality sets in. The program may seek features or uses that the site cannot provide. Conversely, the site may have been found suited to activities that the clients had not even considered. In either case, the client's program may be modified at this point to bring a closer match between the client's needs and the site's potential for development. Once the fitting of the program to the site has been accomplished, the designer proceeds to develop specifics of the design, first in preliminary form for review by the client, and later in final form. Once the design is in final form and has been approved by the client, the designer may proceed to the development of specifications, construction drawings, and other support materials needed to get the project constructed.

Diagrams and Drawings Within the Design Process

The diagrams and drawings of the design process are also sequential. Each is increasingly more specific and detailed than the one that preceded it.

Functional diagrams begin the arrangement of the client's program on the site, Figures 5-3 and 5-4. Some designers refer to these as "bubble diagrams," because they use loosely drawn freeform shapes to represent the use areas or spaces that will accommodate the client's program features. More concerned with relationships between the spaces than with the specifics of how those spaces will be developed, functional diagrams help the designer make important logical decisions concerning layout of the site, size requirements of each use area, circulation patterns between use areas and throughout the total site, potential conflicts of use or circulation, and the relationship of off-site features to on-site areas.

Preliminary designs break the bubbles to reveal the designer's first draft vision of how each area of the landscape will be shaped. Development of the outdoor rooms of the landscape (as described in the next chapter) begins at this stage. Using the functional diagram as the basis for location and intended uses of each room, the designer gives the landscape its form. Decisions are made regarding the types of materials to be used to create the

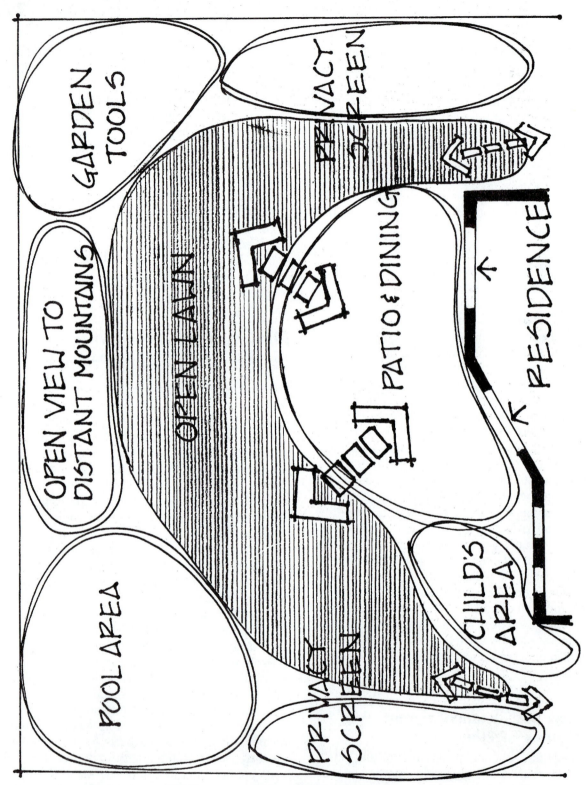

Figure 5-3 A preliminary functional diagram

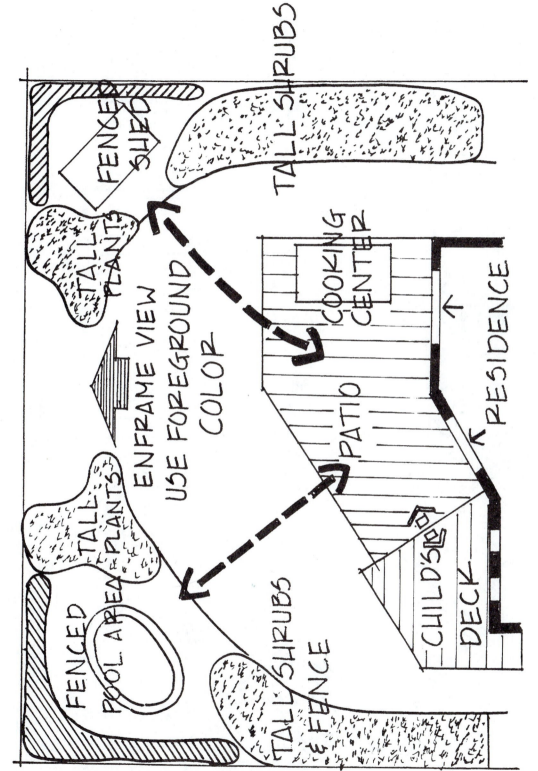

Figure 5-4 A refinement of the functional diagram

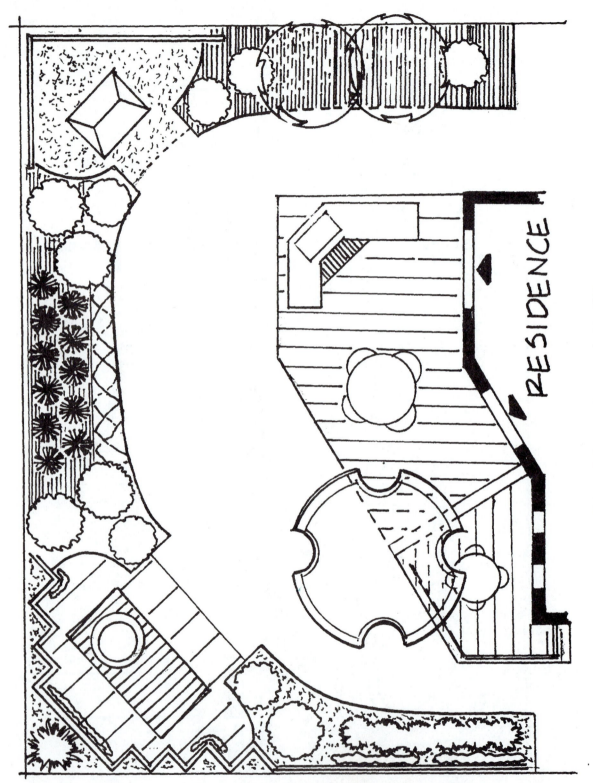

Figure 5-5 A preliminary design plan

RESIDENCE

screenings, ceilings, and surfacings. The principles of design (as discussed in Chapter 8) are applied to the use areas. The designer begins to think in three-dimensional terms, planning not only the horizontal relationships between objects and areas, but the vertical relationships as well, Figure 5-5. While the graphic delineation of the preliminary design is simplistic and detailed specifics are lacking, the preliminary design encapsulates the concepts intended by the designer. It is a suitable drawing to present to the client for review. To carry the design any further without getting feedback from the client would be foolish, since few preliminary designs pass the all-important client test without modification. With difficult projects (or difficult clients) there may be a number of preliminary designs created before the project advances to the final stage.

Final plans incorporate all of the suggestions and reactions of the client into a master drawing that is graphically detailed and completely specific in its intent for the landscape. Plant and hardscape materials are precisely identified. Paving patterns and enrichment features are explained and diagrammed. Graphic styling is designed to impress and to aid the client's visualization of how the completed landscape will appear. Accompanying the final plan may be a series of construction drawings or interpretive drawings to aid the client and/or the contractor in understanding how the design is to be built.

By properly and patiently working through the design process, a landscape architect or landscape designer can create an imaginative and workable solution to almost any project challenge.

Achievement Review

A. Respond to the following statement of opinion. "I would not want to be a landscape architect or designer because there are only a limited number of properties to develop, and once they are designed there will be no more work. I would eventually put myself out of a job!"

B. List sequentially the steps typical to the development of a landscape project.

C. List sequentially the steps involved in the landscape maintenance process.

D. Explain why site analysis and program analysis are done separately during the initial stages of the landscape design process.

E. Indicate if the following features are most typical of

    a. functional diagrams
    b. preliminary designs
    c. final plans

1. They provide specific details of plant selections and hardscape material choices.
2. They represent general design decisions, such as where plant masses will be located, what shape the patio and paved areas will be.
3. They display the most sophisticated graphics of the three types of drawings.
4. They are most concerned with relationships between the proposed use areas of the landscape.
5. This is usually the drawing shown to the client as the first draft of the designer's proposal.

Suggested Activities

1. Have students role play to accomplish a residential client interview.

2. Have them do similar role playing, but do it as a nonresidential project. Let the clients be a panel, perhaps portraying a park commission or a cooperative tenants' association.

3. Have the class brainstorm the program needs and desires of a perfect school campus. Then walk the school property and determine which ones could be accommodated on their site and which ones would have to be eliminated . . . and explain why.

Site Analysis Checklist

_____ Plan and develop objectives of the site.

_____ Relationship of the site to the local environment.

- Orientation, microclimate, and noise sources
- Contaminated soils and filled areas

_____ Relationship of the site to adjoining properties.

- Buildings
- Views to and from the site
- Access and connection points
- Fences, boundaries, and easements
- Major trees and vegetation on adjoining properties

_____ Physical characteristics of the site.

- Contours
- Existing vegetation including significant trees
- Drainage and services

_____ Identify the key influences of the design.

_____ Examine how the proposed site will relate to the immediate surroundings.

- Location and height of walls built to the site boundary
- Characteristics of any adjacent or nearby fencing or gardens
- Building levels
- Heritage considerations

_____ Identify issues to consider in the design process.

_____ Identify the best methods to facilitate preparation of the site plan.

CHAPTER 6

The Outdoor Room Concept

Objectives:

Upon completion of this chapter, you should be able to

- identify indoor and outdoor use areas
- list and define the features of the outdoor room

Modern American homes vary in size from three or four rooms to several dozen. Regardless of the simplicity or complexity of the home, the rooms are usually divided into four different categories of use, Figure 6-1. The **public area** of the home is the portion that is seen by anyone coming to the house. The public area includes the entry foyer, reception room, and enclosed porch. The **family living area** includes those rooms of the house that are used for family activities and for entertaining friends. Rooms such as the living room, dining room, family room, and game room fall into this category. The **service area** includes the rooms of the house that are used to meet the family's operating needs. These include the laundry room, sewing room, kitchen, and utility room. The **private living area** of the home is used only by the members of the family for their personal activities. The bedrooms comprise the major rooms of this area. Dressing rooms, if present, are also included.

Outdoor Use Areas

Just as a house and most other buildings are divided into areas that are commonly described by function, so too can landscapes. Many people identify areas of the landscape by *location*, such as the front yard or the backyard. Landscape designers should instead identify areas of a property by the *function* each is to fulfill for the users of the landscape. Those functional areas should correspond as closely as possible to the use areas of the home or other building and should be physically linked to those areas as closely as practical, Figure 6-2. As many as four *use areas* may be assigned to a residential property.

- **Public Area** Most nondesigners refer to this by location as the front yard. Instead, now

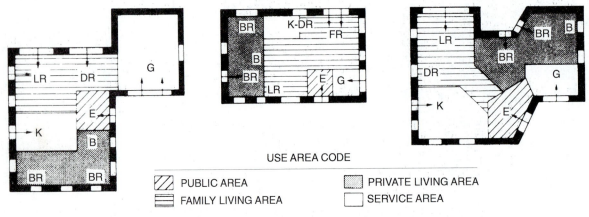

USE AREA CODE

▨ PUBLIC AREA		▤ PRIVATE LIVING AREA	
☰ FAMILY LIVING AREA		☐ SERVICE AREA	

Figure 6-1   Examples of use of area categories within the home

begin to regard it by its function. It is the region of the landscape that is seen by everyone who passes the house or chooses to approach it. It has a three-fold function:

1. **Put the house into an attractive setting**. Since a home is the most expensive thing that most people will purchase in their lifetime, it is logical that it should be displayed in a manner that gives recognition to its value. With the house positioned as the focus of the viewer's attention, the landscape of the public area should create a picture frame type of setting that will enhance the architecture, not hide it, and integrate the home with the adjacent neighborhood (in an urban locale) or the surrounding natural landscape (in a rural locale).

2. **Identify the point of entry**. Persons wanting to enter the home from the street or from the driveway need to recognize the front entry door. Since some homes have secondary points of entry, the main entry must be highlighted by the design of the public area.

3. **Provide access to the entry**. Whether approaching on foot or by car, visitors to the home need an easy way to get from the public sidewalk or their car in the drive to the entry door.

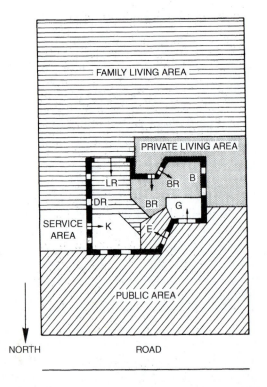

Figure 6-2   Indoor and outdoor use areas should be matched in location as closely as possible. For example, notice that in this drawing, the outdoor service area adjoins the kitchen, the indoor service area.

Linkage to the House By meeting the house at the front door, the landscape's public area connects with the front porch, entry foyer, or reception room, and usually an attached garage.

Size of the Area The size of the public area varies with the overall size of the lot, but in most cases it needs to be only large enough to satisfy the above functions. In many locations, there is a common setback required of all buildings, so there may be no opportunity for the client or landscape advisor to have input into the size of the area. In general, the public area can be regarded as a passive-use area—to make it too large impacts on the amount of space that can be given over to more active use areas of the landscape.

Sun Orientation The public area is usually not as sun-sensitive as other areas of the landscape, so the direction that it faces is only significant in terms of the amount of sunlight allowed the remaining use areas of the landscape.

■ **Family Living Area** The residents of the home carry their social lives into the landscape through the family living area. It is here that a designer will plan for the patio, outdoor dining, swimming pool and spa, lawn games, children's play, and assorted other uses that permit the members of the household to interact together and with their family, friends, and neighbors.

Linkage to the House The family living area, also termed the *general living area*, should link with the house in a way that allows for uninterrupted continuation of the residents' social activities as they move back and forth between the indoors and outdoors. Ideally the family living area will adjoin the living room, family room, and/or dining room of the house. The most important views from the house will usually look into the family living area, so it is common for the home's family area rooms to have large expanses of glass thereby making indoors and outdoors seem like mere opposite sides of a single pane of glass.

Size of the Area The family living area is usually the largest use area of the residential landscape, due primarily to the number and size of activities that it must accommodate. While there are differences in the ways that a large, young family and an elderly retired couple will use their family living areas, on each site the family area will still usually require the greatest space in comparison to other use areas on that site.

Sun Orientation The objective of the designer is to maximize the number of hours each day and the number of days per year that this area can serve the residents. Comfort is the key, and for properties in the temperate areas of the country, that means that a southern exposure is the ideal. The second best sun orientation is facing west, since the afternoon sun is the warmest. An eastern orientation would leave parts of the area cool in the afternoon due to shadows cast by the house, trees, and other structures. A northern exposure is always without sunlight in those areas directly adjoining the house, making it difficult to have a patio or pool directly off the home that can be used comfortably in cool weather. Conversely, if the landscape is being developed in a tropical or subtropical zone, the preferred orientation for a family area may be the opposite of above. Where the summer sun is excessively hot and the season lengthy, an eastern orientation, and sometimes even a northern one, may be preferred.

■ **Service Area** The function of the service area is utilitarian. It may house the storage shed for garden tools, a dog yard for the family pet, a garden for vegetables or cut flowers, a clothesline or compost pile, or provide storage for a camper trailer.

Linkage to the House The area should be placed as close to the kitchen and laundry room as possible when garbage, trash, and/or wet laundry must be moved from the house to the area as easily as possible. There may be a need for more than one service area. For a use such as wood storage, it may require the service area to be closer to the place where it will be used most often, such as near the family room fireplace. A hobby greenhouse may require placement in a location where it will get the most sun, with linkage to the house being of less importance. Because the service area(s) are not developed for beauty, they should not be viewed into from major rooms of the house.

Size of the Area The possible uses of the service area are so varied that no general statement of correct size can be made. For some properties, the area(s) required is extensive, while for others it may be little more than space for a few trash cans. A dog run will require some space, and while a clothesline seems very linear, there must be enough room to permit the clothes to blow back and forth without brushing against a fence or building wall.

Sun Orientation Obviously tool storage doesn't require concern for the sun, nor does a pile of wood. However, for clothes to dry or vegetables to grow, there must be sunshine. To prevent a pet from overheating, there must be shade. In short, whether the location of the service area has relevance to the amount of sunlight depends upon its use(s).

- **Private Living Area** This area typically adjoins the master bedroom of the home and provides an outdoor room for adults to use in the morning and at other times during the day when they desire a place that provides total privacy. Semipassive activities such as reading the morning paper with a cup of coffee or relaxing in the hot tub commonly occur in the private living area. Sometimes the area is separate from the rest of the landscape, while at other times, it may be a screened area of the larger family living area that has its own access off the bedroom(s).

Size of the Area The private living area is seldom large. It does require screening from out-siders' viewing, and it is usually developed aesthetically. Thus, enough area is needed to accomplish those objectives while also providing enough space to accommodate the intended use by the residents.

Sun Orientation Since the area is more often used in the morning than at other hours of the day, the private living area is ideally oriented to the east side of the house. That allows the sun to reach the area during the hours when it is in use. Lacking an eastern orientation, the second best orientation is to the south where it can pick up the morning sun almost as well.

While all landscapes can have different use areas assigned based upon the activities that occur within those areas, it may not be necessary or possible to have all four of the above areas on every property. The public area and the family (general) living areas are nearly always found in residential designs. The service area may be reduced to a corner of the garage if the residents have no other needs for an outdoor space or if someone other than the residents maintains the property. Where there is no access off the bedrooms or where there is inadequate space for development, the private living area may not be possible. In other cases, even where there is adequate space or opportunity, the clients may not want a private living area.

Figure 6-3 shows several different houses and lots with use areas assigned. Note the relationships between indoor and outdoor areas and the opportunities or limitations that different lots allow. Figures 6-4 and 6-5 illustrate a public area and a family living area. Even in these photos the differences in their use are so apparent that describing the areas by location (front yard and backyard) is not as logical as describing them by their function.

The Outdoor Room Concept

Once the use areas have been assigned to the property, the next step is to design and develop those areas into livable outdoor spaces. To accomplish this, it helps to think of the outdoor space as an outdoor room. In this way, students can apply

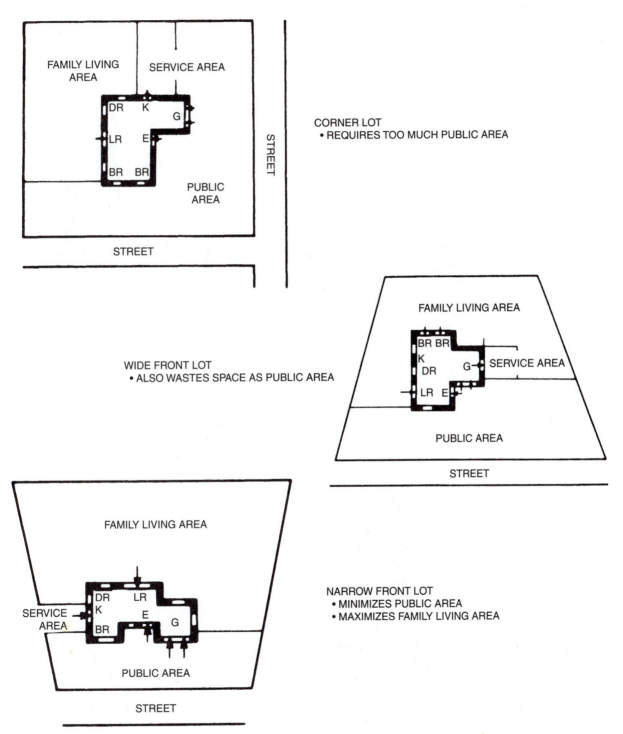

Figure 6-3 Assigning landscape use areas is easier on some lots than on others.

Figure 6-4 The Public Area of the landscape displays the building in an attractive and inviting manner, while explaining how to approach the entry.

some of what they know about indoor rooms to the outdoors.

The composition of the indoor room is the same whether it is simple or ornate, consisting of walls, a ceiling, and a floor. While the materials may vary from room to room, the basic structure remains the same.

The outdoor room also has walls, a ceiling, and a floor, Figure 6-6. The materials are usually different from those used indoors, but they accomplish the same purpose. The **outdoor wall** defines the limits or size of the outdoor room. It can also slow or prevent movement in a certain direction. The walls of the outdoor room determine the vertical sides of the room in the same manner as the walls of an indoor room. Thus, outdoor wall materials should not be placed in the middle of the lawn, where a wall would not logically be located. Materials used to form outdoor walls may be natural (shrubs, small trees, ground covers, and flowers) or man-made (fencing and masonry).

The **outdoor floor** provides the surfacing for the outdoor room. The materials used for the outdoor floor might be natural surfacing, such as grass, ground covers, sand, gravel, or water. They might also be man-made surfacings, such as brick, concrete, patio blocks, or tile.

The **outdoor ceiling** defines the upper limits of the outdoor room. It may offer physical protection, such as an awning or aluminum covering, or merely provide shade, such as a tree. In temperate regions, a deciduous tree is an ideal ceiling material for placement near the home. It gives shade from the hot summer sun, and then drops its leaves in the winter to allow the sunlight through, which warms the house.

Figures 6-7 and 6-8 illustrate two use areas around a home landscape. Carefully analyze the development of these areas as outdoor rooms, noting the different materials used for the walls, ceilings, and floors. Also examine the corresponding landscape plan for each area.

Figure 6-5 The Family Living Area provides a setting that enables users to pursue their recreational interests outdoors.

Figure 6-6 A typical outdoor room, containing wall, ceiling, and floor components

Figure 6-7A Example of a Public Area

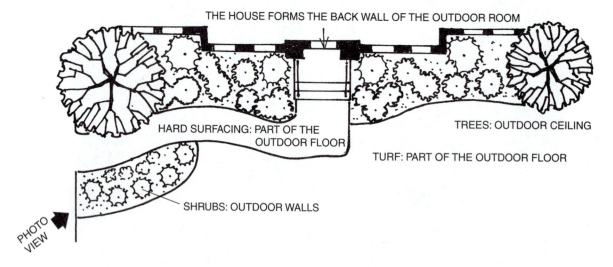

Figure 6-7B Plan view of the Public Area

Figure 6-8A Example of a Family Living Area

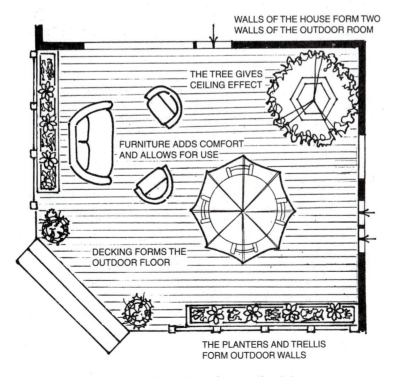

WALLS OF THE HOUSE FORM TWO
WALLS OF THE OUTDOOR ROOM

THE TREE GIVES
CEILING EFFECT

FURNITURE ADDS COMFORT
AND ALLOWS FOR USE

DECKING FORMS THE
OUTDOOR FLOOR

THE PLANTERS AND TRELLIS
FORM OUTDOOR WALLS

Figure 6-8B Plan view of a Family Living Area

Practice Exercises

A. Figure 6-9 shows three individual houses on their own lots. All windows, doors, and rooms are marked. Also provided is information on the family living in each house. Trace each house and its lot on tracing paper. Using a straightedge and pencil, divide each lot into the three or four use areas that seem appropriate for the family. With a scale instrument, determine the approximate square footage of each area you lay out. Is each area a practical size? Are the shapes such that rooms could later be developed if necessary? Is the family living area going to receive a good deal of sunshine? (It should be located on the south or west side of

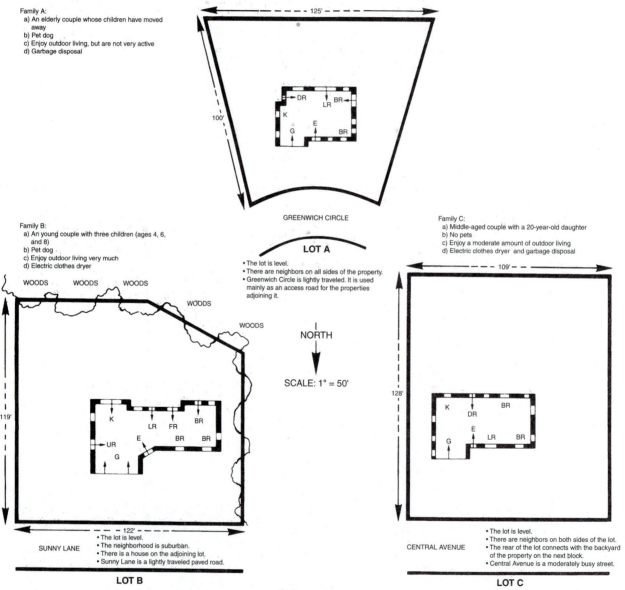

Family A:
a) An elderly couple whose children have moved away
b) Pet dog
c) Enjoy outdoor living, but are not very active
d) Garbage disposal

GREENWICH CIRCLE

LOT A

• The lot is level.
• There are neighbors on all sides of the property.
• Greenwich Circle is lightly traveled. It is used mainly as an access road for the properties adjoining it.

NORTH

SCALE: 1" = 50'

Family B:
a) An young couple with three children (ages 4, 6, and 8)
b) Pet dog
c) Enjoy outdoor living very much
d) Electric clothes dryer

SUNNY LANE

• The lot is level.
• The neighborhood is suburban.
• There is a house on the adjoining lot.
• Sunny Lane is a lightly traveled paved road.

LOT B

Family C:
a) Middle-aged couple with a 20-year-old daughter
b) No pets
c) Enjoy a moderate amount of outdoor living
d) Electric clothes dryer and garbage disposal

CENTRAL AVENUE

• The lot is level.
• There are neighbors on both sides of the lot.
• The rear of the lot connects with the backyard of the property on the next block.
• Central Avenue is a moderately busy street.

LOT C

Figure 6-9

the property if possible.) Is the private living area receiving morning sun from the east? If it is intended as a morning use area, western afternoon sun is of little value.

B. Figures 6-10 and 6-11 illustrate two partially completed landscapes. Trace each landscape onto paper. Applying your knowledge of symbolization and the outdoor room concept, complete the landscape plans. As you insert symbols for trees, shrubs, and construction materials into the plans, be certain that each feature is playing the role of a wall, ceiling, or floor element.

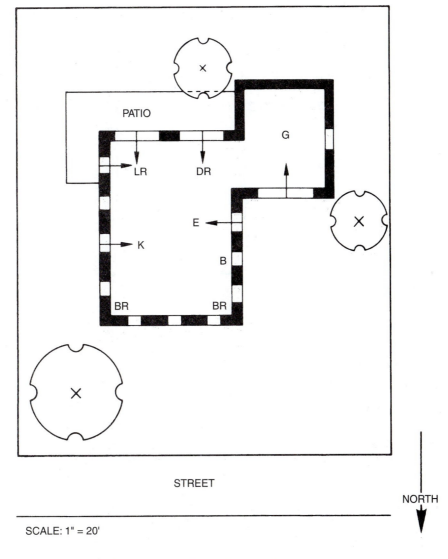

SCALE: 1" = 20'

Figure 6-10 Partially completed landscape

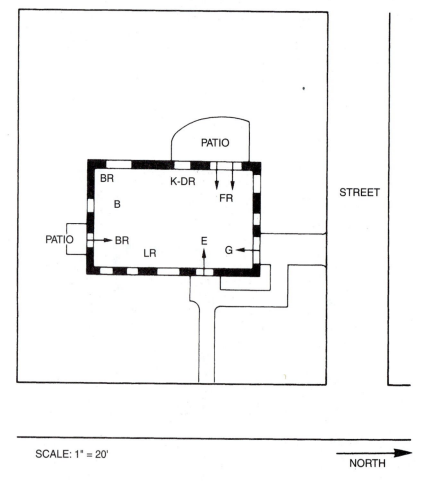

SCALE: 1" = 20'

NORTH

Figure 6-11 Partially completed landscape

Achievement Review

A. Indicate in which use area the following activities occur.

Activity	Public Area	Family Living Area	Service Area	Private Living Area
picnicking				
welcoming guests				
hanging laundry				
breakfast, coffee				
badminton				
pitch and catch				
trash can storage				

B. Indicate if the following materials are used for outdoor walls, ceilings, or floors.

Material	Outdoor Wall	Outdoor Ceiling	Outdoor Floor
brick wall			
shrubs			
crushed stone			
high-branching tree			
turf grass			
fencing			

C. Designating use areas based on how a space functions within the landscape has application to all types of landscapes, not just residential properties. Consider the following settings and indicate which functional use area would be assigned.

  a. public area
  b. general use area
  c. service area
  d. private living area

1. In a shopping mall, it's the area where deliveries are made to the stores.
2. In an industrial park, it's the area where employees park their cars.
3. On a college campus, it's the area where the dining hall and bookstore are located.
4. It's the view of a medical center as seen by people driving by.
5. In a state park, it's the changing rooms for people going to the beach.
6. It's the main entrance to a national museum.
7. Within a city park, it's the area where families gather for picnics.
8. It's the main entrance to a university, leading from the highway to the admissions office.
9. In a shopping center, it's the maintenance complex.
10. In a high school, it's the athletic fields.

Suggested Activities

1. Study various types of house floor plans. (These can be found in numerous magazines.) Which ones provide the best access to the outdoors through doorways? Which ones have only visual connection between the house and garden through windows?

2. Walk down a neighborhood street. Evaluate the public areas of the homes. Have the houses been placed in attractive settings? Are the wall elements where they should be or do they project into the center of the outdoor room?

3. Evaluate the property on which your school is located. Can different use areas be seen? Does the landscape appear to be a series of connecting outdoor rooms?

4. Prepare a photo study of your home property. Take photographs of each area of the property that seems to have a different purpose. Mount the photos on a sheet of poster board and label each photo as one of the four use areas described in this chapter. Then, identify the exact materials that serve as wall, ceiling, and floor components. Finally, make a judgment of how well the property does or does not reflect the outdoor room concept and why.

CHAPTER 7

Plant Selection

Objectives:

Upon completion of this chapter, you should be able to

- describe the origins and forms of the plants used in landscapes
- explain plant nomenclature
- describe the factors relevant to proper plant selection

Of all the features in the designed landscape, none is likely to attract as much attention from the client as the selection of the plants that will become all or part of the walls, ceiling, and floors of the outdoor room. Logic may or may not prevail at this stage of design development. Sentimentality, pretentiousness, or economics may outweigh horticultural reality in determining what plants will be specified, purchased, installed, and maintained in the landscape. Every landscape designer and landscape architect must learn the plants of the region where he or she practices if clients are to be served most professionally. Regrettably, too few schools of landscape architecture give adequate emphasis to this important training detail, causing many graduates of these programs to believe that their site planning and engineering expertise is all that they owe to their clients. Add to this problem clients who have strong but unqualified opinions about plants, and the resulting plant specifications for a landscape design can be limited, inappropriate, dated, or disastrous.

The Origins of Landscape Plants

Plants have found their way into American landscapes from the farthest reaches of the world. The diversity of our national flora is far greater today than it was when the *Mayflower* dropped anchor at Plymouth. As people from across the world have converged on America, they have brought with them seeds, plants, and propagative stock to re-create in this country's fertile soils the favored plants of their homelands. Thus, the plant palette from which modern landscape designers can select includes literally thousands of choices.

Native plants have evolved naturally within a certain geographic location over a long period of time. Examples are the eastern white pine of the northeastern United States and the Douglas-fir of the Pacific Northwest.

Exotic plants have been introduced into an area by some means other than nature. For example, many of the junipers and yews common to American landscapes actually are native to China and Japan. Brought back by plant collectors either directly or as imports via other countries, many exotics have adapted very well to growing conditions in the United States and constitute an important group of plants under production in our nurseries and greenhouses.

Naturalized plants entered the country as exotics, but have adapted so well that in many areas they have escaped cultivation and are often mistaken for native plants. They now are commonly found both in and out of landscaped settings. The bird of paradise is such a plant. Native to South America, it grows like a native plant in the desert of the southwestern United States.

Plant Forms Available

The nursery industry works closely with the landscape industry to produce the plant materials required for the country's private and public landscapes. While some plants are collected in the wild and others may be acquired from private sources, most landscape plants are produced for sale in field nurseries and greenhouses. Inventoried, described, and priced in catalogues provided by wholesale nurseries, the plants are itemized according to whatever system best categorizes the material grown. Some catalogues will distinguish the plants as coniferous (cone-bearing) or flowering; others may separate trees from shrubs and groundcovers from vines. Herbaceous (non-woody) plants may be listed as annuals or perennials, depending upon whether they live only one growing season or will survive the winter to bloom again the following year.

Trees, shrubs, vines, and many groundcovers are known collectively as woody plants. Other groundcovers may be **herbaceous**, as described above, a term which also applies to some vines,

grasses, and nearly all of the plants used to create floral displays. The common term for this latter group of herbaceous landscape plants is simply, flowers.

How Landscape Plants Are Sold

Woody plants grown in nurseries are sold in one of three forms: bare rooted, balled and burlapped, or containerized. Each form has certain advantages and disadvantages that landscapers should be aware of before selecting plants for installation.

A **bare-rooted plant** is one that has been dug from the nursery field, then had the soil washed away from the root system, Figure 7-1. To prevent excessive drying of the roots until the plant is purchased and installed, the root system may be wrapped in damp moss and inserted into a plastic bag, or the roots may be dipped in wax. Plants sold in bare-rooted forms are usually deciduous species that are small and are dormant at the time of harvest.

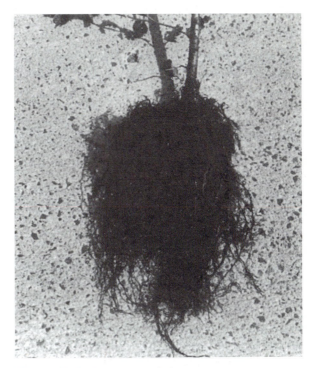

Figure 7-1 Bare-rooted plant

BARE-ROOTED PLANTS

Advantages	Disadvantages
• Lightweight	• Harvesting technique reduces the root system and slows new growth
• Least expensive	• Transplanting season is limited to the early spring and late fall

Balled and burlapped (or B & B) plants are dug at the nursery with a ball of soil intact around the root system, Figure 7-2. The size of the ball is determined by the size of the plant and is further governed by standards established by The American Nursery and Landscape Association. These industry standards assure that the plants will have the greatest opportunity for successful transplanting and protect the landscaper from inadvertently receiving plants with too little root system to survive the transplant. Once the root ball is dug, it is wrapped with burlap and tied or pinned to prevent the soil ball from breaking apart. A more recent harvest trend, especially suited for larger plants, replaces or supplements the burlap with a wire basket around the root ball, Figure 7-3. This gives added strength to assure that the soil ball

Figure 7-2 Balled and burlapped plants

remains intact around the roots. Plants sold in B & B form are heavier and more cumbersome to handle, but the technique allows plants of nearly any size and type to be successfully transplanted. B & B soil balls are nearly essential for the successful transplant of all types of evergreens and large deciduous woody plants.

BALLED AND BURLAPPED PLANTS

Advantages	Disadvantages
• Less disturbance to root system	• Added weight makes handling more difficult
• Allows a longer season for transplanting	• More costly
• Allows large plants to be moved and installed	

Containerized plants are grown and sold in containers such as metal cans, plastic buckets, bushel baskets, plastic bags, and wooden boxes,

Figure 7-3 This tree ball has been placed directly from the field into a burlap-lined wire basket, then tied at the top with twine.

Figure 7-4. Since they are sold in the same containers they have been grown in, there is no loss of roots. Therefore they transplant readily, with little or no transplant shock and no growth time lost. Containerized landscape plants are young plants, seldom more than three or four years old. Although usually smaller than balled and burlapped plants, containerized plants can also be installed at almost any time of the year since their entire root system is intact.

CONTAINERIZED PLANTS

Advantages	Disadvantages
• Usually small and easy to handle	• Large plants seldom available
• Entire root system is intact. No harvest injuries to impede transplanting success	• Plants become rootbound if in container too long
• Few restrictions on time of transplant	

Herbaceous plants are sold in a variety of forms, some of which are best suited to amateur gardeners, while others are more appropriate for professional landscape installations. Commonly known simply as flowers these plants may be started from seeds, which is often the way that home gardeners obtain their plants. It is the least expensive way to produce the colorful annuals and perennials that they desire for their gardens, Figure 7-5. However, landscape contractors and landscape management professionals are more likely to use flowers that are already growing and ready to begin blooming when they are installed. In this form, the flowers are collectively termed **bedding plants**. The plants are started in commercial greenhouses in late winter and are ready for sale to landscapers by early spring. The containers may be plastic or pressed peat moss. The peat container is biodegradable and allows the plant and its container to be installed directly into the soil without injury to the root system. The plant is able to continue growing

almost without interruption. Non-biodegradable containers must be removed before the bedding plants are installed.

Bulbous perennials are available to both amateurs and professionals as propagative structures known collectively as **bulbs**. Homeowners may purchase bulbs in small quantities at their nearby garden centers, while professional landscapers usually acquire theirs in larger quantities from wholesale suppliers. Bulbous perennials may be either **tender** or **hardy**, depending upon their time of planting, time of bloom, and ability to survive the winter. Tender bulbs, such as gladiolus and begonia, will not survive the winter in most parts of the country, so they are planted in late spring, bloom in the summer, and are dug up and stored indoors until the next year. Hardy bulbs, such as tulip, crocus, and daffodil, will

Figure 7-4 Containerized plant

Figure 7-5 Packaged seeds are an easy and inexpensive way to start a flower garden.

survive most American winters, so they are planted in the fall and flower the following spring. They do not need to be dug up and brought indoors at any time.

Other herbaceous plants, including vines and ornamental grasses, are grown by wholesale nurseries and made available to the landscape industry in assorted containers that may include any of those described above.

Plant Nomenclature

Most plants have two names. One is a **common name**, by which the plant is known within a country or region of a country. Winged euonymus or grey dogwood are examples of common names. The other name is the **botanical name** of the plant, a name that is used and recognized internationally. The botanical name is expressed in Latin and is assigned to the plant by a taxonomist (a person who specializes in the classification of plants

and animals). Plant taxonomy is governed by the International Rules of Botanical Nomenclature, which are established by an international botanic congress. Botanical names are recognized in all countries, regardless of native language. In the examples above, the botanical name of winged euonymus is *Euonymus alatus*, and that of grey dogwood is *Cornus racemosa*.

In *plant nomenclature* (naming of plants), each plant in the world has a unique name. Each identical plant sharing that two-part name, termed a *Latin binomial*, is referred to as a *species*. The first part of the name (such as *Cornus*) is called the genus. The second part is called the epithet (such as *racemosa*).

NOTE: Many texts and individuals use the word *species* to mean *epithet* as well as using the term to refer to the complete binomial.

A genus (plural is *genera*) name may be applied to several closely related types of plants. For example, all pines are in the genus *Pinus*. The epithet eliminates all other types but one and creates the unique species. Thus, *Pinus sylvestris*, *Pinus resinosa*, *Pinus nigra*, and *Pinus strobus* are four distinctly different plants.

Landscapers should know both the common and botanical names of the plants with which they work. The common name is most recognizable to clients. However, to assure that the nursery provides the correct plant for installation, the landscaper must specify it by its botanical name. This is necessary because many common names are localized in their usage, and because a single plant may have numerous common names, even within one locality. Conversely, the botanical name of a plant is the same around the world.

Selecting the Proper Plant

Even after the landscape designer has defined the general type of plants needed for the landscape, the selection of exact species remains. Too often the designer abbreviates this important phase of design development or simply does it incorrectly. Plants may be selected for reasons of sentimentality, easy availability, low price, or trendiness. While those reasons may have occasional validity if the chosen plants are appropriate in other more important ways, it is unlikely that plants will serve the landscape properly if selected for only those reasons.

Selection Factors

Plants have great diversity, and as such they must be matched against criteria established by the designer if they are to fill their specified roles in the landscape. While not intended to be a complete listing, here are some common factors that must be considered when selecting a specific plant for the landscape.

Role Factors. What function will it serve in the landscape?

a. Is it to be an architectural element serving as a wall, ceiling, or floor component of the outdoor room? Will it totally or partially control the views into and out from the landscape? Will it add privacy? Will it create an enframed setting for the house or other buildings, Figure 7-6?

b. Is it to be an engineering element that will aid in the solution of problems such as erosion control, traffic control, diverting winds, retaining moisture, or blocking glare, Figures 7-7 and 7-8?

WITH FENCING . . .
TO PROVIDE SCREENING AND SECURITY

WITH TREES . . .
TO SCREEN AN UNSIGHTLY AREA

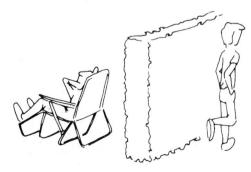

AS A HEDGE FOR PRIVACY

Figure 7-6 Various ways to screen with shrubs

Figure 7-7   This hedge, bordering the sidewalk, prevents pedestrians from cutting across the corner of the lawn.

c. Is it to serve in climate control by filtering or blocking sunlight, or softening the impact of a prevailing breeze, Figure 7-9?

d. Is its role primarily aesthetic, contributing one or more sensory qualities such as color, fragrance, texture, or taste, as shown in the color photo section of this text?

Hardiness Factors.   Will it survive the winter in the region of the country where the landscape is located? One of the most reliable measurements of a plant's potential for survival is its hardiness rating. The United States Department of Agriculture has prepared, and periodically updates, a Hardiness Zone Map, Figure 7-10, which shows the average annual minimum temperatures of the 50 states and much of Canada and Mexico. It divides the continent into 11 hardiness zones. On the most detailed maps, every county within the 48 adjacent mainland states is shown, along with its hardiness zone rating.

Each hardiness zone has an average annual minimum temperature variation of 10 degrees Fahrenheit. As the hardiness zone number increases, so does the temperature minimum. For example, northern Kansas temperatures (Zone 5) drop to between -10 and -20 degrees F in the winter, but southern Kansas (Zone 6) drops only to between 0 and -10 degrees F during the same months. Zones 1 and 11 represent the coldest and warmest regions of the country and are the least common zones within the United States. On color coded and detailed Hardiness Zone Maps, Zones 2 through 10 are further divided into *a* and *b* zones. The *a* zone represents the colder half of the 10-degree variance, and the *b* zone represents the warmer half of the 10-degree variance.

NOTE: The *a* and *b* zones do not show on the map in this text.

Figure 7-8 This hedge is set back too far from the sidewalk to function as an outdoor wall. Pedestrians have cut across the corner, destroying the lawn.

GENTLE BREEZE

WIND

Figure 7-9 Shrubs can reduce the velocity of wind, turning a gust into a breeze.

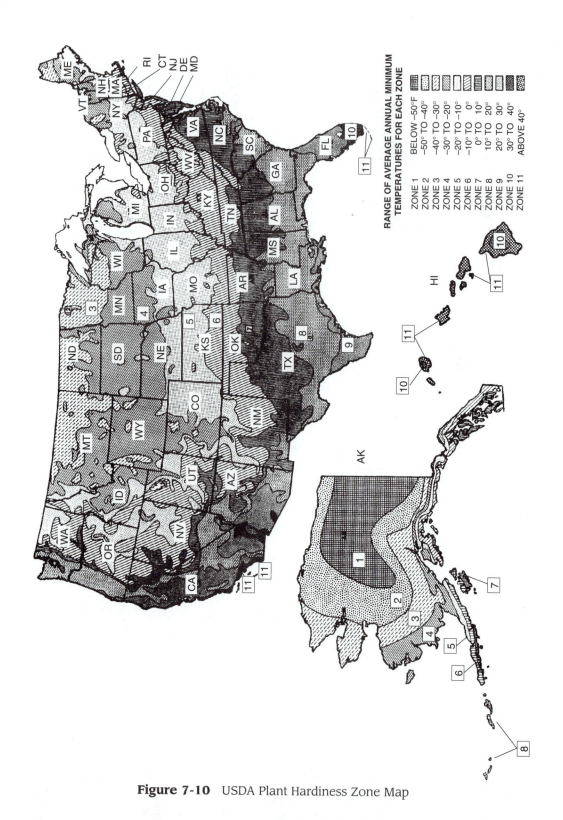

RANGE OF AVERAGE ANNUAL MINIMUM
TEMPERATURES FOR EACH ZONE

ZONE 1   BELOW −50°F
ZONE 2   −50° TO −40°
ZONE 3   −40° TO −30°
ZONE 4   −30° TO −20°
ZONE 5   −20° TO −10°
ZONE 6   −10° TO   0°
ZONE 7   0° TO   10°
ZONE 8   10° TO   20°
ZONE 9   20° TO   30°
ZONE 10   30° TO   40°
ZONE 11   ABOVE 40°

Figure 7-10   USDA Plant Hardiness Zone Map

When a plant is assigned a hardiness zone rating, it means that the plant may survive the winter in that zone or in zones with a higher numbered rating. Thus a tree or shrub rated as a Zone 5 plant may survive the winters in Zones 5 through 11, but not in Zones 4 through 1. However, there are also upper temperature limits to a plant's hardiness. Certain plants cannot survive or grow normally in climates that are too warm. That is why there are no apple trees in the Florida Keys. Still other plants can have unreliable hardiness zone ratings when grown in **microclimate** areas, regions with atypical growing conditions. For example, some plants will survive the winter because they are buried under heavy snow and escape the severe cold or drying winds that would otherwise kill them. If a winter fails to provide the protective snow cover, the plants would be unable to survive.

Heat Zone Considerations. While hardiness zone ratings for plants have been and continue to be studied and adjusted for many years, in response to the earth's natural climatic changes, there has been little research into the impact of excessive warmth on plant growth and survival. In the mid-90s, the American Horticultural Society began the development of data that tracked the number of days that various regions of the United States experienced temperatures over 86 degrees Fahrenheit (30 degrees Celsius). Termed a *heat day*, it represents the point at which a plant will begin to experience physiological injury because of the temperature. Using a county-specific map, similar to the USDA Hardiness Zone Map, the AHS has divided the nation into 12 *heat zones*. A heat zone represents the average number of heat days per year for that region. The Plant Heat Zone Map divides the country into 12 zones, with Zone 1 having no heat days and Zone 12 averaging 210 heat days. Plants can then be described not only on the basis of the coldest hardiness zone in which they will survive, but the warmest heat zone that they can tolerate without injury.

Several factors complicate the current use of the AHS Plant Heat Zone Map. First, it does not have the popular recognition and longtime familiarity that the Hardiness Zone Map has, so many people are simply unaware of it. Second, far fewer plants are coded for heat tolerance than for cold tolerance, so finding the appropriate heat zone rating for a specific species can be difficult. Eventually there will be ample information, but for now it is uneven. Third, the current trend toward global warming, whether short-term or long-term, seems certain to change the configuration of the Heat Zone Map even before it reaches its desired level of public and professional appreciation. Finally, the heat zone coding assumes that the plants are growing with adequate water supplies and with no factors that will create stress or provide an artificially supportive microclimate. Given the variances that are introduced into a region by periodic drought, elevation differences, disturbances to the soil from construction, compaction, or other causes, changes in the natural vegetation, or the addition of buildings and other structures, it is predictable that plants will often be asked to grow in conditions that are more or less favorable than those envisioned when they were coded.

A few texts on heat zone gardening are beginning to appear, and their number can be predicted to increase in the next few years. As more plants are coded for their heat zone tolerances, that information will appear in the popular and trade publications, giving proper and overdue recognition to the role that heat plays in plant survival. However, ratings for cold and heat tolerances will always be generalizations, so it is wise to avoid selecting plants that are at the limits of their tolerance range. To view or purchase a copy of the Heat Zone Map, visit the Web site of the American Horticultural Society at <http://www.ahs.org>.

Physical Factors. Closely associated with the role of a plant are its physical features, which enable it to fulfill the role or perhaps make it unsuitable. Some of the features of plants that determine their suitability for particular uses include

a. mature size and rate of growth.

b. density of the canopy, with and without leaves.

c. branching habit.

d. thorns.

e. shape or silhouette.

f. leaf size and thickness.

g. leaf pubescence.

h. type of root system.

i. flowering characteristics.

j. seasonal variation.

Unless the landscape designer or landscape architect is also a knowledgeable horticulturist, some plants selected may grow to a mature size unsuitable for their place in the landscape. Just as puppies begin life small, but some grow to be toy poodles while others become German shepherds, so too do nursery plants grow to vastly different sizes. The shrub that matures to a height of eight feet and a width of ten feet is incorrectly used beneath a residential picture window. Yet it may be the perfect plant for creating a high, dense screen to block an unsightly view. A tall tree can seem threatening if planted too close to a low, single-story house. However, if the intent is to create a cathedral ceiling effect with a high canopy for the outdoor room, then that same tall tree may fit perfectly, Figure 7-11. Also to be considered when comparing the mature size of a possible plant choice is its rate of growth. Depending upon the species, the length of the growing season, and the growing conditions, some species will attain full size in only a few years. Others may require a century to reach maturity. The designer must match a plant's rate of growth with the expected time that the plant is to be of significance in the landscape. For example, most oaks are very large at maturity, but they grow much more slowly than ash trees, which are also large at maturity. For a residential estate, the ash tree might be a preferred choice if the design called for a large tree, simply because it would be large sooner. However, if the site is a large park or cemetery, where a stronger, longer-lived species is preferred and where there is not the same need for rapid effect as in a residential setting, then the oak might be the proper selection.

Canopy density results from a combination of leaf size and the extensiveness of the branches.

PROPERLY SIZED TREES PROVIDE A VISUALLY APPEALING FRAME FOR HOMES.

OVERSIZED TREES DWARF AND DEHUMANIZE A RESIDENTIAL LANDSCAPE.

LARGE TREES PROVIDE AN EXCELLENT HIGH CEILING FOR THE RECREATIONAL LANDSCAPE.

Figure 7-11 Size is an important factor in tree selection.

Obviously, more branches with large leaves will create a denser canopy than will smaller leaves and/or more open branching. Denser canopies allow less sunlight and water to reach the surface of the ground. That may or may not be desired, so it must be considered by the landscape designer or landscape architect at the time of plant selection. It would not be appropriate for shrubs and flowers requiring dense shade to be planted beneath trees that offered filtered sunlight. Canopy density also affects both what can and cannot be seen through a planting, and the amount of air that can pass through the planting. When screening or wind control is the role of the plants selected, canopy density becomes an important selection factor. While the canopy density of evergreens is consistent year round, deciduous trees and shrubs change with the seasons, becoming less dense as their leaves fall. That may mean a reduction in the amount of privacy provided by a planting. With deciduous trees, the change in canopy density resulting from the change of seasons may mean that the shade provided to the house during the hot summer becomes

the welcomed warming sunshine of winter, Figure 7-12.

The **branching habit** of plants is a combination of the number of branches, the average size of the branches, the flexibility of the branches, and the vertical or horizontal direction of their growth. This plant characteristic becomes important when traffic control is the role of the species being selected. Species such as the spireas have small, thin branches that bend easily and are readily trampled. They are inappropriate for traffic control roles. However, plants such as winged euonymus with dense, thick, inflexible branches stand up better against invaders and are good choices in such situations. Another example where branching habit must be considered is when species are being used as formal hedges. The shaping and shearing used to create and maintain such plantings dictates the need for species that mass well and that have some vertically oriented branches. Horizontal branching species may work well as informal, unsheared hedges, but they do not do well in formal roles.

Figure 7-12 A deciduous tree off the southwest corner of a building provides summer shade while allowing winter warmth.

The presence of thorns makes certain species suitable for certain roles in the landscape and equally unsuitable for other roles. When traffic control is the role, the presence of thorns may be a positive supplement to the branching habit of potential species. If safety or ease of litter cleanup is a concern, then the thorniness of a species otherwise well suited may be a negative factor.

The shape or **silhouette** of a plant may also determine its suitability for use in the landscape. Figures 7-13 and 7-14 illustrate some of the shapes that trees and shrubs may take. Some of the more common species, their characteristics, and recommended uses are also listed. It is better to select plants that will provide the desired silhouette naturally than to select species that must be pruned and trained to do the job required in the landscape. Plants such as pin oaks and willows have low, sweeping branches as they mature, Figure 7-15. They make graceful lawn specimens where they can spread their branches fully. However, it is difficult to walk, park, or play beneath them, so they are unsuited for playgrounds or parking lots. When human activity is to occur beneath trees, a high-branching species should be selected. For example, small, vase-shaped trees (25 to 30 feet tall) make excellent outdoor ceilings for home patios. They provide shade while keeping the outdoor ceiling low and intimate. Large vase-shaped trees (50+ feet tall) function well as street tree plantings and in parking lots where cars can pass easily beneath them. Another example could be the formal approach to a commercial site where the intent of the design is to convey dignity and impress visitors. Lining both sides of a long entry drive with evenly spaced columnar forms of a single species might accomplish that design goal. With some species, the shape of the plant may suggest its best use in the design or how well it will fare during certain seasons. Many shrubs, especially needled evergreens, have rigid, geometric shapes. As such, they function well in the role of specimen or accent plants and must be used cautiously. Other shrubs are softer and less defined in their silhouettes. They mass easily and quickly lose their individual identity. Where heavy snow

loads on top of plants are predictable, the most appropriate plants may be those that are more prostrate (with horizontal branches) since the branches can bend rather than break beneath the weight of the snow.

Leaf size affects the density of the canopy of trees and shrubs. It also determines the foliage texture of plants, with large leaves creating coarse textures and small leaves creating fine textures. Leaf thickness contributes to the texture of the foliage. In addition, thicker leaves tend to absorb sound and thereby can aid in noise reduction. When noise control is a function of the plants being selected for the landscape, the designer needs to know which plants will perform most satisfactorily. As noted before, that necessitates a familiarity with the plant palette of the region.

Pubescence is the presence of fine hairs on the surface of leaves. Some species are noted for having visible, tangible amounts of pubescence. Such leaves are more likely to collect dust, pollen, and other fine particulate matter from the air than leaves without pubescence. In dusty or sooty locations, leaf pubescence might make one plant species more suitable than others if a contribution to air quality was one of the roles assigned to the selected plant.

The **root systems** of plants range from total tap root systems, with a large single root growing straight down into the soil, to full fibrous systems, which have thousands of fine, hairlike roots spreading out in all directions from the center of the plant. Fibrous root systems are useful for trapping and holding soil particles. That is an attribute that plants with tap roots do not have. If the role of the plant being selected was to reduce soil erosion due to wind or water runoff, only fibrous rooted plant species would logically be considered.

Depending upon whether the plant is woody or herbaceous, perennial or annual, cone bearing or non-cone bearing, and numerous other differences, the flowering characteristics of landscape plants vary greatly. Some plants, especially woody plants, flower once each year in the spring or summer. The flower display may be bright and showy or comparatively inconspicuous, depending upon

Silhouette and Examples	Characteristics	Possible Landscape Uses	Silhouette and Examples	Characteristics	Possible Landscape Uses
wide-oval  Flowering crabapple Silk tree Cockspur hawthorn Flowering dogwood	• spreads to be much wider than it is tall • often a small tree • horizontal branching pattern • branches low to the ground	• focal point plant • works well to frame and screen • can be grouped with spreading shrubs beneath	**round** Shinyleaf magnolia Cornelian cherry dogwood American yellow wood Norway maple	• width and height are nearly equal at maturity • usually dense foliage • if the tree is large, a heavy shade is cast	• lawn trees • mass well to create grove effect • larger growing species may be used for street plantings • smaller growing species can be pruned and used for patio trees
vase-shaped 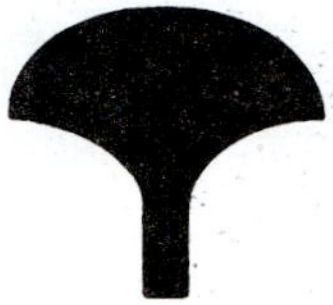 American elm	• high, wide-spreading branches • majestic appearance • usually gives excellent shade • an uncommon tree shape	• excellent street trees • allows human activities underneath • frames structures • use above large shrubs or small trees *note:* the American elm is easily killed by Dutch elm disease; this limits its use	**columnar**  Columnar Norway maple Columnar Chinese juniper Fastigiate European birch	• somewhat rigid in appearance • much taller than wide • branching strongly vertical	• useful in formal settings • accent plant • group with less formal shrubs to soften its appearance • frames views and structures
pyramidal  Pines Fir Spruce Hemlock Filbert Sweetgum Pin oak Sprenger magnolia	• pyramidal evergreen trees are geometric in early years • pyramidal deciduous trees are less geometric • pyramidal shape is less noticeable as the trees mature	• accent plant • large, high-branching trees allow human activity beneath • older trees may be valued for their irregular shapes *note:* avoid planting large trees near small buildings	**weeping** Weeping willow Weeping hemlock Weeping cherry Weeping beech	• very graceful appearance • branching to the ground • easily attracts the eye • grass or other plants cannot be grown beneath them	• focal point plant • screens • attractive lawn trees *note:* avoid grouping with other plants

Figure 7-13   Typical tree silhouettes, characteristics, and landscaping uses

Shrub Silhouette and Examples	Characteristics	Recommended Landscape Uses
pyramidal Upright yew Pyramidal junipers False cypress Arborvitae	• taller than it is wide • rigid and stiff • attracts attention • geometric shape • usually evergreen	• accent plant • focal point • use to mark entries and at incurves • group with less formal spreading shrubs
upright and loose Lilac Smoke bush Rose of Sharon Rhododendron	• taller than it is wide • loose, informal shape • usually requires pruning to prevent leggy growth	• closely spaced for privacy • use to soften building corners and lines • useful for screening and framing views
columnar Hicks yew Italian cypress Arizona cypress	• width is about half the height • geometric, flat topped, and dense	• accent plant • foundation plantings • closely spaced for hedges • mass closely when a solid wall is desired
globular 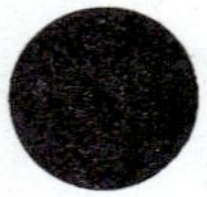 Brown's yew Globe arborvitae Burford holly Globosa red cedar	• as wide as it is tall • geometric shape • attracts attention • does not mass very well	• accent plant • use several with a single pyramidal shrub for strong eye attraction • avoid overuse
low and creeping Andorra juniper Bar-Harbor juniper Cranberry cotoneaster Prostrate holly	• low growing • much wider than it is tall • masses well • irregular shape • loose, informal shape	• use to edge walks • cascades over walls • controls erosion on banks • grown in front of taller shrubs
spreading Hetz junipers Pfitzer junipers Spreading yew Mugo pine	• wider than it is tall • medium to large shrub • masses well • usually dense foliage	• use at outcurve • place at corners of buildings • useful for screening, privacy, and traffic control
arching Forsythia Beautybush Vanhoutte spirea Large cotoneaster	• wider than it is tall • prevents the growth of other plants beneath itself • graceful silhouette • usually requires yearly thinning	• provides screening and dense enclosure • softens building corners and lines • background for flowers, statuary, fountains

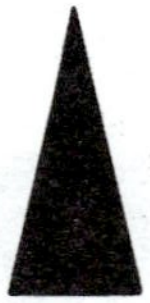

Figure 7-14 Shrub silhouettes

Figure 7-15 Trees with weeping growth habits are attractive to view, but difficult to walk beneath. They should not be used as street trees.

the species. Herbaceous plants may bloom throughout the growing season, as with annual bedding plants; or they may bloom for several weeks, then wane, as with bulbs and other perennials. Some plants produce sterile flowers, incapable of producing fruit. That may or may not be important. With sterile flowers, there is no crushed fruit to stain sidewalks, but also no food to attract wildlife to the garden. Some flowers have desirable fragrances, making them ideal for certain settings. Others have no fragrance or even an unpleasant one, making them undesirable in certain locations at certain times. Some flowers attract bees and other insects, while others will repel certain insects. Some plants require certain soil conditions or temperatures or day lengths to induce flowering. Others will thrive under a wide range of conditions. In short, flowering characteristics are diverse and can contribute positively or negatively to the landscape's design, depending upon how a particular species matches the role assigned it by the designer.

Seasonal variation includes the presence or lack of flowers, but involves other plant attributes as well. It addresses a concern for consistency. The selected plant species must meet its obligation as a component of the outdoor room throughout all of the seasons of use, not just for part of the year. If year-round privacy screening is required, it is unacceptable for the plants to drop their leaves and permit viewing through the planting. If a flower border is to offer color enrichment throughout the growing season, then the plants used must come into bloom at different times, rather than coming and going at the same time. If the role of a plant is to be a strong focal point, a two-week color display is not good enough; it needs to catch the eye every day of the year. Conversely, if changing focal points are planned for an area, then the short-term effectiveness of individual plants is desirable.

Cultural Factors. The requirements for the growth and maintenance of plantings are their cultural requirements. Some plants establish easily, require minimal maintenance, and grow satisfactorily in a variety of locations and under assorted conditions. Other plants can only be successfully transplanted within a narrow window of time each year. Still others have very exacting requirements for growth and/or maintenance if they are to make their maximum contribution to the landscape. The more soil conditioning, spraying, pruning, thinning, weeding, fertilizing, or other requirements that are necessary to sustain a planting, the more costly it is to the client. The cost may appear at the time of installation or it may not become apparent until after a season of growth, but it is an expense in the form of materials and labor.

To be certain that the cultural requirements of the plants selected for a landscape are appropriate for the site and the client, the designer should seek answers to questions such as these:

■ Will the plants attain and hold the size and shape desired without the need for extensive pruning and training?

■ How frequently do the plants need pruning to prevent legginess or to induce fullness or to eliminate water sprouts and sucker growth?

A Guide to Landscape Trees

Tree		Evergreen	Deciduous	Maximum Height		
Common Name	**Botanical Name**			10'–25'	25'–60'	60' and up
Almond	Prunus amygdalus		•	•		
Amur Corktree	Phellodendron amurense		•		•	
Apples	Malus species		•		•	
Apricot	Prunus armeniaca		•		•	
Arborvitae	Thuja occidentalis	•			•	
Ash Arizona Green White Flowering	Fraxinus species F. velutina F. pennsylvanica F. americana F. ornus		 • • • •		 • • •	 •
Beech American European	Fagus species F. grandifolia F. sylvatica		 • •			 • •
Birch Canoe European Sweet	Betula species B. papyrifera B. pendula B. lenta		 • • •		 • 	 • •
Cherry	Prunus padus		•		•	
Chestnut, Chinese	Castanea mollissima		•		•	
Crabapple Flowering Fruiting	Malus species 		 • •	 • •		
Crape Myrtle	Lagerstroemia indica		•			
Cypress Italian Monterey Sawara false	 Cupressus sempervirens Cupressus macrocarpa Chamaecyparis pisifera	 • • •				 • • •
Dogwood Flowering Kousa	Cornus species C. florida C. kousa		 • •	 • •		
Douglas Fir	Pseudotsuga menziesi	•				•
Elm American Chinese Pioneer Smoothleaf	Ulmus species U. americana U. parvifolia U. carpinifolia 		 • • 		 • 	 • •
Fig	Ficus carica	•			•	
Fir Balsam White	Abies species A. balsamea A. concolor	 • •				 • •
Fringe tree	Chionanthus virginicus		•	•		
Ginkgo	Ginkgo biloba		•		•	
Goldenchain	Laburnum watereri		•	•		
Golden Rain Tree	Koelreuteria paniculata		•	•		

Time of Flowering			Fruiting Time			Good Fall Color	Hardiness Zone Rating	Comment
Early Spring	Late Spring	Early Fall	Late Summer	Early Fall	Late Fall			
	•	•					8	Edible fruit
	•			•			4 to 8	Corky bark
•			•				4 to 8	Does not fruit in warmer zones
•			•				5 to 5	Edible fruit
							3 to 7	Many types in wide size range
	•			•		•	5 to 9	Seedless forms
	•			•			3 to 9	are recommended
	•			•			3 to 9	
	•			•			5 to 9	
							5 to 9	
						•	3 to 9	Low-branching. Beeches
						•	3 to 9	do poorly in city air
							5 to 9	
•						•	5 to 9	Often short-lived because
•						•	3 to 9	of certain insect
•						•	3 to 7	damage
•			•				4 to 8	Edible fruit; attracts wildlife
	•			•			4 to 8	Edible fruit; disease resistant
	•			•		•	4 to 8	Showy flowers and fruit
	•			•		•	4 to 7	Edible fruit
		•					7 to 10	Difficult to transplant
							7	Pyramidal growth habit
							7	
							3 to 8	
•						•	5 to 9	Good patio tree
•						•	5 to 8	
							4 to 7	Dense foliage; pyramidal
	•			•			3 to 9	Vase shaped; disease prone
	•			•			5 to 9	Good for residential use
	•			•			5 to 8	Disease resistant
•			•				6 to 10	Edible fruit
							4 to 8	
							4 to 8	
	•			•			4 to 8	May also be used as a shrub
						•	4 to 8	Use only male trees
	•						5 to 7	Somewhat short-lived
		•			•		5 to 9	Coarse texture

A Guide to Landscape Trees *(continued)*

Tree		Evergreen	Deciduous	Maximum Height		
Common Name	**Botanical Name**			**10'–25'**	**25'–60'**	**60' and up**
Hawthorn	Crataegus species					
English	C. oxyacantha		•	•		
Green	C. viridis		•	•		
Washington	C. phaenopyrum		•		•	
Hemlock	Tsuga canadensis	•				•
Holly	Ilex species					
American	I. opaca	•			•	
English	I. aquifolium	•				•
Honeylocust, Thornless	Gleditsia triacanthos inermis		•			•
Hornbeam	Carpinus species					
American	C. caroliniana		•	•		
European	C. betulus		•		•	
Larch	Larix laricina		•			•
Linden	Tilia species					
American	T. americana		•			•
Little leaf	T. cordata		•		•	
Silver	T. tomentosa		•		•	
Magnolia	Magnolia species					
Bigleaf	M. macrophylla	•			•	
Saucer	M. soulangeana		•	•		
Southern	M. grandiflora	•			•	
Star	M. stellata		•	•		
Sweetbay	M. virginiana		•			
Maple	Acer species					
Amur	A. ginnala		•	•		
Hedge	A. campestre		•		•	
Japanese	A. palmatum		•	•		
Norway	A. platanoides		•		•	
Red	A. rubrum		•			•
Sugar	A. saccharum		•			•
Mountain Ash, European	Sorbus aucuparia		•		•	
Oak	Quercus species					
Live	Q. virginiana	•			•	
Pin	Q. palustris		•		•	
Red	Q. rubra		•			•
Scarlet	Q. coccinea		•			•
White	Q. alba		•			•
Peach	Prunus persica		•		•	
Pear, Bradford	Pyrus calleryana Bradford		•		•	
Pecan	Carya illinoinensis		•			•
Pine	Pinus species					
Austrian	P. nigra	•				•
Loblolly	P. taeda	•				•
Red	P. resinosa	•				•
Scotch	P. sylvestris	•			•	
White	P. strobus	•				•

Time of Flowering			Fruiting Time			Good Fall Color	Hardiness Zone Rating	Comment
Early Spring	Late Spring	Early Fall	Late Summer	Early Fall	Late Fall			
	•			•			4 to 7	Thorny
	•			•			4 to 7	
	•			•		•	3 to 8	
							3 to 7	Grows best in partial sunlight
					•		6 to 9	Male and female plants are needed for fruit set; pyramidal
					•		6 to 9	
				•			4 to 8	Good city tree; several varieties
	•		•			•	3 to 8	Shade tolerant
	•		•			•	4 to 8	Conical growth habit
						•	2 to 6	A deciduous, needled conifer
	•		•				3 to 8	Good street trees
	•		•				3 to 8	
	•		•				4 to 8	
	•						6 to 9	Very large leaves and flowers
				•			5 to 9	
•	•	•					7 to 9	Also usable as a large shrub
•		•				•	5 to 9	
	•						5 to 9	
•			•			•	3 to 8	Good fall foliage color
•			•				4 to 7	They make good lawn trees
•			•			•	5 to 8	
•			•				3 to 7	
•			•			•	3 to 9	
•			•			•	3 to 8	
	•		•			•	3 to 8	Susceptible to borer insects
							7 to 9	Strong trees; used widely as lawn and shade trees
						•	4 to 9	
						•	4 to 8	
						•	5 to 9	
							3 to 9	
•			•			•	5 to 8	Edible fruit
	•			•		•	4 to 9	Symmetrical and formal
	•			•			5 to 8	Nuts only mature in warmer areas
							5 to 7	Good for use as windbreaks
							6 to 9	Can be massed for grove effects
							2 to 7	
							2 to 7	
							3 to 7	

A Guide to Landscape Trees *(continued)*

Tree		Evergreen	Deciduous	Maximum Height		
Common Name	Botanical Name			10'–25'	25'–60'	60' and up
Plum Fruiting Purple flowering	Prunus species P. domestica P. cerasifera pissardi		• •	• •		
Redbud, Eastern	Cercis canadensis		•		•	
Russian Olive	Elaeagnus angustifolia		•	•		
Sapodilla	Achras zapota	•			•	
Sik Tree	Albizia julibrissin		•	•		
Spruce Blue Colorado Golden White Norway White	Picea species P. pungens glauca P. pungens P. glauca aurea P. abies P. glauca	• • • • •			• •	• • •
Sweet Gum	Liquidambar styraciflua		•			•
Sycamore	Platanus occidentalis		•			•
Tulip tree	Liriodendron tulipifera		•			•
Walnut Black English	Juglans species J. nigra J. cinerea		• •			• •
Willow Babylon weeping Corkscrew Pussy Thurlow weeping	Salix species S. babylonica S.matsudana tortuosa S. discolor S. elegantissima		• • • •	• 	• • •	
Zelkova, Japanese	Zelkova serrata		•			•

Time of Flowering			Fruiting Time			Good Fall Color	Hardiness Zone Rating	Comment
Early Spring	Late Spring	Early Fall	Late Summer	Early Fall	Late Fall			
•			•				4 to 8	Edible fruit
•			•				4 to 9	Deep red leaf color
•							4 to 9	Attractive, delicate flowers
	•			•			4 to 8	Silver foliage
•		•					10 to 11	
	•						6 to 9	
							2 to 7	Rigid, dense conifers; they make
							2 to 7	excellent lawn trees, but give
							2 to 7	them plenty of room to spread
							2 to 7	
							2 to 7	
				•		•	5 to 9	Excellent fall color; mixed tones
	•				•		3 to 9	White, peeling bark
	•		•				4 to 9	Needs room to grow and spread
	•			•			4 to 7	Edible nuts; detrimental to the
	•			•			6 to 8	growth of nearby plants
•						•	4 to 9	
•							4 to 9	Twisted stems; specimen plant
•							4 to 9	Willows grow quickly, have weak
•						•	4 to 9	wood, and thrive in wet areas
						•	4 to 8	Often a subsitiute for American elm

A Guide to Landscape Shrubs

Shrub		Evergreen	Deciduous	Mature Height		
Common Name	**Botanical Name**			**3'–5'**	**5'–8'**	**8' and up**
Almond, Flowering	Prunus glandulosa		•	•		
Azaelas	Rhododendron species					
Gable	R. poukanense hybrid		•	•		
Hiryu	R. obtusum Hiryu		•	•		
Indica	R. indica	•			•	
Kurume	R. obtusum Kurume	•		•		
Mollis	R. kosterianum		•	•		
Torch	R. calendulaceum		•	•		
Barberry	Berberis species					
Japanese	B. thunbergi		•		•	
Redleaved	B. thunbergi atropurpurea		•		•	
Wintergreen	B. julianae	•		•		
Bayberry	Myrica pennsylvanica	semi				•
Boxwood	Buxus species					
Common	B. sempervirens	•			•	
Little leaf	B. microphylla	•			•	
Camellia	Camellia species					
Japanese	C. japonica	•				•
Sasanqua	C. sasanqua	•				•
Coralberry	Symphoricarpos orbiculatus		•	•		
Cotoneaster	Cotoneaster species					
Cranberry	C. apiculata		•	•		
Rockspray	C. horizontalis	semi		•		
Spreading	C. divaricata		•		•	
Deutzia, slender	Deutzia gracilis		•	•		
Dogwood	Cornus species					
Cornelian Cherry	C. mas		•			•
Grey	C. racemosa		•			•
Red twig	C. stolonifera		•			•
Firethorn	Pyracantha species					
Scarlet	P. coccinea	semi				•
Formosa	P. koidzumii	•				•
Forsythia	Forsythia species					
Early	F. ovata		•		•	
Lynwood Gold	F. intermedia Lynwood		•			•
Showy border	F. intermedia spectabilis		•			•
Gardenia	Gardenia jasminoides	•			•	
Hibiscus	Hibiscus species					
Chinese	H. rosa sinensis	•				•
Shrub Althea	H. syriacus		•			•
Holly	Ilex species					
Chinese	I. cornuta	•				•
Inkberry	I. glabra		•			•
Japanese	I. crenata convexa	•			•	

Semi-evergreen indicates that the plants retain their leaves all year in warmer climates, but drop them during the winter in colder areas.

Season of Bloom**			Light Tolerance			Good Fall Color	Zone of Hardiness	Comment
Early Spring	Late Spring	Early Fall	Sun	Semi-Shade	Heavy Shade			
•			•				4 to 8	Very showy blooms
			•	•			6 to 9	Requires acidic soil condition and often iron chelate fertilizers
•	•			•			7 to 9	
	•			•			8 to 9	
	•			•			6 to 9	
	•			•			6 to 9	
	•			•		•	6 to 9	
	•		•	•		•	4 to 8	Good plants for traffic control; thorny
	•		•	•		•	4 to 8	
•			•	•			5 to 8	
			•	•			2 to 6	Fragrant leaves and berries
			•	•			6 to 9	Prunes well; good for hedges
			•	•			5 to 10	
•				•			7 to 9	Fragrant
		•	•	•	•		7 to 9	
		•		•		•	2 to 7	Good for erosion control
	•		•	•		•	4 to 8	Fall color comes from bright red fruit
	•		•	•		•	4 to 8	
	•		•	•		•	5 to 8	
	•		•	•			5 to 8	Delicate foliage; white blossoms
•			•	•		•	4 to 8	Also used as a small tree
•			•			•	4 to 8	Good for erosion control
•			•			•	2 to 8	
	•		•	•		•	6 to 9	Fall color comes from brightly colored fruit
	•		•			•	7 to 10	
•			•				4 to 9	Bright yellow flowers; cascading branching
•			•				5 to 9	
•			•				5 to 9	
	•	•	•	•			8 to 10	Fragrant flowers
	•		•				9 to 10	Also called Rose of Sharon
		•	•	•			5 to 9	
			•	•			6 to 9	Fruit is dark blue and not as showy as on the tree hollies
			•	•			3 to 7	
			•	•			5 to 8	

**Where no rating is given, flowers are either not produced or are not of importance.

A Guide to Landscape Shrubs *(continued)*

Shrub		Evergreen	Deciduous	Mature Height		
Common Name	**Botanical Name**			3'–5'	5'–8'	8' and up
Honeysuckle	Lonicera species					
Blue leaf	L. korolkowii zabel		•			•
Morrow	L. morrowii		•		•	
Tatarian	L. tatarica		•			•
Hydrangea	Hydrangea species					
Hills of Snow	H. aborescens grandiflora		•	•		
Oak Leaf	H. quercifolia		•	•		
Pee Gee	H. paniculata grandiflora		•			•
Jasmine	Jasminum species					
Common White	J. officinale	semi				•
Florida	J. floridum	•		•		
Winter	J. nudiflorum		•	•		
Juniper	Juniperus species					
Andorra	J. horizontalis plumosa	•		•		
Hetz	J. chinensis hetzi	•			•	
Japanese Garden	J. procumbens	•		•		
Savin	J. sabina	•			•	
Pfitzer	J. chinensis pfitzeriana	•				•
Lilac	Syringa vulgaris		•			•
Mahonia	Mahonia species					
Leatherleaf	M. bealei	•			•	
Oregon-grape	M. aquifolium	•		•		
Mockorange	Philadelphus virginalis		•		•	
Nandina	Nandina domestica	•			•	
Ninebark	Physocarpus opulifolius		•			•
Oleander	Nerium oleander	•				•
Photinia	Photinia species					
Chinese	P. serrulata	•				•
Red	P. glabra	•				•
Red-tip	P. fraseri	•				•
Pieris (Andromeda)	Pieris species					
Japanese	P. japonica	•			•	
Mountain	P. floribunda	•			•	
Pine, Dwarf Mugo	Pinus mugo compacta	•		•		
Poinsettia	Euphorbia pulcherrima	•				•
Pomegranate	Punica granatum		•			•
Potentilla (Cinquefoil)	Potentilla fruticosa		•	•		
Privet	Ligustrum species					
Amur	L. amurense		•			•
California	L. ovalifolium	semi				•
Regal	L. obtusifolium regelianum		•		•	
Waxleaf	L. lucidum	•				•
Quince, Flowering	Chaenomeles species					
Common	C. speciosa		•		•	
Japanese	C. japonica		•	•		

Semi-evergreen indicates that the plants retain their leaves all year in warmer climates, but drop them during the winter in colder areas.

Season of Bloom**			Light Tolerance			Good Fall Color	Zone of Hardiness	Comment
Early Spring	Late Spring	Early Fall	Sun	Semi-Shade	Heavy Shade			
	•		•	•			5 to 8	
	•		•	•			4 to 8	
	•		•				3 to 8	Fragrant
	•		•				4 to 9	Coarse leaf texture; showy
		•	•			•	4 to 9	blossoms
		•	•	•		•	3 to 8	
	•		•				7 to 9	
	•		•	•			7 to 9	
•			•	•			6 to 10	
			•				3 to 8	Junipers grow well in hot, dry
			•				4 to 9	soil; many are tolerant of
			•				4 to 9	salted soils and heavy snow
			•				3 to 7	loads.
			•				4 to 9	
	•		•				3 to 7	Large, fragrant flowers
•				•	•		6 to 8	Holly-like foliage
•				•	•		5 to 8	Bluish, grape-like fruit
	•		•				4 to 9	Creamy white fragrant flower
•			•	•		•	6 to 9	Both flowers and fruits attractive
	•		•			•	2 to 7	
	•		•	•			8 to 10	All parts are poisonous if eaten
•			•	•			7 to 9	Rapid growing; prone to
•			•	•			7 to 9	fungal diseases
•			•	•			7 to 9	
•				•			6 to 8	
•				•			4 to 8	
			•	•			2 to 7	Slow-growing
		late fall	•				9 to 10	Long-lasting blooms
	•		•	•		•	8 to 9	Colorful both spring and fall
	•	•	•				2 to 7	Produces flowers all summer
	•		•	•			3 to 8	Prunes well;
	•		•	•			5 to 8	popular hedge plants
	•		•	•			3 to 8	
		•	•				7 to 10	
•			•	•			4 to 8	Densely branched; thorny
•			•	•			4 to 8	Good for traffic control

**Where no rating is given, flowers are either not produced or are not of importance.

A Guide to Landscape Shrubs *(continued)*

Shrub		Evergreen	Deciduous	Mature Height		
Common Name	Botanical Name			3'–5'	5'–8'	8' and up
Rhododendron	Rhododendron species					
Carolina	R. carolinianum	•			•	
Catawba	R. catawbiense	•				•
Hybrid	R. hybrida	•			•	
Rose, Hybrid tea	Rosa species		•	•		
Spirea	Spiraea species					
Anthony Waterer	S. bumalda Anthony Waterer		•	•		
Bridal wreath	S. prunifolia		•		•	
Billiard	S. billardi		•		•	
Frobel	S. bumalda Froebelii		•	•		
Thunberg	S. thunbergi		•	•		
Vanhoutte	S. vanhouttei		•		•	
Viburnum	Viburnum species					
Arrowwood	V. dentatum		•			•
Black Haw	V. prunifolium		•			•
Cranberrybush	V. opulus		•			•
Doublefile	V. plicatum tomentosum		•			•
Fragrant	V. carlcephalum		•			•
Japanese Snowball	V. plicatum		•			•
Leatherleaf	V. rhytidophyllum	•				•
Sandankwa	V. suspensum	•			•	
Wax Myrtle	Myrica cerifera	•				•
Weigela	Weigela florida		•			•
Winged Euonymus	Euonymus alatus		•		•	
Wintercreeper	Euonymus fortunei vegetus	•		•		
Yew	Taxus species					
Spreading Anglo-Japanese	T. media	•				•
Upright Anglo-Japanese	T. media hatfield	•				•
Spreading Japanese	T. cuspidata	•				•
Upright Japanese	T. cuspidata capitata	•				•
English	T. baccata	•				•
Canada	T. canadensis	•		•		

Semi-evergreen indicates that the plants retain their leaves in warmer climates, but drop them during the winter in colder areas.

Season of Bloom**			Light Tolerance			Good Fall Color	Zone of Hardiness	Comment
Early Spring	Late Spring	Early Fall	Sun	Semi-Shade	Heavy Shade			
	•			•			6 to 8	Showy plants; require well-drained, acidic soil
	•			•			4 to 8	
	•		•	•			4 to 8	
	•	•	•				varies	Very diverse group of plants; large blooms; high maintenance
•	•	•	•	•		•	4 to 9	Attractive when flowering; most are resistant to insects and diseases
	•		•	•			4 to 9	
			•	•			4 to 9	
•			•			•	5 to 8	
	•		•	•		•	4 to 8	
	•		•	•		•	2 to 8	Attractive spring flowers; good fall color; many provide good wildlife food
	•		•	•		•	3 to 8	
	•		•	•		•	4 to 8	
	•		•	•		•	5 to 8	
	•		•	•		•	5 to 7	
•	•		•	•		•	5 to 8	
	•				•		5 to 8	
	•		•	•			9 to 10	
			•	•			7 to 9	Tiny, waxy grey berries
		•	•				4 to 8	Blooms late
			•	•		•	3 to 9	Crimson fall color
				•		•	5 to 9	
			•	•			4 to 7	Excellent for foundation plantings; prunes well; long lived; will not tolerate poorly drained soil
			•	•			4 to 7	
			•	•			4 to 7	
			•	•			4 to 7	
			•	•			6 to 7	
				•			2 to 6	

**Where no time of bloom is given, the flowers are either not produced or are inconspicuous.

A Guide to Groundcovers

Groundcovers									
Common Name	Botanical Name	Evergreen	Deciduous	Height	Optimum Spacing	No. Needed to Plant 100 sq. ft.	Light Tolerance	Zone of Hardiness	Flower or Fruit Color and Time of Effectiveness
Aaron's Beard	Hypericum calycinum	•		18"	18 inches	44	full/partial sun	5 to 8	Yellow flowers in late spring
Ajuga or Bugle	Ajuga reptans		•	5"	6 inches	400	sun or shade	4 to 9	Blue or white flowers in summer
Bearberry	Arctostaphylus uvamursi	•		10"	12 inches	92	sun or shade	2 to 6	Pink flowers in spring
Bigleaf Wintercreeper	Euonymus fortunei radicans	•		18"	3 feet	14	sun or shade	4 to 9	Orange fruit in the fall
Cast iron plant	Aspidistra elatior	•		18"	10 inches	144	shade	6 to 10	Not noticeable
Candytuft, Evergreen	Iberis sempervirens	•		12"	12 inches	92	full/partial sun	4 to 9	White flowers in spring and summer
Cotoneaster, Creeping	Cotoneaster adpressa		•	12"	4 feet	10	sun	4 to 8	Pink flowers/spring; red berries in fall
Cotoneaster, Rockspray	Cotoneaster horizontalis	semi		18" plus	4 feet	10	sun	4 to 8	Pink in spring/red berries in fall
Himalayan Sweet Box	Sarcococca hookerana humilis	•		15"	18 inches	44	shade	6 to 9	Insignificant
Honeysuckle, Creeping	Lonicera prostrata		•	12"	3 feet	14	sun	5 to 9	Yellow flowers/spring; red berries/fall
Hosta	Hosta plantaginea		•	18"	18 inches	40	shade	3 to 9	White flowers in late summer
Ivy, Baltic English	Hedera helix baltica	•		8"	12 inches	44	shade	5 to 9	Insignificant
Juniper Blue Rug	Juniperus horizontalis wiltonii	•		6"	3 feet	14	sun	3 to 9	Insignificant

A Guide to Groundcovers (continued)

Groundcovers

Common Name	Botanical Name	Evergreen	Deciduous	Height	Optimum Spacing	No. Needed to Plant 100 sq. ft.	Light Tolerance	Hardiness Zone Rating	Flower or Fruit Color and Time of Effectiveness
Juniper Jap Garden	Juniperus procumbens	•		18"	3 feet	14	sun	4 to 9	Insignificant
Juniper Shore	Juniperus conferta	•		18"	3 feet	14	sun	5 to 9	Insignificant
Liriope Variegated	Liriope muscari variegata	•		15"	18 inches	40	shade	6 to 9	Lavender flowers in late summer
Mondo grass	Ophiopogon japonicus	•		10"	10 inches	144	partial shade	7 to 9	White or pink flowers/ late summer
Myrtle (Periwinkle)	Vinca minor	•		6"	12 inches	92	shade	4 to 9	Blue flowers in spring
Oyster plant	Tragopogon porrifolius		•	12"	12 inches	92	sun or shade	9	Insignificant
Pachysandra	Pachysandra terminalis	•		8"	12 inches	92	shade	4 to 8	White flowers in spring
Phlox (Moss Pink)	Phlox subulata	•		5"	6 inches	400	sun	3 to 9	Varied pastel colors in spring
Purpleleaf Wintercreeper	Euonymus fortunei coloratus	•		18"	3 feet	14	sun or shade	5 to 9	Insignificant
Wandering Jew	Tradescantia albiflora	•		6"	12 inches	92	shade	9 to 11	Purple flowers spring/summer
Weeping lantana	Lantana montevidensis	•		18" plus	24 inches	25	sun	9 to 11	Lavender flowers all year
Yellowroot	Xanthorhiza simplicissima		•	18" plus	18 inches	44	sun	5 to 7	Brown-purple flowers in spring

A Guide to Vines

Common Name	Botanical Name	Broad Leaf Evergreen	Deciduous	Height	Clinging	Twining or Tendrils	Light Tolerance	Hardiness Zone Rating	Flower or Fruit Color and Time of Effectiveness
Actinidia, Bower	Actinidia arguta		•	30'		•	full/partial sun	4 to 7	White flowers in spring
Actinidia, Chinese	Actinidia chinensis		•	30'		•	full/partial sun	7 to 9	Insignificant
Akebia, Fiveleaf	Akebia quinata	semi		30'		•	full/partial sun	5 to 8	Purple flowers in spring
Ampelopsis, Porcelain	Ampelopsis brevi-pedunculata		•	20'		•	semi-shade	4 to 7	Multicolored fall fruit
Bignonia (Crossvine)	Bignonia capreolata	•		60'		•	full/partial sun	6 to 9	Orange-red flowers in spring
Bittersweet, American	Celastrus scandens		•	20'		•	full/partial sun	2 to 8	Yellow-red fruit in fall/winter
Boston ivy	Parthenocissus tricuspidata		•	35'	•		sun or shade	4 to 8	Insignificant
Bougainvillea	Bougainvillea glabra	•		20'	•		sun	7 to 10	Multicolored in summer
Clematis	Clematis hybrida		•	varies		•	semi-shade	4 to 8	Spring color varies by species
Fig, Climbing	Ficus pumila	•		40'	•		sun or shade	8 to 9	Insignificant
Honeysuckle, Trumpet	Lonicera sempervirens		•	20'		•	full/partial sun	4 to 9	Orange flowers/summer
Hydrangea, Climbing	Hydrangea, anomala petiolaris		•	60'	•		full/partial sun	4 to 8	White flowers in summer
Ivy, English	Hedera helix	•		70'	•		semi-shade	5 to 9	Insignificant
Jasmine, Star	Trachelospermum jasminoides	•		15'		•	shade	8 to 9	White flowers in late spring
Lace vine, Silver	Polygonum aubertii		•	25'		•	sun or shade	4 to 9	White flowers in summer
Monks hood vine	Ampelopsis aconitifolia		•	20'		•	semi-shade	4 to 7	Yellow-orange fruit in fall
Rambling roses	Rosa multiflora hybrida			10' to 20'	Tied		sun	5 to 8	Multicolored in summer
Trumpet vine	Campsis radicans		•	30'	•		sun	5 to 9	Orange flowers/summer
Virginia creeper	Parthenocissus quinquefolia		•	40'	•		sun or shade	3 to 9	Insignificant
Wisteria, Japanese	Wisteria floribunda		•	30'		•	full/partial sun	5 to 9	Purple flowers in spring
Woodbine, Chinese	Lonicera tragophylla		•	50'		•	shade	5 to 8	Yellow flowers/summer; red fruit/fall

- Are the soil and drainage characteristics of the site suitable for good plant growth?

- Will these species transplant easily and at the time of year that this project will be installed?

- Are the plants susceptible to insects and diseases common to this region?

- Will spraying or dusting be necessary to keep the plants pest-free?

- Do all of the plants used in this section of the landscape have similar cultural requirements?

- If replacement becomes necessary, are similar plants commonly available from nearby sources?

- What are the clients' attitudes about garden maintenance? Are they hobby gardeners who will enjoy giving special attention to their plants, or are they more interested in the lowest maintenance possible?

Only after the cultural requirements of the possible plant choices are matched against these questions can the proper plants be selected for the landscape.

Computer-Aided Plant Selection

Several computer software programs have been developed to aid landscape designers and landscape architects in their selection of plants for projects. These programs allow the designer to enter into the computer the specific plant features required; then the program presents a selection of plants that meet those requirements. Some of the programs even provide full-color photos of the plants, turning the computer screen into a garden catalogue. While such technology is welcomed and used increasingly by designers, there is some danger that good plant choices may be overlooked because they are not programmed into the software. There is also the potential for overlooking microclimate variations unique to sites. Computer-aided plant selection should be regarded as a tool for use by landscape architects and designers, but not a substitute for experience and knowledge of the horticultural flora of the region.

Achievement Review

A. Define the following terms.

1. native plant
2. exotic plant
3. naturalized plant
4. bare root plant
5. B & B plant
6. containerized plant
7. Latin binomial
8. genus
9. hardiness
10. microclimate

B. Of the following characteristics, indicate which ones are typical of plants sold as (BR) bare rooted, (B & B) balled and burlapped, or (C) containerized.

1. most common with very large plants
2. the root system is totally intact
3. may be cumbersome to handle
4. more common with small deciduous plants
5. limited season for planting
6. roots may bind if kept too long in this form
7. heavy
8. lightweight
9. usually the least expensive
10. usually the most expensive

C. Explain why the use of botanical names is important when selecting plants for the landscape.

D. Identify the major categories of factors that must be evaluated when selecting specific plants for a particular project.

E. Indicate if the following statements are true or false.

1. A hardiness zone map and a heat zone map duplicate temperature information.
2. A heat zone is a measure of warm temperatures, whereas a hardiness zone is a measure of cold temperatures.
3. Heat zones are less susceptible to microclimatic variations than are hardiness zones.
4. Heat days are beneficial to plant growth and survival.
5. Apple trees will not grow in tropical climates. It is likely that the reason is related to the high number of heat days.
6. Palm trees will not grow outdoors in Maine. It is likely that the reason is related to the low number of heat days.
7. A plant species can have difficulty growing for reasons unrelated to either hardiness zones or heat zones.
8. The time of year can affect the probability of a plant surviving transplanting.
9. Plants that are hardy and heat tolerant are always more resistant to pests.
10. The root form of a plant is more related to its *suitability* for the landscape than to its *survivability* in the landscape.

Suggested Activities

1. Consult the Hardiness Zone Map to determine:
   a. the hardiness zone rating of your home county.
   b. other regions of the country having winter temperatures similar to those of your home county.

2. Compare the Hardiness Zone Map with a geographic map. Determine what can influence winter temperatures in an area besides the north-south location.

3. Determine why the coast of Maine has a warmer hardiness zone rating than upstate New York, even though Maine is farther north.

4. If a light meter is available, take readings beneath several different types of trees. Be sure that the day selected for taking the readings is clear and sunny. Temperature readings can be taken as well. Rate the foliage density of the trees tested.

5. Make a collection of tree flowers, fruits, or colored leaves. Rate them on the basis of how colorful they are.

6. Collect a number of plants in the same botanical genus, but of different species (Example: different kinds of pines or dogwoods or palms). Make a listing of all of the characteristics that are common to all of them. Then look for one characteristic of each species that makes it different from all of the others in the genus. This is how taxonomists separate plant species.

   NOTE: Many of the distinctions are not easily noticed by the untrained eye.

7. Visit a nursery and compare the plants for sale based upon the form of their root systems. Compare the plant types with their methods of harvest and sale. Compare them on the basis of size, weight, price, suggested time of transplant, and any other factors that seem to illustrate differences based upon the form of the root system.

CHAPTER 8

The Principles of Design

Objectives:

Upon completion of this chapter, you should be able to

- describe the basic principles that lead to good design
- explain how these principles are applied to landscape designing
- describe features of plant materials, hardscape materials, and architecture that make them useful as design elements

The Foundation for Designing

Many texts address the subject of design principles, and each does it in a slightly different way. Since the names and number of principles vary from book to book, the reader might begin to wonder if any one source is complete or totally correct. While each author may present the subject in a slightly different way, by chapter's end most have explained the principles fully even if the terminology fails to match completely. As discussed in this chapter, **principles** refer to standards by which designs can be created, measured, discussed, and evaluated. Because design is at times very personal, it can be difficult to evaluate objectively. Whether someone likes or dislikes a certain plant selection or paving pattern does not necessarily

credit or discredit the design. Only when the design can be shown to be in compliance with or in violation of the principles that guide all design can it be judged as good, bad, or in between. Only when a judgment can be offered that the design is good or bad *because* it applies or fails to apply one or more of the principles of design does it become a judgment or critique that is understandable or justifiable to the designer.

Here then are six principles that artists have been using in the fine and applied arts for centuries. Whether the art form is oil painting or flower arranging, the same principles of design can be applied during the creation of the work and can be used to describe a viewer's reaction to the work. The reassuring thing about these principles is that they were not created centuries ago by artists. Rather, they stem from an inherent visual sense possessed by most people, whether they regard themselves as creative or not.

Balance

Balance is a state of being as well as seeing. We are physically uncomfortable when we are off balance. Whether while held hostage on top of the see-saw by a heavier playmate when we were children, or tipping over in a canoe when the contents shifted unexpectedly, we experience a lack of balance at various times in our lives, and we do not like it. It follows that we are most appreciative of and comfortable in landscape settings that are visually balanced.

There are three types of balance: symmetric, asymmetric, and proximal/distal. Symmetric balance is the balance of formal gardens. One side of

the composition is a reflective mirror image of the opposite side, Figure 8-1. While it can be somewhat stiff, especially if the plants used are geometric or sheared forms, symmetric balance is acceptable to a great many people because it is comfortable and easily understood. It is not necessary that landscape composition be stiff, though, in order to be symmetric. The combination of materials used may be loose and casual, yet as long as the shapes, colors, and specific materials match on both sides of center, the balance will be symmetric. Asymmetric balance is informal balance. The visual weight on opposite sides of the composition is the same, but the materials used and their placement may vary, Figure 8-2. Asymmetric balance has the potential

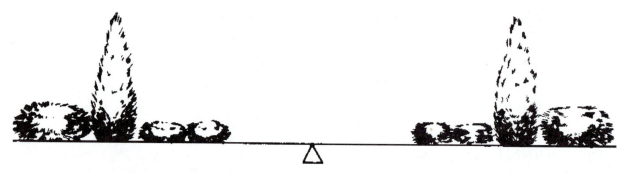

Figure 8-1 Symmetric balance. One side of the outdoor room is a mirror image of the opposite side.

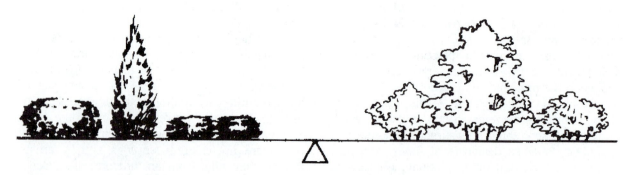

Figure 8-2 Asymmetric balance. One side of the outdoor room has as much interest as the opposite side, but does not duplicate it exactly.

to be more visually interesting to the viewer because there are two sides to observe and explore. With symmetric balance that is not true. Proximal/distal balance is asymmetric but carries it further by dealing with depth in the field of vision. In addition to balancing the left/right relationship in the composition of the landscape, there is a need to balance the near/far. This becomes a concern more frequently than might appear at first. The landscape designer is seldom able to design without concern for off-site features. Those features may be pleasant and desirable to include within the composition, but once included they must be factored into the balance of the setting. Tall buildings, distant mountains, large, off-site trees, and other such features can affect the balance of a design that looks good on paper, but would be off-balance once installed, Figure 8-3.

Focalization of Interest

Anything that is designed well has a focal point, one place within the composition where the viewer's eye is first attracted. Whether it be the enigmatic smile on the face of the Mona Lisa,

the largest, brightest flower in a table arrangement, or the jeweled pin on an evening gown, there will be something placed by the designer to command the attention of the viewer. Everything else in the composition will serve to complement that feature. **Focalization of interest** is the principle of design that selects and positions visually strong items into the landscape composition. Focal points can be created using plants, hardscape items, architectural elements, color, movement, texture, or a combination of these and other features, Figure 8-4. Some focal points are predestined. For example, in the public area of most residential landscapes, the entry to the home is understood to be the most important focus of attention. All of the design decisions made in that area serve to support that focal point. In other areas of the landscape, the designer usually has more opportunity to select those areas to be highlighted as focal points. Beginning designers often must learn restraint in applying the principle of focalization of interest. There is often a tendency to overuse focal points, which creates complexity and visual confusion within the landscape composition.

Figure 8-3 Proximal/distal balance. The foreground tree counterbalances the distant mountain.

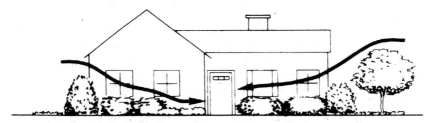

Figure 8-4 Focalization. Plants are arranged in an asymmetrical step-down manner to move the viewer's eye toward the entry, which is the intended focal point.

Simplicity

As with the principle of balance, **simplicity** seeks to make the viewer feel comfortable within the landscape. Few people are happy when exposed for any length of time to settings that are cluttered or fussy. Complexity is not always the opposite of simplicity, at least where landscape design is concerned. Landscapes may involve buildings with complex and intricate architecture. Other projects may be technically complex, with extensive lighting, water features, sound systems, circulation patterns, or security systems. If such things are present, they are probably important to the client or to the site; but the landscape into which they are placed can still be simple and comfortable for the user. Simplicity does not imply simplistic, boring, or lack of imagination. It does avoid the use of too many species, too many colors, too many textures, too many shapes, curves, and angles within an area or within a project.

Rhythm and Line

Rhythm and Line may be regarded separately or collectively as a principle of design. When something repeats enough times with a standard interval between repetitions, a rhythm is established. In landscape designing, the interval is usually a measured space. What repeats may be something structural, such as lamp posts or benches. It may be something patterned, such as a sidewalk pattern that is replicated every 50 feet from one end of a mall to the other end. As a user experiences the landscape, moving at a fairly even rate of speed,

these repeating design features establish a rhythm within the user's experience. The lines of the design establish the very shape and form of the landscape. Lines are created where different materials meet, such as where turf meets pavement or where turf meets the mulched planting bed. The merger of the two edges of the materials creates the line. Lines are also created for the eye when paved areas or walls cut through lawns, or when patterns are created within paved areas. Nature too provides lines, in the crest of surrounding hills or at an edge where a lake or river meets the land. When enough lines are parallel to each other, and all can be seen by a viewer at the same time, then a rhythm of lines is established, bringing the two concepts together as one, Figure 8-5. No landscape should be without a rhythm to the layout of its major line-making features.

Proportion

Proportion is concerned with the size relationships between all of the features of the landscape. That includes both vertical and horizontal relationships as well as spatial relationships. Much of our perception of vertical proportion is influenced by the height of the viewer and in particular by the viewer's eye level, which varies between standing, seated, and reclining. Shorter people perceive the vertical space of some landscapes differently than do taller people. Children have needs for certain types of landscapes that are vastly different from those of their parents if they are to feel comfortable within the space. The concern for proper propor-

Figure 8-5 The repeating parallel arcs of turf, pavement, and flowers add rhythm and line to the design.

Figure 8-6 Proportion. Each component of the landscape must be in the proper size relationship with all other elements. The large tree is appropriate with the multi-storied house B, but is too large to be so close to the single-storied house A.

tion extends to building size, lot size, plant size, the relationship between the areas of mass and void, and the human users of the landscape, Figure 8-6.

Unity

As a principle of design, **unity** is the easiest to measure if the other five principles have been applied properly and comprehensively to the design. A unified design is one in which all of the separate parts contribute to the creation of the total design. The color schemes and textural choices support each other rather than demanding individual attention. In a painting, the background details match the foreground features. In a flower arrangement a suggestion of each color is distributed evenly throughout the design. In a fabric pattern, the same shapes and colors repeat at regular intervals throughout the cloth. In each case, while individual components are valued and appreciated, they collectively create a single overall design. The same principle applies to landscape designing. Each component of the design, whether it is the plant materials, the shape of the planting beds, the choice and use of paving materials, the color selections, the lighting plan, or any other component of the outdoor rooms within a project, is obligated to be a part of the whole.

Applying the Principles of Design to Landscape Designing

Some references have already been made to ways that the six principles of design are used in designing and evaluating landscapes. Beginning design-

ers may have to make a conscious effort to assure that all of the principles are applied to a specific design. However, as designers and landscape architects become more experienced, using the principles becomes part of their creative thought processes, and the application becomes nearly automatic.

It is important to understand that the principles of design must be applied at all of the levels that people will experience the landscape. While the designer may be able to see the entire project as it is spread out across the drawing board, that is not the way human users will actually experience the landscape. They will move through it one outdoor room at a time and at varying rates of speed and attentiveness. At any one time, users of the landscape will only see what falls within their cone of vision, Figure 8-7. Thus, a focal point is only effective when it falls within that cone of vision. Balance is only apparent when the viewer is able to see both sides at the same time. As people move back and forth through a landscape, their perception of it continually changes. As a designer works to

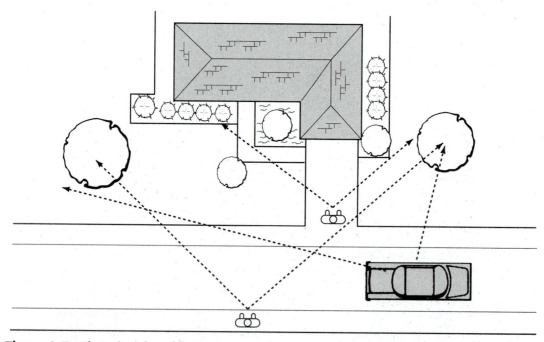

Figure 8-7 The principles of design must apply even when the position of the viewer changes.

develop a plan that is responsive to those who will be using the landscape, it is important to envision the design from every logical vantage point, realizing that only birds will see the property as an overhead panorama, and birds are not the clients. The human user will experience it as an ongoing series of visual images, and no two people will link those images together in the same way.

Applying the Principle of Balance

Balance must be addressed from several ranges. At **macro-range**, the viewer sees the landscape from the most distant vantage point, such as the opposite side of the street. This extends the cone of vision and often permits the viewer to see the entire house or building plus the entire adjacent landscape. Frequently houses are not centered within the lot, and predictably there will be a driveway connecting the street to the garage, creating a barren, no-plant space on one side of center that

does not exist on the opposite side. Visible from behind the house may be large trees or utility poles that affect the balance of the setting. In addition, immediate off-site features, such as the neighbors' houses and trees, are often visible when viewed from curbside, making them a site feature to factor into the consideration of balance. It is almost mandatory that the designer take a series of photographs during the site analysis to be certain that none of these factors is overlooked when designing the public area of a landscape. It is the only way to assure that the curbside view of the landscape is balanced, Figure 8-8.

The same concern for macro-range balance must be applied to all other use areas of the landscape. For example, when viewing the family living area or the private living area from inside the house or from the patio, how much of the view is on-site and how much is off-site? What must be done on-site to bring the view into balance? Correct design solutions may require the addition of trees

Figure 8-8 A balanced public area as viewed from curbside

Figure 8-9 The design is balanced when the viewer is close to the house (A), but the balance is lost when viewed from a distance that widens the cone of vision (B).

or the thinning or removal of existing ones to bring the site into balance with the immediate off-site setting.

At closer range, the viewer's shortening cone of vision becomes a major concern. For example, as the person approaches the entry of a building, certain features such as the masses of trees that enframe the building move outside the viewing range and no longer contribute to the viewer's perception of balance. Instead, the placement of shrubs on either side of the entry assumes greater importance. The viewer's eye will make a judgment about the type of balance and the success or failure of the principle's application, Figure 8-9. Similarly, the viewer will pass judgment on whether the plantings on each side of an important window are or are not balanced, or if the trees lining the drive are as densely planted on one side as on the other.

The viewer will decide if the shrub planting at one end of the patio has sufficient mass to counterbalance the stone grill constructed at the other end.

The designer must envision dozens of viewing stations, looking across every section of the landscape at ground level as well as from upper story windows and decks. Further, the designer must envision a dividing line down the center of each view to determine if the design balances from each vantage point.

Applying the Principle of Focalization of Interest

As discussed earlier, focal points are places in the landscape where attention either is naturally drawn or where the designer elects to direct it. In the public area, focusing attention on the entry is just common sense, because the first question a visitor to any building asks is "How do I get into this building?" See Figure 8-4. To place anything else in the public area that will compete with the entry area design would be a violation of the principle of focalization of interest. Both beginning and experienced designers frequently violate this principle. Among the most common mistakes are the use of flowers or night lighting in places removed from the entry. Brightly colored flowers are strong magnets for the eye. If used at all in the public area, they should be clustered at the entry and not spread across the front of the building or tucked between the shrubs bordering the lawn. Likewise, night illumination of a tree at the end of the house or in the center of the lawn will shift attention away from the entry. In the public area, night lighting should be used to reinforce the entry design so that the focal point is not lost when darkness falls.

In other use areas of the landscape, the designer often has a choice for placement of the focal points. Depending upon the size of the property and the type of design (private or public, residential or commercial), it may be suitable to have one or multiple focal points. In a typical residential family living area, a single focal point is usually appropriate, Figure 8-10, unless the area is large enough to develop additional ones without conflict. Again, the viewer's cone of vision plays a major role in deciding what works. If the property is so small that two focal points can be seen at the same time, there may be a visual conflict. As long as the focal points do not compete with each other, then more than one can be appropriate.

Many things can function as focal points in the landscape. Some of them will fulfill that role continuously and without changing. Statuary and other garden ornaments exemplify such items. Plants can also serve as focal points when they are unusually colorful or have striking growth habits. Their time of service can vary immensely. Many plants that are visually powerful when in spring flower or fall color become uninteresting during the rest of the year. By carefully selecting and positioning the focal features of the landscape, the landscape designer can introduce shifting focal points,

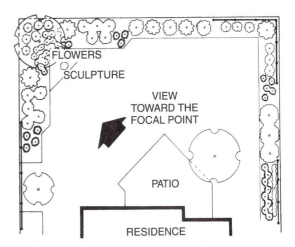

Figure 8-10 Focalization draws the eye of the viewer to one key feature within the composition.

Figure 8-11. This is an opportunity that is unique to landscape design. No other artist is able to do that.

The Corner Planting. One of the most natural places to position a focal point is in the corner of the outdoor room. When bedlines converge at the corner from two directions the eyes of viewers follow willingly to the point of convergence. Termed the **incurve** of the planting, the point of convergence easily accepts a focal feature such as a specimen plant (highly attracting to the eye) or an accent plant (less attracting than a specimen plant, but sufficiently different from other plants around it to be distinctive). The incurve plant is usually taller than the other plants extending from it to the outer reaches of the corner bed, the **outcurve**, Figure 8-12. The placement of plants into the corner arrangement should be done so that when the plants are mature, the viewer's eye will step up from the outcurves to the incurve, Figure 8-13.

Many variations are possible for the corner planting, Figure 8-14. A garden ornament, bench, or fountain might be used instead of a plant as the focal feature at the incurve. Flowers could be used directly at the incurve in front of the focal feature to lend some seasonal variation to the design.

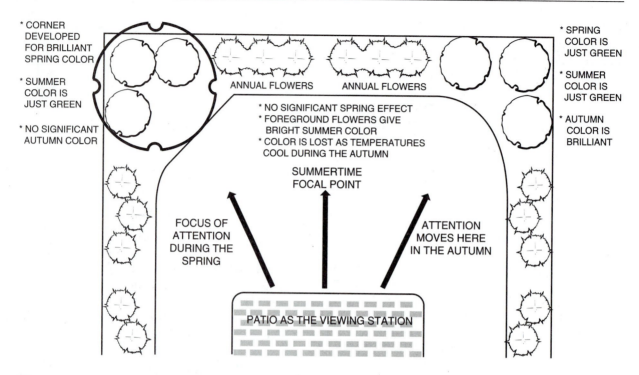

Figure 8-11 By selectively locating seasonal plants, the designer can introduce shifting focal points into the landscape composition.

Applying the Principle of Simplicity

The provision for simplicity in landscape design is the exclusion of unnecessary change. The key word here is *unnecessary*. Each time that there is a change of shape or color, complexity is introduced. Each time a bedline changes direction, complexity can result. Too many garden ornaments, too many different kinds of furniture, too many different types of wall or paving materials, can spoil a design that has everything else done correctly. It is not that change, diversity, or variety are wrong or unwarranted. A good design contains all of that. It is when the introduction of that change or variation does not advance or improve the design that it becomes a violation of the principle of simplicity. Unfortunately no magic number establishes a threshold that tells a designer when to stop. How many different colors or materials become too many? At what point does something cross the line

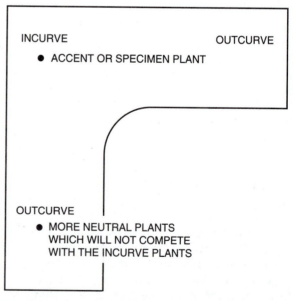

Figure 8-12 Parts of the corner planting bed: the incurve and the outcurves

from tasteful to garish? There are no clear answers, and there is a great deal of subjectivity in the interpretation of this principle. However, here are a few guidelines and suggestions for designing an imaginative, but simple landscape.

a. Avoid using too many angles in the layout of planting beds and paved areas. Use wide arcs in designing curvilinear features such as planting beds, and match concave arcs with convex ones of closely comparable size, Figure 8-15.

b. Limit the number of different species of plants in any use area. While no exact number can be given because use areas can vary in size so greatly, do not create an arboretum.

c. Massing plants eliminates the need to look at each one individually.

d. Arrange plants so that they create a silhouette that flows smoothly and is not choppy, Figure 8-16.

e. Avoid paving patterns and shapes that introduce independent lines, angles, or colors into an area.

f. When planning for color, high color contrasts are appropriate at the focal points of the design, but elsewhere it is best to use more subtle color schemes. If the architecture or nearby off-site features already contain numerous colors, the landscape should only repeat the existing colors rather than introducing new ones.

g. Avoid the overuse of geometrically shaped or sheared plants. Pyramidal forms are especially eye-attracting, so only use them where a specimen or accent plant is appropriate.

Applying the Principle of Rhythm and Line

The establishment of rhythm and line in the landscape has been compared to the creation of frozen motion. Drawing upon strong existing lines (and sometimes objects) and replicating them to

Figure 8-14 Three variations of the corner planting. Notice that in each example, the eye is drawn from the ends of the planting to the center and from the front to the rear.

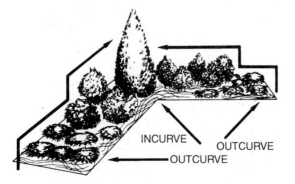

Figure 8-13 In a corner planting, attention is drawn from the outcurves to the incurve by stairstepping plants.

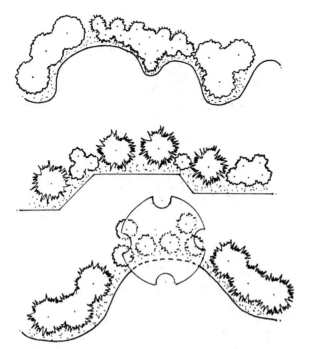

A. CHOPPY SILHOUETTE – NO MASSING

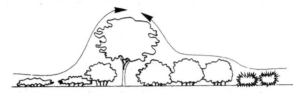

B. FLOWING SILHOUETTE – SPECIES MASSED

Figure 8-16   Simplicity in plant arrangement. Grouping B is better than A because it permits the viewer's eye to move smoothly along the planting.

Figure 8-15   Bedline design: Top bedline is fussy. It has too many arcs that do not counterbalance each other. The middle bedline is simple, but impractical. The sharp angularity appears stiff and unnatural. Unless retained with edging material, the angles would become rounded after several mowings. The lower bedline is well designed. It features simple curves with a good balance of concave and convex arcs.

the extent that the person using the landscape is consciously or subconsciously aware of them is the way this principle of design is applied. As an example, consider a typical residential landscape. The strongest and most dominant lines are the lines of the house. Architecture provides an excellent springboard from which to begin. It is seldom wrong for a landscape designer to repeat the lines and angles of the house in the lines and angles of the landscape. It's a technique that nearly always works. In other situations, the designer may choose to repeat the lines of some other visually powerful item, such as a swimming pool or a patio, Figure 8-17.

The principle of rhythm and line also can be used to evoke certain emotional responses from the users of the landscape. Strongly rectilinear designs suggest greater formality than those which are mostly curvilinear. One may seem stiffer to the viewer, while the other more casual and leisurely. One might be ideal for the design of a corporate headquarters, and the other better suited to a city park. In neither case is the use of the principle of rhythm and line incorrect. Should the lines of the landscape not be sufficiently replicated to establish at least a suggestion of rhythm, then the lines become erratic. That will also evoke an emotional reaction from the user of the landscape, but it will not be a pleasant one. Erratic lines introduce complexity and confusion.

The Line Planting.   Concern for the vertical linearity of the landscape is as important as concern for the horizontal linearity. Off-site features such as distant hills or the spire of a church might offer a strong design line that could be replicated on-site. Other times, the planting design must first fill a functional role such as blocking an unsightly view or providing some degree of privacy, Figure 8-18. Still, the line created by the silhouette of the mature

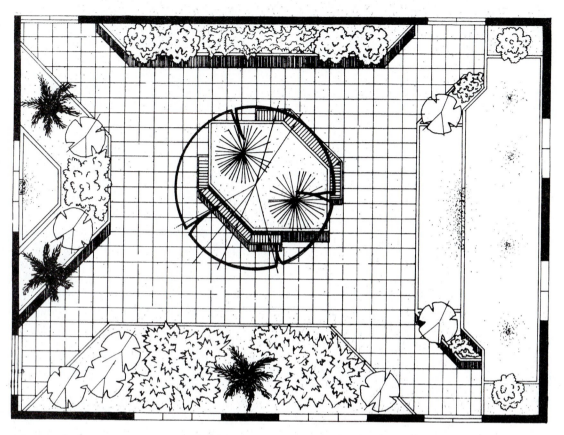

Figure 8-17 Rhythm and line: The repeating horizontal and vertical lines and repeating 45-degree angles create a rhythm of lines in this design.

plants should be envisioned by the designer at the time of plant selection, and its appropriateness to the overall design should be considered.

Positioning plants in the line planting must also be done with care at the time the design is created. If not done carefully, the plants can create a vertical line, which will not match the shape intended for the outdoor wall, Figure 8-19. While the bedlines appear very prominent on the drawn plan view, the vertical linearity created by the plants and other materials is responsible for the creation of the shapes of landscape spaces. The plants need to be positioned so that they replicate the lines of the beds. Where smaller plants are used in front of taller plants, or where plants are staggered for

shadow effects, both ranks of plants should replicate the bedlines, Figure 8-19.

The Foundation Planting. One of the most prominent line plantings is the **foundation planting**. It has been passed down from an era when houses were built on top of high, unattractive foundations. The planting was designed to screen the foundation. Even though modern houses often have no exposed foundations, the foundation planting still is used; however its functions have been broadened. The foundation planting includes the entry design, and as such must focus attention on the front door of the residence. That was discussed in detail earlier. Being so close to the building, the modern foundation planting also serves as the single most important element connecting the man-made building with the more natural landscape. In applying the principle of rhythm and line to the design of the foundation planting, a designer may use the lines and angles of the house as described previously, but with some added objectives. Where space permits, the foundation planting should reach outward from the ends of the building, not stopping just because the building has ended. It should also reach forward from the entry

rather than sticking snugly against the building. In both places, the intent is to free the planting from the constraints of the building and, by extending the lines toward the surrounding landscape, cause the house to interface more comfortably with its setting, Figure 8-20. Figures 8-21 and 8-22 illustrate contemporary foundation plantings.

Applying the Principle of Proportion

Maintaining the proper size relationships within the landscape first requires that the designer clearly understand what those relationships are to be. Anyone who has read *Alice in Wonderland* or seen the animated motion picture can understand how the scale of the landscape, in comparison to the human users, evokes predictable reactions from the users. The designer is in a position to control those reactions, assuming they are anticipated in advance. If the ceiling of the outdoor room is low, the feeling

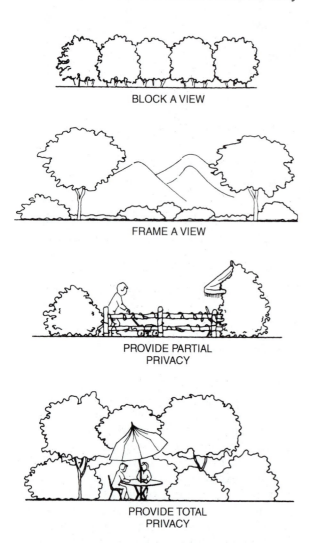

Figure 8-18 Some functions of the line planting

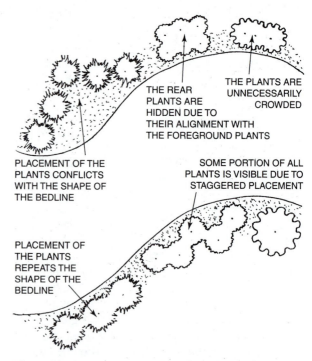

Figure 8-19 Correct and incorrect placement of plants within a planting bed

Figure 8-20 The foundation plants reach outward from the corner of the house, forward from the entry, and stay below window levels.

Figure 8-21 A modern foundation planting displaying a welcoming entry court

is more intimate. That is frequently desirable when developing a residential patio. If the trees loom high overhead, people feel smaller by comparison. Even small buildings appear out of scale and threatened when the nearby trees are disproportionately large. The higher the wall and ceiling elements become in the outdoor room, the more dehumanizing the landscape gets. People who walk the canyons of New York City, Chicago, or similar cities learn to appreciate the important role that street trees play in bringing the bigger-than-life cityscape back into human scale. Children enjoy play areas where the equipment and the plantings are scaled to their heights, not that of adults. Small, low-growing plants should not be planted immediately adjacent to large, tall plants. Instead, it is preferable to bring the canopy down in stages from the tallest plants to intermediate heights, and finally down to the lowest growing forms.

Landscapes need proper proportional relationships between

- buildings and people.
- buildings and plants.
- plants and people.
- plants and plants.
- masses and voids.

The need to balance masses and voids changes the concern from putting things in proportion to putting spaces in proportion. If the horizontal open areas of the landscape, the **voids**, are much larger or smaller than the solid vertical areas of plantings, buildings, walls, or land forms, the **masses**, and if that size difference is apparent to the viewer, then the principle of proportion will be missing from the design. For example, if a use area such as the family living area has an unusually large central lawn (the void) and adjoins another equally large void such as a field, the planting bed (mass) separating the two voids must be sizable enough to truly separate them without looking weak and apologetic. It needs sufficient width, height, and density to match their importance.

Applying the Principle of Unity

The principle of unity must be applied within each use area of the landscape to tie together the individual components of those areas. It must also be applied to the total landscape to integrate all of the separate use areas into a single design. It can even be applied to larger areas, including residential or commercial developments or even entire geographic regions. As with the other principles, unity can be applied at both the macro and micro levels of design.

Figure 8-22A Before: The foundation planting not yet installed (Courtesy of Lied's, Inc., Sussex, WI)

Figure 8-22B After: The completed foundation planting and entry design (Courtesy of Lied's, Inc., Sussex, WI)

Repetition is the key to unity. As people move through the landscape, they recognize the oneness of the design when certain familiar themes are repeated often enough. Here is a partial listing of techniques often used to bring unity to a landscape design.

- Repeat prominent colors inside and outside the house.

- Repeat the construction materials of the house in the garden's constructed features.

- Continue the design themes of the interior rooms into the outdoor rooms.

- Use large expanses of glass to visually connect indoor and outdoor use areas.

- Raise patios, porches, decks, and landings to door level, rather than requiring users to step up or down when moving between the indoors and the outdoors.

- Repeat plant species between use areas, particularly at prominent points where the repetition will be noticed.

- Standardize the style of lighting fixtures and furniture.

- Standardize signs throughout a project or region.

- Maintain a consistent architectural style among buildings.

- If there is a strikingly prominent feature, such as a distant mountain or a city skyline, allow it to be seen from as many areas in the landscape as possible.

Material Characteristics Supportive of the Design Principles

All of the components of the landscape have characteristics that can support the application of the six design principles as long as the designer uses them correctly. If misused, certain materials may result in the weakening of one or more of the design principles.

Architecture

While there are many different styles of architecture, some of them historic and some of them contemporary, most styles can be clearly described as symmetrical or asymmetrical. Regardless of whether it is a private residence or a corporate structure, small or enormous, compact or rambling, it will be either a symmetric or asymmetric structure. Symmetric structures such as the Cape

Cod frame and Federal or Georgian buildings easily accept symmetrically balanced landscape designs, Figure 8-23. Houses and other buildings that are not symmetric are often better suited for more informal, asymmetric landscape settings. However, it shouldn't be concluded that buildings are pre-destined for a certain type of design based solely upon the architecture. That would relegate landscape design to a paint-by-numbers status. However, some building styles do support the application of certain principles of design more obviously than do others.

Building doorways can make the focalization of interest at the entry much easier at times. Georgian doorways are traditionally ornate, making it nearly impossible for the viewer to mistake how to get into the building. Contemporary entries may feature bright colors, ornate light fixtures, double widths, or other devices to engage the viewer's eye.

Other architectural features can add to the difficulty of applying some of the design principles. For example, the wide, gaping door of an attached garage is strong competition for the viewer's eye. Bay windows, shutters, or rambling porches can all compete for attention with the front entry.

Plant Materials

Earlier reference has been made to the distinctiveness of certain plants in developing focal points within a design. To elaborate on what has already been written, specimen plants are the most satisfactory for use in focalization. Due to one or more extraordinary features, such as color or growth habit, specimen plants attract more attention than any other plants around them. If used in the public area, they need to be placed at or near the entry so that they support the intended focal point rather than competing with it. If used elsewhere, the duration of their eye appeal is important for the designer to consider. A specimen plant whose only contribution is one week of showy flowering in the

Figure 8-23 Symmetric architecture and symmetrically balanced landscape plantings fit together comfortably.

early spring is not very significant as a focal point when compared to an evergreen with year-round contortion or highly glossy foliage.

While most specimen plants are naturally occurring species, it is possible to create unusual forms through special pruning or training.

Accent plants were also discussed briefly before. These are plants that can be used to counterbalance a specimen plant without competing against it for the viewer's attention. They create lesser, secondary focal points at the corners of planting beds, at secondary entries, at intersections, and other places where the design needs emphasis but not dominance. Examples of accent plants would be the upright forms of yew or juniper or holly. When a single upright plant is placed within a mass of spreading forms of the same genus, the upright gives emphasis to the planting. It becomes the accent plant.

Most landscape plants are neither specimen plants nor accent plants. They are the workhorses of the landscape, serving as **massing plants**. They fill large amounts of space both on the ground and in the air. As such they are most likely to be the plants used when the principle of balance is being applied.

Hardscape Materials

This is a broad category of components, and a more in-depth presentation is given in Chapter 11; therefore the suggestions made here are general and abbreviated. As a design element, hardscape materials can make no contribution until the designer decides how to use them. Through the merger of design imagination with construction fabrication, hardscape materials can be transformed into landscape features that attract the eye, add mass and weight to the composition, create themes, and add pleasure to the landscape. If sufficiently unusual or attractive, they aid in application of the principle of focalization. Their mass and visual weight make them a factor to use or contend with in applying the principle of balance. When formed into a shape or structure that repeats in all areas of the landscape, hardscape contributes to the principle of unity.

Hardscape materials, such as pavings, fencing, and walls, frequently create strong lines in the landscape. The viewer's eye will willingly follow the line to its end. Therefore it is important that the line lead somewhere, warranting the attention. When the lines created by hardscape materials need to be understated, as when they lead to the garage door rather than to the entry, the color of the material used should be restrained. A white concrete walk running to the front door of the house has a better chance of focusing attention on the entry if the driveway is paved with a less noticeable color. In turn, the drive should never be lined with flowers or other materials that would serve to outline and emphasize it.

Practice Exercises

A. Using the proper instruments and a drawing board, duplicate the corner planting bed shown in Figure 8-24, drawn to the scale of 1" = 10'. Design a planting for a viewer looking in the direction of the incurve. Select plants and dimensions from Chart A. Draw to scale and label all species used.

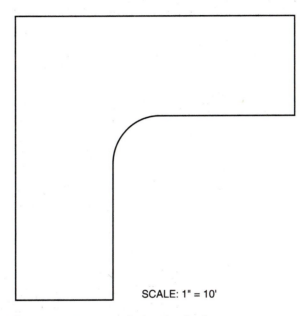

SCALE: 1" = 10'

Figure 8-24   Corner planting bed

Chart A

Species		Width	Height
Redbud tree	(D)	20 ft.	25 ft.
Viburnum	(D)	10 ft.	12 ft.
Forsythia	(D)	10 ft.	9 ft.
Cotoneaster	(D)	5 ft.	5 ft.
Spirea	(D)	5 ft.	4 ft.
Grape holly	(BLE)	3 ft.	3 ft.
Andorra juniper	(NE)	3 ft.	1½ ft.
Myrtle	(G)	Vining	1 ft.

D: Deciduous   BLE: Broadleaved Evergreen
G: Groundcover   NE: Needled Evergreen

B. With the proper equipment, duplicate the planting bed in Figure 8-25. Assume that the viewer is located south of the bed and that there is an attractive mountain scene north of the bed. Design a planting that frames but does not block the view. Select plants from Chart A. Draw to the scale of 1" = 10'.

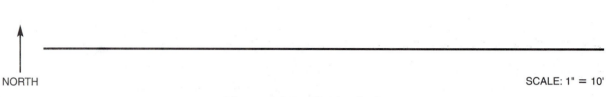

NORTH   SCALE: 1" = 10'

Figure 8-25   Planting bed

C. Figure 8-26 illustrates the entry portion of a house. The garage and driveway are also shown. With the proper instruments, design the public area portion of the foundation planting. Use the plants in Chart A. Design to the scale of 1" = 10'.

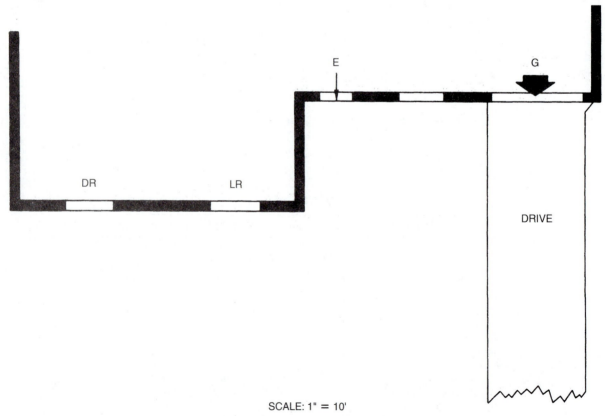

SCALE: 1" = 10'

Figure 8-26   Entry portion of house

Achievement Review

A. Indicate which principle of design is being applied in the following situations.

1. The view toward the house from the front street is framed by large masses of trees in the woodlands that border both sides of the property.
2. A dark, recessed entry door is illuminated with a bright floodlight each evening as part of the night lighting design.
3. At the point of entry into a private residential complex there is a distinctive logo affixed to a custom designed stone column. That column and logo are also featured prominently at the community pool, the clubhouse, and on all signs throughout the complex.
4. The sharp, erratic angles in the planting beds of a property being renovated are replaced with gently curvilinear beds.
5. A matching planting is placed on each side of the entry to a home.
6. A piece of garden sculpture is placed into a line planting so that it can be seen from the client's bedroom window throughout the year.
7. A house is designed to feature 90-degree and 45-degree angles prominently in its construction. The landscape designer uses those same angles in the design of the patio and pergola that adjoin the house.
8. To re-create the sense of being in a large park setting, a designer specifies that very tall tree species be selected for the canopy of a private residential family living area.
9. At regular intervals of 100 feet, a designer plans for a pedestrian mall to be lined with identical planters filled with flowers of the same color. There are a total of 20 replications of the planter from one end to the other.
10. In the above example, at the same mall, similar planters and color selections are found in the nearby parking areas, in the food court, and down the side alleys.

B. Name and explain the three types of balance.

C. Explain how the principles of design are affected by the cone of vision of the viewers.

D. Describe the parts of a corner planting.

E. What are the functions of modern foundation plantings?

F. Distinguish between the following types of plants.

1. specimen plant
2. accent plant
3. massing plant

Suggested Activities

1. Working from photographs of unlandscaped new homes or from slides of the same projected onto drawing paper, draw in different possibilities for foundation plantings. Try creating both symmetrical and asymmetrical designs for the same house. Have someone else critique the proposals using the principles of design as the basis for judgment.

2. Prepare a written critique of an existing landscape. Explain why you do or do not approve of the design. Avoid all subjective opinions and use only the principles of design to justify your statements.

3. Working from a book of house plans that show the houses in elevation and perspective views, select an array of different styles. With a pencil or marker, circle and note features of the architecture that would make it easier or more difficult to design foundation plantings that would focus attention on the entries.

CHAPTER 9

Flowers

Objectives:

Upon completion of this chapter, you should be able to

- describe the uses and limitations of flowers in a landscape design
- explain the differences between annual, perennial, and biennial flowers
- list the characteristics of hardy and tender bulbs
- explain the difference between a flower bed and a flower border
- design a flower planting
- plan a naturalized bulb planting

The Uses and Limitations of Flowers in the Landscape

Flowers are one of the most desired and difficult elements of the landscape, whether the landscape is residential or commercial. While the motivation for using flowers varies, their manner of use varies little. A residential client may desire flowers to attract certain birds to the backyard, while a commercial developer may desire flowers to attract tenants to a business park. In each case, the use of flowers is driven by desire. It then falls to the designer to use the flowers in ways that complement and enhance the design rather than marring it.

Very few flowers have much impact on a viewer when used singly; however, mass several dozen, or several hundred, together and they become one of the most powerful magnets for the eye that can be placed in a landscape composition. Placed correctly, such flower plantings can reinforce a strong design, strengthen a weak one, and make unique contributions to the design. Placed incorrectly, such flower plantings can weaken or destroy an otherwise good design for all or part of the year.

In some cases, flowers are the only plants that clients know or care about. They may not know a maple from a palm tree, but they know petunias and geraniums, and how those plants are used in the landscape will be of great interest to them. A designer who knows less than clients about specific flowers and their proper use and culture will have difficulty satisfying the client's desires or guiding the client toward a proper design use of flowers.

Using Flowers Properly

In most regions of America, flowers are the most fragile of the landscape's plants in that their time of effectiveness is limited. Flowers are also the plants most likely to evoke emotional responses from people. Winter-weary residents of the northern states cheer when the first crocus breaks through the snow. Chrysanthemums evoke memories of autumns past and high school football games. Home gardeners across the country eagerly await the arrival of seed catalogues each spring to see what new varieties of flowers are being introduced. In short, flowers have many important roles to fill in the landscape:

- Their bright colors create strong focal points.

- They often possess fragrances that few other plants can match.

- Their colors and fragrance may attract birds and butterflies.

- They can soften harsh lines of the architecture and areas of the landscape such as pavement lines.

- Specific flowers suggest the changes of the seasons.

The Limitations of Use

Despite their many contributions, flowers can be overused or misused, so landscape architects and designers must be cautious as well.

- If misplaced, the focal points created by flowers can detract from the intended focal point.

- In addition to the birds and butterflies, flowers may attract bees, which could be troublesome for children, pets, or people with sensitivities to bees.

- In most regions of the country, flowers die back after a frost, and abandon their role in the landscape. If used in key areas of the landscape, the design could suffer.

- Flowers are expensive to purchase and time-consuming to maintain.

Different Flowers Have Different Life Spans

Annuals

An **annual** flower is one that completes its life cycle in one year. That is, it goes from seed to blossom in a single growing season and dies as winter approaches. Generally, annuals are most commonly used in summer landscapes. They bloom during the months when the days are long and warm. They offer color, especially in northern regions, when many bulbous perennials are past blooming. Hundreds of annuals are commonly used throughout the country. Some examples of well-known annuals are the petunia, marigold, salvia, and zinnia.

Landscapers obtain annuals by two different methods. In one method, they are directly seeded into the ground after the danger of frost is past. Packages of annual seeds, Figure 9-1, are available at most garden centers in the spring. Seeds can also be obtained from mail-order supply houses.

Direct seeding of annuals (the placement of seeds in the ground) is the least expensive way to place flowers in the landscape. The major limitations of direct seeding are: (1) the young plants usually require thinning; (2) the plants require more time than the other method to reach blooming age; and (3) it is difficult to create definite patterns in the flower planting.

The other method used to obtain annuals is from bedding plants. These are plants that were started in a greenhouse and are already partially grown at the time they are set into the garden, Figure 9-2. Often the plants are grown in a pressed peat moss container, which can be planted directly into the ground. This creates no disturbance for the annual's root system; thus, the flower planting is well established from the beginning.

Figure 9-1 Packaged seeds are an easy and inexpensive way to start a flower garden. (Courtesy of National Garden Bureau, Inc., Sycamore, IL)

Bedding plants are more expensive than seeded annuals. However, there is no need for thinning, and definite patterns can be easily created.

Perennials

A **perennial** flower is one that does not die at the end of its first growing season. While it may become dormant as cold weather approaches, it lives to bloom again the following year. (When a plant is **dormant**, it is experiencing a period of rest in which it continues to live, but has little or no growth.) Most perennials live at least three or four years; many live for much longer.

Nearly all early spring flowers are perennials. Some grow from bulbs; others do not. Many special autumn flowers are also perennials. There are numerous summer perennials that act with annuals to add color to rock gardens and border plantings. Some examples of perennials are the hyacinth, iris, daffodil, tulip, poppy, phlox, gladiolus, dahlia, and mum.

Some perennials are available in seed form. However, most appear on the market as bedding plants or reproductive structures such as bulbs. Since they do not die at the end of the growing season, most perennials reproduce themselves and may eventually cover a larger area of the garden than was originally intended. This tendency to propagate may be a side benefit or a maintenance nuisance, depending upon the situation.

Bulbous Perennials. The very popular bulb comprises a large number of perennials. Bulbs survive the winter as dormant fleshy storage structures known to botanists as tubers, corms, rhizomes, tuberous roots, and true bulbs. In the landscape trade, they are usually simply called **bulbs**.

Most bulb perennials bloom only once, in the spring, summer, or fall. There are a few exceptions, however; these may bloom repeatedly. Bulbs may be classified as hardy or tender.

Hardy bulbs are perennials that are able to survive the winter outside and therefore do not require removal from the soil in the autumn. The only time they must be moved is when they are being thinned. Hardy bulbs usually bloom in the spring. Examples are the hyacinth, iris, daffodil, and tulip.

Tender bulbs are perennials that cannot survive northern winters and must be taken up each fall and set out each spring after the frost is gone. These bulbs usually bloom during the summer months. Examples are the canna, gladiolus, caladium, and tuberous begonia. Refer to Figures 9-3 and 9-4 for planting instructions.

NOTE: Some bulbs that are tender in the north are hardy in the south where they can be left in the ground throughout the year.

Figure 9-2 Bedding plants are available from the greenhouse ready to install in the garden. They give color faster than direct seeding, but at greater cost.

Biennials

Flowers known as **biennials** complete their life cycles in two years. They produce only leaves during their first year of growth and flower the second year. After they have bloomed, they die. Biennials include the English daisy, foxglove, Japanese primrose, and pansy.

Flower Beds and Flower Borders

A **flower bed** is a freestanding planting made entirely of flowers. It does not share the site with shrubs or other plants. As Figures 9-5 and 9-6 illustrate, flower beds are effective as focal points where sidewalks or streets intersect. They also work well in open lawn areas where they do not conflict with more important focal points and where the lawn is not used for activities that could be damaging to the beds.

Flower beds should never be planted in the public area of a residential landscape. When this is done, the beds attract more attention than the entry to the home, which is the most important part of any public area.

Flower beds must be designed so that they can be viewed from all sides, and must be planted accordingly. Because of this requirement, flower beds contain no woody plants to provide a backdrop for the blossoms. This has given the flower bed the reputation of being the most difficult flower planting to design. Probably for this reason, it is used much less often than the flower border.

The **flower border** is a planting that is placed in front of a larger planting of woody shrubs. The foliage of the shrubbery provides a background to set off the colors of the blossoms, Figure 9-7. Since the flower border can only be viewed from one side, it is more easily controlled by the designer. Figure 9-8 illustrates the difference between the viewing perspectives of beds and borders.

Figure 9-3 Bulbs are spaced before installation to assure proper color patterns and to prevent crowding.

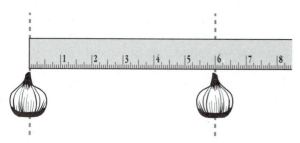

Figure 9-4 Each type of bulb has a recommended planting depth and spacing. As a general rule, bulbs should be spaced the same distance apart as their planting depth.

Modern landscapers are more likely to use flowers in borders than beds for two reasons: they are easier to design, and the strong visual attraction of the flowers is more easily controlled.

Designing Flower Plantings

Flowers are high-cost, high-maintenance items, so they should be used where they will give the best return for the investment. Flowers filling a role in the landscape's design cannot be cut and taken into the house. Neither should they be wasted in side yards or stuck beneath windows where they will be looked over rather than being looked at. Regardless of whether the landscape project is large or small, commercial or residential, flowers should be used in high-profile locations. They also lend themselves to uses in novel ways, such as container plantings or hanging pots. A visit to any major theme park will display the many creative ways that flowers can be used to enrich landscapes.

Landscape designers planning the flower effects for a landscape must know the plants well. Here are some of the factors to be considered.

- Are the plants annuals, perennials, biennials, or a combination?

- Do the plants that are grouped together have similar cultural requirements? Are they compatible with the cultural requirements of the woody plants in the planting bed?

- Do all plants bloom at the same time? If not, what plants will be in bloom together at different times during the growing season? Incompatible colors can coexist within the same flower planting as long as they do not bloom at the same time.

- How will the colors of flowers blooming simultaneously blend? Will there be subtle blends, or strong, sharp contrasts, or uncomplementary clashes of colors or tones?

- Are blossom and foliage textures compatible?

- Dark, vibrant colors attract the eye strongly and should be used sparingly.

- Pale, pastel colors attract the eye less and can be used in the greatest quantities.

- Rigid, spike-form flowers have the strongest eye appeal and should be used in limited numbers at the center of the flower bed or at the incurve of the flower border.

- Flowers with round blossoms can be used in greater numbers at the outcurves of flower borders and as foreground plants in flower beds.

- In between the spike forms and the round forms fall the non-typical flower shapes, such

Figure 9-5 This freestanding flower bed acts as a traffic divider while adding beauty to the area.

Figure 9-6 Walks on all sides create various points from which this flower bed can be viewed. Despite its size, only two species were used (salvia and dusty miller).

Figure 9-7 The flower border has a background, unlike the flower bed. Here, tall shrubs provide a dark contrast to the brighter flower blossoms.

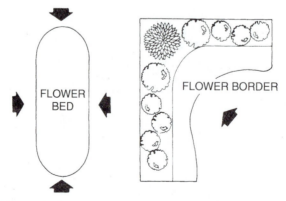

Figure 9-8 Flower beds are viewed from all sides and have no background foliage. Flower borders have a more limited viewing opportunity and have background foliage.

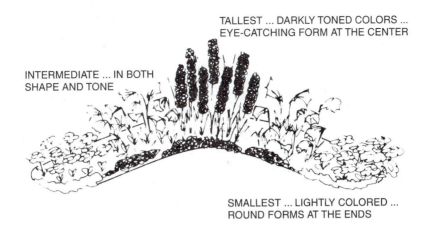

INTERMEDIATE ... IN BOTH
SHAPE AND TONE

TALLEST ... DARKLY TONED COLORS ...
EYE-CATCHING FORM AT THE CENTER

SMALLEST ... LIGHTLY COLORED ...
ROUND FORMS AT THE ENDS

Figure 9-9 One example of how to arrange flower shapes, colors, and sizes in the flower border

as tubular blossoms, which are collectively termed the *intermediate blossom shapes.* They bridge the gap between the spike forms and the round forms of flowers. See Figure 9-9 for an example of how to arrange blossom shapes within a flower border.

If an all-season flower planting is being designed, species must be selected and arranged so that there will always be a vibrantly colored, spike-form species blooming at the center or rear of the planting and larger quantities of shorter intermediate and round forms in complementary colors distributed evenly throughout the bed or border.

Naturalized Plantings

In landscapes that are designed to appear more park-like or pastoral, neither the flower bed nor the flower border may be suitable. Instead, flowers may be used in a way suggestive of how nature

would distribute and multiply flowers. Bulbous flowers such as daffodils are especially well suited to naturalized plantings, because once installed, they can overwinter in the ground and will reproduce freely. Thus the planting becomes more realistically natural each year.

To accomplish a naturalized bulb planting, bulbs are tossed gently onto a sunny slope at the appropriate season of the year, then planted wherever they happen to land, Figure 9-10. If enough bulbs are installed, when they bloom they will create the effect of large drifts of wildflowers sweeping across the lawn. Such plantings require no formal beds, and, once they go dormant, they disappear beneath the turf until the next spring. It is important, however, that the foliage be left uncut long enough to allow the bulb to gain strength for the next year. Naturalized plantings within lawns should not be used in landscapes where the sight of uncut grass filled with spent bulbs will upset the viewers.

Figure 9-10 To naturalize a bulb display, toss them rather than space them, and plant them where they land.

Achievement Review

A. If you were a designer discussing the use of flowers with a client, what advice would you offer in response to the following client comments:

1. "I love lots of flowers. I want them everywhere in my yard."
2. "I think flowers spread all across the front of the house would be pretty."
3. "I would like to use extensive flower plantings to separate the use areas of my landscape."
4. "I hate yardwork, so let's use more flowers than trees and shrubs, since flowers don't have to be pruned and I won't have so many leaves to rake."
5. "Flowers are so expensive and so short-lived. Wouldn't it be wiser for me to spend my money on longer-lived trees and shrubs?"

B. What are the differences between annual, biennial, and perennial flowers?

C. Indicate if the following characteristics typify

a. hardy bulbs   c. both
b. tender bulbs   d. neither

1. Perennials
2. Planted in the spring
3. Bloom in the spring
4. Commonly started from seed
5. Usually bloom once per year
6. Planted in the fall
7. Bloom in the summer
8. Dug and stored over winter
9. Annuals
10. Fleshy storage organs

D. Describe the technique commonly used to create naturalized plantings of bulbs within lawn areas.

E. Distinguish between flower beds and flower borders.

F. Match what is known about the eye appeal of certain flower shapes with what is known about the eye appeal of certain color tones and explain how both can contribute to the design of flower plantings.

G. If you wanted to use flowers as a means of applying the principle of simplicity to the design of the landscape, how could it be done?

Suggested Activities

1. Write to seed companies and request catalogs. The gardening section of your newspaper may have some addresses of suppliers. Posting the pictures around the room will help you become familiar with common flowers in the area.

2. With drafting tools, design a flower bed or border. It should cover an area of at least 40 square feet. Design it so that there are plants in bloom from early spring to late fall.

3. Start annual flowers from seed. The seed can be purchased at a local garden center, supermarket, or hardware store. Plant them in greenhouse flats or flowerpots and place in a greenhouse or near a sunny window. When the weather is warm enough, set them outside around the school building. As an added activity, design a planting for the building first, using the flowers being grown in class.

4. Practice blending flower colors. Cut out patches of colored paper to represent different types and quantities of flowers. Arrange them so that the brightest, most attracting colors are used where the attraction is desired. Pastel colors should be used in greater quantities and placed away from the central color. Follow the suggestions given in the text for arranging flower colors.

5. To understand better the anatomy of bulbs, select several different types for dissection. Carefully peel away each layer. Try to identify those tissues that will become the roots, stem, leaves, and flowers of each bulb. A botany text may be consulted for assistance and exact identification of all parts.

6. Study areas where wildflowers are common. Note how in the natural world flowers are grouped together in plant masses, not randomly distributed. This can help you attain natural effects when arranging landscape flowers.

CHAPTER 10

Xeriscaping

Objectives:

Upon completion of this chapter, you should be able to

- define and describe the purpose of Xeriscaping
- list the economic, environmental, and aesthetic benefits of Xeriscaping
- explain the seven principles of Xeriscaping
- apply the principles of Xeriscaping to landscape designing

Xeriscape: Water-wise Landscaping

Earth's water supply is finite. What is here now is all that will ever be available no matter how many new needs we have in the future. For generations of Americans, the distribution of our national water resources dictated where the majority of the people would live, where most of our manufacturing would be centered, and where the most pleasurable recreation would occur. In each case, the abundance of water was the key. The northeastern states grew in population and industrial prominence due largely to the generous water supplies provided by the lakes and rivers of those states. Those same lakes and rivers and the nearby ocean also gave the northeastern states a recreational advantage over other states.

Eventually technology and inventiveness permitted other states to share in the bounty. Southern California learned how to tap into the snow melt of far distant mountains and divert it to the cities of the state, thereby permitting their growth and development. Manufacturers discovered that they could obtain the water they needed in the southern states by mining the fresh water of swamps and the Everglades. The landlocked midwestern states opened international ports on the Great Lakes and linked them to the ocean via the St. Lawrence Seaway. Today our national water supply is more interconnected than it has ever been. It is also experiencing a greater level of stress due to overuse and misuse than ever before, and there is no reason to believe that the stress load will ever diminish. Cities continue to sprawl and grow even though there is insufficient water to support the growth. Underground aquifers are being depleted faster than they can be recharged by natural forces.

Our national hunger for new products and new synthetic materials is filling many of our water bodies with artificial waste by-products that cannot be decomposed by naturally occurring organisms.

Today, despite a generally favorable public awareness of the need for water conservation and water quality improvement, our national water supply is in jeopardy. Government at all levels is attempting to address the problem. Some of the legislation is helpful and long overdue. Some of it is short-term, knee-jerk reaction to an immediate crisis and is quickly forgotten once the crisis subsides. Landscapers are significant users of water and are frequently caught in the middle of a community's water crisis.

In the arid regions of the country the declining availability and quality of water were first apparent. Arizona, Colorado, New Mexico, Texas, and parts of California, Nevada, Oklahoma, and Utah all have regions where water quality or quantity is limited due to insufficient rainfall or other environmental factors such as drying winds or mountainous terrain. While the native plants and animals of these regions adapted long ago to their dry environments, Table 10-1, the people who have moved into these regions of the country to retire, seek recreation, or work have not adapted. Instead they have tried to bring their lifestyles with them, causing New England and midwestern houses and gardens to spring up in areas once home to the desert. Grassy lawns grow on desert sand, and eastern shrubs bloom from planting beds filled with carefully blended soil mixtures. As the population of the Southwest has grown, the demand for water has grown disproportionately faster.

Other areas of the country are now beginning to experience water stress as well. Due to the demands of a water-greedy population and the environmental mistakes of past generations, fresh water supplies are beginning to fail in every geographic region of the nation. The resulting reactions generally take two forms. One form of action/reaction is legislative. Governments go after polluters, forcing them to stop further polluting activity and clean up their past mistakes whenever possible. The other reaction is usually to impose water use restrictions on the population, which is where landscapers get hit the hardest. Plants that can't be watered following installation will usually die. Plants that are established will also die or be stunted if not given sufficient water to maintain them.

Xeriscaping is the term used to describe techniques of landscaping that conserve water. *Xeriscape* is a trademarked term, spelled with a capital *X*, and pronounced *zeer-escape*. It was first used in 1981 when the Denver, Colorado, Water Department and the Associated Landscape Contractors of Colorado joined together to create a program enlisting public cooperation to make landscape water use more efficient. It was an idea whose time had come. It was quickly adopted by other arid states, and in 1986 the National Xeriscape Council was created. They hold the trademark on the name *Xeriscape*. Currently cities in 42 states have public education programs in Xeriscaping, and interest continues to grow, both nationally and internationally. Xeriscaping, if done correctly, should not be perceived as punitive or restrictive, but as an opportunity to return the designed landscape to a more natural arrangement and relationship of plants and materials. In the process, a great deal of water can be saved.

The Benefits of Xeriscaping

Economic Benefits

Simply stated, water costs money. To the community that supplies, treats, and recycles it, it is at best a break-even responsibility of local government. To those in the landscape industry, it costs to use water, and it costs to apply it. If over-applied, the water goes to waste as it goes down the drain. If under-applied and plants die, the water was of no benefit but there was still a cost. If water-deprived plants must be replaced, there is additional cost.

Comparatively, landscape designs that replace water-requiring plants and expansive lawn areas with hardscape surfacings save water costs and yield higher profit returns to the landscape contractor. Xeriscaped landscapes are also still new

Table 10-1 A guide to selected Southwestern plants

Plant		Growth Habit	Mature Height		
Common Name	Botanical Name		1' or less	2'–5'	6'–9'
Ash					
Arizona	Fraxinus velutina	T			
Modesto	Fraxinus velutina var. glabra	T			
Citrus trees	Citrus sp.	T			
Coral tree	Erythrina sp.	T			
Crabapple, Flowering	Malus sp.	T			
Cypress, Arizona	Cupressus arizonica	T			
Elderberry, Desert	Sambucus arizonica	T			
Elephant tree	Bursera microphylla	T			
Elm					
Chinese	Ulmus parvifolia	T			
Siberian	Ulmus pumila	T			
Eucalyptus	Eucalyptus sp.	T			
Hackberry, Netleaf	Celtis reticulata	T			
Honeylocust					
Shademaster	Gleditsia triacanthos var. inermis 'Shademaster'	T			
Sunburst	Gleditsia triacanthos var. inermis 'Sunburst'	T			
Thornless	Gleditsia triacanthos var. inermis	T			
Ironwood, Desert	Olneya tesota	T			•
Jujube, Chinese	Ziziphus jujuba	T			
Locust					
Black	Robinia pseudoacacia	T			
Idaho	Robinia pseudoacacia 'Idahoensis'	T			
Pink flowering	Robinia pseudoacacia 'Decaisneana'	T			
Magnolia, Southern	Magnolia grandiflora	T			
Mesquite					
Honey	Prosopis glandulosa var. glandulosa	T			
Screwbean	Prosopis pubescens	T			
Mulberry, White	Morus alba	T			
Olive, European	Olea europaea	T			
Pagoda tree, Japanese	Sophora japonica	T			
Paloverde					
Blue	Cercidium floridum	T			
Little leaf	Cercidium microphyllum	T			
Mexican	Parkinsonia aculeata	T			
Pine					
Aleppo	Pinus halepensis	T			
Digger	Pinus sabiniana	T			
Italian stone	Pinus pinea	T			
Japanese black	Pinus thunbergii	T			
Pinyon	Pinus cembroides	T			

KEY: T = Trees   S = Shrubs   V = Vines   G = Groundcovers   NS = Flowers are not showy.

Mature Height				Season of Bloom					Special Use in the Landscape
10'–15'	15'–30'	30'–50'	Over 50'	Early Spring	Late Spring	Summer	Fall	Winter	
		•		NS					shade tree
		•		NS					shade tree
•				varies with the variety					excellent for containers
	•			•					brilliant flowers
	•				•				specimen plant
		•		NS					screens and windbreaks
	•			NS					screens and windbreaks
	•			NS					
		•		NS					shade tree
		•		NS					windbreak
		•		varies with the variety					many species / prized for flower and/or foliage
		•		NS					shade tree
	•			NS					good in dry, desert conditions
	•			NS					
		•		NS					
	•				•				specimen tree
									very salt tolerant
		•			•				frequent pruning makes these attractive flowering trees
		•			•				
		•			•				
			•			•			lawn tree
									shade trees and windbreaks
	•			NS					
	•			NS					
		•		NS					shade tree
	•			NS					good multi-stemmed tree
	•					•			lawn tree
									specimen trees
	•				•				
	•				•				
	•				•				
		•		NS					grows well in poor soil
		•		NS					specimen plant
			•	NS					good in desert conditions
	•			NS					good in planters; prune well
•				NS					multi-stemmed effects

Table 10-1 A guide to selected Southwestern plants *(continued)*

Plant		Growth Habit	Mature Height		
Common Name	**Botanical Name**		**1' or less**	**2'–5'**	**6'–9'**
Pistache, Chinese	Pistacia chinensis	T			
Poplar					
Balm-of-Gilead	Populus balsamifera	T			
Bolleana	Populus alba var. 'Pyramidalis'	T			
Cottonwood	Populus fremontii	T			
Lombardy	Populus nigra 'Italica'	T			
White	Populus alba	T			
Silk tree	Albizia julibrissin	T			
Smoke tree	Dalea spinosa	T			
Sycamore					
American	Platanus occidentalis	T			
Arizona	Platanus racemosa 'Wrightii'	T			
California	Platanus racemosa	T			
Tamarisk					
Athel tree	Tamarix aphylla	T			
Salt cedar	Tamarix parviflora	T			
Umbrella tree, Texas	Melia azedorach 'Umbraculiformis'	T			
Willow					
Babylon	Salix babylonica	T			
Globe Navajo		T			
Wisconsin	Salix x blanda	T			
Zelkova, Sawleaf	Zelkova serrata	T			
Abelia, Glossy	Abelia x grandiflora	S			•
Apache plume	Fallugia paradoxa	S		•	
Arborvitae, Oriental	Thuja orientalis	S			
Barberry					
Darwin	Berberis darwinii	S			•
Japanese	Berberis thunbergii	S		•	
Bird of Paradise	Caesalpinia gilliesii	S			•
Brittlebush	Encelia farinosa	S		•	
Butterfly bush	Buddleia davidii	S			•
Cherry laurel, Carolina	Prunus caroliniana	S			
Cotoneaster, Silverleaf	Cotoneaster pannosus	S			•
Crape myrtle	Lagerstroemia indica	S			
Creosote bush	Larrea tridentata	S			•
Firethorn, Laland	Pyracantha coccinea 'Lalandei'	S			•
Hibiscus					
Chinese	Hibiscus rosa-sinensis	S			
Rose of Sharon	Hibiscus syriacus	S			
Holly					
Burford	Ilex cornuta 'Burfordii'	S			•
Wilson	Ilex wilsonii	S			•
Yaupon	Ilex vomitoria	S			

KEY: T = Trees   S = Shrubs   V = Vines   G = Groundcovers   NS = Flowers are not showy.

Mature Height				Season of Bloom					Special Use in the Landscape
10'–15'	15'–30'	30'–50'	Over 50'	Early Spring	Late Spring	Summer	Fall	Winter	
			•	NS					good patio tree; good fall color
		•		NS					narrow columnar form windbreaks
		•		NS					
			•	NS					
			•	NS					
		•		NS					
		•				•			showy shade tree
•					•				
									excellent street trees
			•	NS					
			•	NS					
			•	NS					
		•				•			wind, drought, and salt resistant
	•					•			
		•			•				shade tree
		•		NS					
			•	NS					
		•		NS					
		•		NS					windbreak
						•			
				•					
•				NS					
									barrier plantings
				NS	•				
						•			
					•				
						•			vigorous growth
	•			•					screens and hedges
					•				wind screen
	•					•			very colorful flowers
						•			screens and hedges
				•					espaliers well
		•				•			
						•			
•									
				NS					Wilson and Yaupon clip and shade well
				NS					
	•			NS					

Table 10-1 A guide to selected Southwestern plants (continued)

Plant		Growth Habit	Mature Height		
Common Name	Botanical Name		1' or less	2'–5'	6'–9'
Hopbush	Dodonaea cuneata	S			
Jojoba	Simmondsia chinensis	S		•	
Juniper					
Armstrong	Juniperus chinensis 'Armstrongii'	S		•	
Hollywood	Juniperus californica	S			
Pfitzer	Juniperus chinensis 'Pfitzeriana'	S			•
Lysiloma	Lysiloma sp.	S			
Myrtle	Myrtus communis	S		•	
Ocotillo	Fouquieria splendens	S			
Oleander	Nerium oleander	S			
Photinia	Photinia glabra	S			
Privet					
California	Ligustrum ovalifolium	S			
Glossy	Ligustrum lucidum	S			
Japanese	Ligustrum japonicum	S			
Texas	Ligustrum japonicum texanum	S			•
Rose, Floribunda	Rosa x floribunda	S		•	
Silverberry	Elaeagnus commutatus	S			
Sugar bush	Rhus ovata	S			
Bougainvillea	Bougainvillea glabra	V			•
Ivy					
Algerian	Hedera canariensis	G	•		
Boston	Parthenocissus tricuspidata	V			
Jasmine, Star	Jasminum multiflorum	V			
Lavender Cotton	Santolina chamaecyparissus	G	•		
Periwinkle	Vinca minor	G	•		
Trumpet creeper	Campsis radicans	V			
Virginia creeper	Parthenocissus quinquefolia	V			
Wisteria	Wisteria sinensis	V			

KEY: T = Trees   S = Shrubs   V = Vines   G = Groundcovers   NS = Flowers are not showy.

enough and few enough that there may be an economic marketing benefit to the homeowner at the time the property is put up for sale.

Environmental Benefits

Anything that reduces the demand on a region's water supply must be judged as beneficial. That is

a prime objective of Xeriscaping. So many plant choices are inappropriate for their sites, surviving only because of supplemental irrigation, fertilization, and horticultural technology. By eliminating all or most of the plants that require excessive amounts of water and chemicals to sustain them, the local water supply is conserved and the remaining wastewater is less polluted and easier to clean

Mature Height				Season of Bloom					Special Use in the Landscape
10'–15'	15'–30'	30'–50'	Over 50'	Early Spring	Late Spring	Summer	Fall	Winter	
•				NS					screens
				NS					hedges
•				NS NS NS					
•					•				good for transition between garden and natural landscape
						•			prunes and shapes well
•					•				specimen plant
•						•	•		does well in heat and poor soil
	•			•					screens
•					•				all species can be pruned to lower heights
					•				
•						•			
					•				massing effects
•				NS					good for containers
•					•				
						•			very colorful
				NS NS					
	•					•			very fragrant
•						•			effective as edging
					•				
	•					•	•		
	•			NS					
•					•				may be trained as shrubs and weeping trees

and recycle. Also, with world hunger at record levels and insufficient quantities of fertilizer available for agricultural use in many parts of the world, it is offensive to many people that fertilizers are used to keep backyard lawns green and blemish-free. Replacing the lawn-dominant landscape with Xeriscaping can make a contribution to the survival of people throughout the world.

Aesthetic Benefits

For many years Americans applied more horticulture than ecology to their landscapes. They so willingly accepted the trendy suggestions of the national garden magazines that much of the geographic distinctiveness of the country was lost. Xeriscaping, with its emphasis on the use of native plants and others suited to the local climatic conditions, offers the nation an opportunity to regain some of its regional distinctiveness. By combining plants with common water requirements, matching them to the climate of the region, and placing them in settings whose construction materials and design themes reflect the heritage of the region, landscape designers will begin the resurrection of our diverse national landscape. It will be possible to have a dialogue with our past while making a contribution to the nation's future.

The Seven Principles of Xeriscaping

The National Xeriscape Council has prepared a list of seven principles that serve as a guide to the development and maintenance of landscapes that use water wisely, not wastefully.

The First Principle:
Proper Planning and Design

Assuming that everything described in Section One regarding the way to analyze a site, measure a client's needs, and develop outdoor rooms has been done, this important principle of Xeriscaping is applied to the selection and use of plant materials. Plants need to be **hydrozoned**, grouped on the basis of their water needs. Three water-use zones are defined:

Low water use	Requiring little or no supplemental water after transplants become established
Moderate water use	Requiring some supplemental water during hot, dry periods
High water use	Limited areas where plants are given as much water as needed at all times

Of the three zones, high water-use zones are usually the smallest zones and are placed in the landscape in highly visible locations such as at the entry to a building or at the public entrances to commercial properties.

The Second Principle:
Proper Soil Analysis

Considering that the regions where Xeriscaping is now practiced range from the desert of Arizona to the subtropics of Florida and the mountain valleys of the Rockies, it is understandable that there is no such thing as a common soil environment. A soil analysis will give the landscaper information about the existing structure, nutrient content, water retention capacity, and drainage characteristics of the soil of a specific site. In turn, that will guide the designer in selecting plants for the site. With appropriate additives, it may be possible to improve water penetration, retention, and soil drainage. The ultimate objective is to provide the plants with a soil environment that will allow them to send their roots deep into the earth.

The Third Principle:
Appropriate Plant Selection

Plants should be chosen for their suitability and adaptability to the site. All of the factors described in the chapter on Plant Selection apply to Xeriscaping. Additionally, plants should be selected for their appropriateness for the water-use zone in which they will be located. The more low water-use zones there are within a landscape, the fewer will be the plants requiring large amounts of water to sustain them. Also, if the designer specifies small plants with wider spacing at the time of installation, less water will be required to establish the transplants, since smaller plants need less water than larger plants.

The Fourth Principle:
Practical Turf Areas

Turf areas can be the greatest users of water in the landscape. An entire industry has grown up around the irrigation requirements of turf. Too often turf has been the glue that held the other components of the landscape together. It has been used for no reason other than to fill the voids of the design. In Xeriscaping, that is no longer an affordable luxury. Turf should be used where it is the functional plant of choice and where an alternative non-living surface is inappropriate. Grasses thrive or go dormant in reaction to changes in temperature and available water. If kept separate from the other plants in the landscape, turfgrass can survive in varying hydrozones, appearing lush and green when natural water is plentiful and temperatures are optimal, and at other times becoming brown and dormant in response to the conditions of the zone.

The Fifth Principle:
Efficient Irrigation

It is a standard horticultural practice to irrigate plants in a way that encourages roots to grow deep into the soil since shallow rooting increases the plant's vulnerability to drought. In Xeriscaping it is important to select irrigation systems that match the site and the needs of the plants. Sprinkler systems waste water because they deliver too much to the foliage and too little to the root zone. Drip systems and other micro-irrigation systems conserve water by delivering low volumes of water at low pressures to precise areas. They should be the systems specified by landscape architects and designers as they practice water-efficient planning. Efficiency can also be applied to irrigation scheduling. Water applied in the early morning will benefit plants more than water applied during the heat of the day, since less will be lost through evaporation. Only enough water should be applied to replace that lost since the last application. The use of soil moisture sensors can help the landscape manager determine the amount of irrigation water needed.

The Sixth Principle:
Mulching

Mulches offer many benefits, including water retention. In Xeriscapes, the most appropriate

mulches are those that are organic, fine-textured, and that do not develop a water-repelling crustiness over time. Selecting an organic mulch that is native to the local region will contribute to the natural appearance of the Xeriscape. The mulch should be applied to a thickness of 4 inches. Over time it will decompose and should be removed before adding fresh mulch.

The Seventh Principle:
Appropriate Maintenance

Any and all maintenance practices that reduce the water needs of the landscape will contribute to water conservation. Sweeping walks rather than hosing them clean will save water. Spraying plants with anti-transpirant chemicals at the time of transplanting and at other times of high water stress can reduce their water needs. Mowing turf areas often enough to prevent excessive growth, which will pull more water from the soil, will conserve water. Keeping plants disease- and insect-free will assure healthy plants, which will not need replacement. Also, if water used for other purposes, such as cooling or washing, can be used in the landscape rather than merely being poured down a drain, it can save on the consumption and expense of freshly treated municipal water.

Applying Xeriscape Principles to Designing

Landscape architects and designers should regard application of Xeriscape principles as a modern updating of their profession. It brings environmental ethics into the planning process, and this benefits everyone involved. It should not be regarded as a constraint on design creativity. Instead, Xeriscape principles offer new opportunities to freshen and re-create much of the American landscape.

Designers must know the plants of the region where they are working. In addition to knowledge of the plants' physical characteristics, they must know their water requirements if they are to fill each hydrozone correctly. There are comparatively

Table 10-2 Plants Suitable for Xeriscaping

Trees

Common Name	Botanical Name	Hardiness Zone	Hydrozone*
Acacia, Sweet	Acacia farnesiana	9 to 10	M
Ash, Arizona	Fraxinus velutina	5 to 9	L/M
Boxelder	Acer negundo	3 to 9	L
Calabash, Mexican	Crescentia alata	10	M
Cornelian cherry	Cornus mas	4 to 8	M/H
Elm, Chinese	Ulmus parvifolia	5 to 9	L/M
Locust, Black	Robinia pseudoacacia	4 to 8	M
Maple, Florida	Acer barbatum	7 to 9	M
Maple, Trident	Acer buergeranum	5 to 9	L
Oak, Live	Quercus virginiana	7 to 10	M
Olive, Black	Bucida buceras	8 to 10	M
Poplar	Populus deltoides	2 to 9	L
Redbud	Cercis canadensis texensis	4 to 9	M
Serviceberry	Amelanchier species	4 to 9	M
Silver Elaeagnus	Elaeagnus multiflora	5 to 8	L/M
Wax Myrtle	Myrica cerifera	7 to 9	M/H

Shrubs

Common Name	Botanical Name	Hardiness Zone	Hydrozone*
Almond, Desert	Prunus fasciculata	6 to 9	L/M
Ash, Singleleaf	Fraxinus anomala	5 to 9	L/M
Barberry, Japanese	Berberis thunbergii	4 to 8	M
Bluebeard	Caryopteris incana	7 to 9	M
Bougainvillea	Bougainvillea species	8 to 10	L/M
Butterfly bush	Buddleia davidii	5 to 9	L
Cowania	Cowania mexicana	9 to 10	L
Euonymus, Dwarf Winged	Euonymus alatus compactus	3 to 9	M
Firethorn, Leland	Pyracantha coccinea lelandei	6 to 9	M
Juniper (Various)	Juniperus species	3 to 8	L/M
Leadplant	Amorpha canescens	2 to 8	L/M
Ligustrum, Golden Japanese	Ligustrum japonicum Howard	7 to 10	M
Lilac (Various)	Syringa species	3 to 7	M
Mahonia, Chinese	Mahonia fortunei	8 to 9	M
Pine, Pinon	Pinus edulis	6 to 8	L/M
Rose, Floribunda	Rosa floribunda	5 to 8	L/M
Rose of Sharon	Hibiscus syriacus	5 to 9	M
Sage, Big Western	Artemisia tridentata	5 to 8	L/M
Saw palmetto	Serenoa repens	8 to 10	M
Scotch broom	Cytisus scoparius	6 to 8	L/M
Sumac (Various)	Rhus species	varies	L/M
Yucca, Adam's Needle	Yucca filamentosa	4 to 9	L/M

*L = Low water requirements   M = Medium water requirements   H = High water requirements

few reference texts that consistently include reliable water-use information about the plants listed. One of the texts most often referred to is *Xeriscape Gardening*, by Connie Lockhart Ellefson, Thomas L. Stephens, and Doug Welsh, Macmillan Publishers, New York, 1992. Another helpful listing of plants and their water needs is found in a publication developed by the Georgia Water-Wise Council in Marietta, Georgia, entitled *Xeriscape: A Guide to Developing a Water-Wise Landscape*. The publication

Table 10-2 Plants Suitable for Xeriscaping (*continued*)

Palm and Ornamental Grasses			
Common Name	**Botanical Name**	**Hardiness Zone**	**Hydrozone***
Fountain grass, Dwarf	Pennisetum alopecuroides hameln	5 to 10	L/M
Fountain grass, Purple	Pennisetum setaceum rubrum	7 to 10	L/M
Maiden grass	Miscanthus sinensis gracillimus	4 to 9	L/M
Palm, Jelly	Butia capitata	8 to 10	L/M
Palm, Mediterranean fan	Chamaerops humilis	8 to 10	L/M
Pampas grass	Cortaderia selloana	7 to 10	L/M
Plume, grass	Erianthus ravennae	5 to 10	M

Turf Grasses			
Common Name	**Botanical Name**	**Warm or Cool Season**	**Hydrozone***
Buffalograss	Buchloe dactyloides	Warm	L
Blue grama	Boutelova gracilis	Warm	L
Crested wheatgrass	Agropyron cristatum	Cool	M
Tall fescue	Fescue species	Cool	M

*L = Low water requirements   M = Medium water requirements   H = High water requirements

may be available through the Xeriscape council in individual states. Designers working in mountain regions may wish to consult *The Xeriscape Flower Gardener*, by Jim Knopf, Johnson Books Publisher, Boulder, Colorado, 1991. As new texts are written and current texts are revised, water requirements and tolerances should be included into the plant lists. Table 10-2 provides a rudimentary list of plants that can be used in designing a landscape that seeks to apply the principles of Xeriscaping.

In summary, here are some of the ways that landscape designers can take the lead to reduce the water consumption of landscapes:

■ Do not create lawn areas where other options, such as hard surfacings or wildflowers, will serve the clients as well and not consume so much water. When turf is used, specify varieties that require less water, such as tall fescue and Bermudagrass.

■ Regard planting beds as hydrozones, placing plants that share common water requirements into common hydrozones.

■ Use as many low water-use hydrozones as possible, reserving the high water-use zones for the most important, high-profile regions of the design.

■ Select plants that can adapt to the existing soil conditions, rather than trying to condition vast areas of the site to accept inappropriate plant species.

■ Place trees to cast shade where it will be of greatest benefit to understory plants in reducing their water consumption.

■ Specify that water-retention rings be placed around each tree at the time of transplanting.

■ Specify the use of organic mulches rather than inorganic mulches. Inorganic mulches, such as marble or stone chips, can retain solar heat and bake the soil beneath.

■ Design irrigation systems that do not waste water and that deliver the water only where and when it is needed.

■ Use containerized plants to provide spot color and interest in areas where paving replaces turf and ground beds. While potted flowers and other plants require more frequent watering than in-ground plantings, the total water consumption should be considerably less, Figure 10-1.

Figure 10-1 When containerized plantings replace in-ground installations, the water savings can be substantial.

Achievement Review

A. Define *Xeriscaping*.

B. Name the seven principles of Xeriscaping.

C. Explain hydrozoning and its application in designing.

D. Describe the economic benefits of Xeriscaping to the landscaper, the local community, and the client.

E. Why does it matter what irrigation system is used as long as plants receive the water they need?

F. Why are lawn-dominant landscapes inappropriate as Xeriscapes?

G. Why is a proper soil analysis important in Xeriscape designing?

H. Why are organic mulches preferable to inorganic mulches for water retention?

Suggested Activities

1. Place into three pots with the same soil mixture, three different plants: one with low water needs, one with medium water needs, and one with high water needs. All plants should be of comparable size. Apply the same amounts of water to each. Gradually reduce the amount of water and extend the time intervals between waterings until you are able to see visible signs of the plants' differing water requirements.

2. Do a similar comparative study using local turf grasses and wildflowers. As above, gradually reduce the amount of water and increase the intervals between waterings. Compare how each appears as water is reduced.

3. Fill greenhouse flats with typical local soil and place thermometers into each at the same depth. Leave one uncovered, and cover the other flats of soil with different organic and inorganic mulches to a depth of 4 inches. Place all flats outdoors or in a greenhouse where they can receive direct summer sunshine. Apply exact amounts of water to each flat. Compare soil temperatures at different times of the day and night.

CHAPTER 11

Hardscape

Objectives:

Upon completion of this chapter, you should be able to

- explain what is meant by hardscape
- describe how hardscape materials serve in functional and/or aesthetic ways
- explain how hardscape materials are selected
- choose from among many types of enclosure and surfacing materials
- explain how the dimensions of outdoor steps are calculated
- describe the basic steps in the creation of a recirculating water feature
- explain the uses of 12-volt lighting systems in illumination of the landscape

Hardscape: An Umbrella Term

The term *hardscape* is used almost exclusively by the landscape industry, and it may never appear in standard dictionaries. Even among landscapers, the term has only recently entered the industry jargon. Simply stated, hardscape is everything that is part of the landscape composition other than the plant materials. Much of it functions to enclose or surface portions of the landscape, but there are other important functions as well. One point is indisputable: when hard-

scape is part of the composition, a landscape will never be perceived as being totally natural. Whether the landscape is then seen as being marred by human intervention or improved by it depends upon the artistry of the designer and the craftsmanship of the contractor.

Hardscape features may perform an *architectural* role in the design of the landscape. In that capacity, they help determine a viewer's or user's perception and understanding of the area. Fences, walls, retained embankments, and trellises help to articulate the shape of an outdoor room. They also may dictate the degree of privacy provided to an area or establish the ease or difficulty of entry and access, Figures 11-1 to 5.

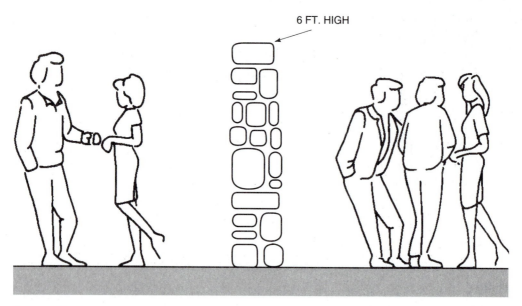

6 FT. HIGH

Figure 11-1 Privacy between two properties on the same level requires a solid enclosure at least 6 feet in height.

Other hardscape features may be used in an *engineering* role to help solve problems in the landscape. Where turf grass would wear away from excess foot or vehicular traffic, a surfacing of hard paving would likely be used. In windy locales, a fence or wall could form a baffle to deflect the wind or to reduce its velocity across a patio area. A steep slope could be retained with a strong wall that would prevent the soil from eroding while allowing water pressure to escape through planned weep holes, Figures 11-6 to 10. Steps or a grade change can be illuminated at night to assure the safety of users.

Still other hardscapes make an *aesthetic* contribution, serving more as enrichment items than as architectural or engineering features. Fountains, reflection pools, seat walls, walk lights, swimming pools, and patios are but a few examples of the sensory contributions that some hardscape features can make to a landscape, Figures 11-11 to 13. Comparisons of hardscape materials commonly used for enclosure and surfacing are shown in the charts at the end of this chapter.

Selection Criteria

The selection process for hardscape should reflect the preferences of the clients, the budget constraints of the project, the concerns of those who will maintain it, and the professional expertise of the landscape designer. These are not always of equal concern, nor are they always ranked in the same hierarchy of importance. Sometimes the clients do not have a preference, or they have one that is impractical. Sometimes the budget is tight and other times it is not. Maintenance concerns may involve cost, time, skill requirements, or other things. As landscapes vary, so do the criteria that determine the suitability of certain hardscape choices for specific projects.

- *Function* The hardscape will have to do something, such as securing a children's play yard, confining a pet, blocking an outsider's view, supporting the weight of a vehicle, or providing protection from the rain. Some materials will be more appropriate than others.

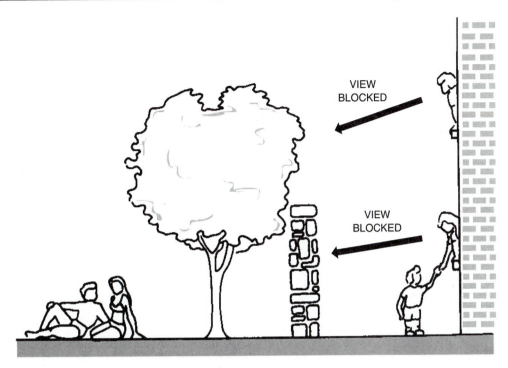

VIEW
BLOCKED

VIEW
BLOCKED

Figure 11-2 The combination of a constructed enclosure and trees gives privacy when the viewer is above ground level.

Figures 11-14 to 16 illustrate the different ways that constructed enclosure functions to create privacy in a landscape.

- *Cost* This is more complex than just a comparison of the initial investments. There are the initial material purchase costs, but there are also the costs associated with the degree and skill of labor required to install the materials. Later maintenance costs must also be considered at the time of initial selection. For example, a brick wall costs more to install than a prefabricated wood fence, but over time the wall may have minimal costs for upkeep, while the fence necessitates annual painting and the replacement of rotted support posts. A concrete surface costs more to install than a seeded lawn surface of the same size, but the concrete does not have to be fertilized, watered, and mowed to sustain its appearance and effectiveness.

- *Strength* The selection of a surfacing material may require a consideration of the weight it will support. What is suitable as a patio is likely to be unsuitable for a driveway. An enclosure material may have to stand up against a strong prevailing wind, the pressure of a retained slope, or the lunging of a large, hyperactive dog.

- *Weather effects* Some hardscape surfacing materials absorb heat, making them uncomfortable to walk on during summer days. Other surfacings may reflect sunlight or create glare requiring that their placement in certain areas be avoided. Painted wood peels and stained wood fades over time. Other materials, many of them new synthetics, are barely affected by exposure to the elements. The intended use and location of the hardscape material may determine the suitability of one material over another.

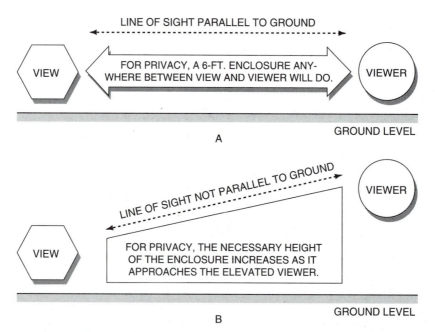

Figure 11-3 The difference in the height of the enclosure necessary for privacy when the line of sight is and is not parallel to the ground

- *Client preference* The choice of hardscape items and materials remains a prerogative of clients. It is a prerogative that they may or may not exercise, or exercise in varying amounts. Some clients will trust their designer or contractor to select the most appropriate materials; others will have something general in mind, but no specific details. Still other clients will not only specify exact items or materials, but may mandate specific suppliers.

- *Integration with other components of the landscape* Hardscape items and materials can have a style or other characteristic that evokes certain attitudes or impressions from viewers. For example, bamboo often suggests the Orient to people. That makes it an excellent material to use in a Japanese garden, but it may not be appropriate somewhere else. A white picket fence may be the perfect choice for enclosure around a Victorian cottage but look out of place in the landscape of a midwestern ranch home. Architecture has

distinctive styling, and many landscapes have thematic styling. The hardscape materials can and should contribute to the development of the total landscape composition, not stand apart as unrelated oddities or in ultrasharp contrast to the other materials. Fencing can be generic in style, for example, solid board or basket weave, or it can be styled to match comfortably with a specific architectural or thematic style, Figure 11-17. Surfacings often can repeat the materials used in the house or in nearby enclosure materials, Figure 11-18.

- *Maintenance concerns* Concern about the cost of upkeep for some hardscape items was described previously. There are a myriad of other concerns that must be measured at the time that hardscape materials are selected to determine their suitability for a particular landscape and for a particular client. Many of the maintenance concerns are item-specific. With items such as fieldstone walls, they can generally be left to age and weather, thus making

Figure 11-4 This architectural use of enclosure combines solid columns and spans of trellis with containerized plant materials.

Figure 11-5 The constructed enclosure components in this riverside setting are varied and imaginative. They contribute greatly to the users' understanding and appreciation of the uniqueness of the area.

Figure 11-6 The raised planting bed adds an interesting effect to this flat rooftop garden. It also offers protection by keeping people away from the edge of the roof.

them nearly maintenance-free. However, water features have to be tended to with some frequency. Pools need their filters cleaned. The pH of the water may need to be monitored with regularity. Decking needs waterproofing and staining. Fountains have to be drained at season's end. Charcoal- or wood-fueled fireplaces need regular ash removal, while gas and electric grills do not. Concrete surfaces can be plowed with less danger of damage than can a cobblestone or brick surface. A designer must not only be knowledgeable of how things are maintained, but he or she has the responsibility to query and/or advise the client before specifying certain hardscape selections.

Changing Levels in the Landscape

When the grade changes in the outdoor room, the hardscape may accompany the change. Some types of enclosure, such as stone walls, are able to parallel a slope. Other types of enclosure, such as panels of fencing, must be stepped down a slope. With surfacing materials, there are numerous design possibilities, but only two technical ways to accommodate the grade change: as steps or as ramps.

Design Concerns

The materials used in the construction of steps and ramps are often the same materials used to surface the levels being connected. The materials used could also mimic those used in nearby enclosure components or even materials used in the facade of the building. Any of those techniques would be a good application of the design principles of simplicity and unity. Sometimes, for the sake of contrast, the steps or ramp may be constructed of a different material. It comes down to the designer's choice and what level of visual attention the grade change is to receive in the design.

A major concern in the design of outdoor steps is that they be designed for safe use. One of the greatest dangers of steps in the landscape is that someone will fall because of them. Ramps reduce that danger and are especially appropriate in pub-

Figure 11-7 What was once a steep slope has been converted into a series of broad, level planting strips connected by steps. The brick retaining walls provide both strength and beauty.

Figure 11-8 Weep holes placed near the bottom of a retaining wall prevent water pressure from building behind the wall.

lic landscapes or in private settings where users may be in wheelchairs or on walkers, or have other physical challenges. Where steps are used, they must be designed to give emphasis to where they begin and to where each step is positioned. On a sunny day, shadow patterns may give emphasis to each step, thereby assuring that users will not miss one and stumble. Those shadow patterns can be duplicated at night with carefully placed night lighting. However, on a cloudy day, with no shadows

Figure 11-9 Timber retaining walls help stabilize this steep slope.

Figure 11-10 The fence effectively blocks the view of this pool apparatus.

Figure 11-11 Aesthetic enclosure. The planter enhances this urban area, doing triple duty as a visual softener, traffic router, and unofficial pedestrian bench.

apparent, steps that are made of the same materials as the surfaces they connect and which give no additional emphasis to distinguish one step from another become potential trip hazards in the landscape. It is helpful to make some type of change at the top and bottom of an outdoor staircase. The change could be one of color, texture, or both. The purpose is to register in the users' consciousness that something is about to happen and they should take notice. It is also helpful to give emphasis to the edge of each step by changing the material or color so that the step separates from the one below it. It becomes similar to a highlight for each step. Figures 11-19 and 11-20 illustrate how an outdoor staircase can be designed for both aesthetics and safety.

Technical Concerns

Steps are composed of two parts: the **riser** and the **tread**, Figure 11-21. In designing outdoor steps,

one of the intents is to permit the user to maintain a close to natural stride pattern while using the steps. That is accomplished by using wider treads and lower risers than are used for indoor stairs. A widely accepted formula for sizing outdoor steps is $T + 2R = 26"$, where T is the tread width and R is the riser height. Furthermore, the tread width must be wide enough to permit a user's foot to rest comfortably and safely on it. That necessitates a 12-inch minimum width. Applying the formula with a 12-inch tread will result in a riser dimension of 7 inches $(12" + [2 \times 7"] = 26")$, Figure 11-22. As the tread width increases, the riser dimension decreases.

Riser	Tread
7 inches	12 inches
6 inches	14 inches
5 inches	16 inches
4 inches	18 inches

The number of steps required to connect two levels is calculated by dividing the elevation by the riser dimension desired. For example, if two levels are 48 inches apart and a 6-inch riser is desired, eight steps are required (48" divided by 6 equals 8 steps), Figure 11-23. The amount of horizontal space required for the steps is determined by multiplying the tread dimension by the number of steps. In the example, the eight steps, with seven 14-inch treads, require 98 inches (8.16 feet) of horizontal space in the landscape for their construction (14" x 7 treads = 98" ÷ 12" = 8.16'). It should be noted that the number of treads is always one less than the number of risers, since the top riser connects with the upper level, not another tread.

Walks, Drives, and Stepping Stones

The choice of materials for walks and stepping stones has both design concerns and utilitarian concerns. A slice of solid pavement cutting across the lawn becomes a strong attraction to the viewer's eye. In most cases, that is not a desirable attribute, so most of the time it is best to select surfacing materials that are repetitive of those used elsewhere in the design and that attract as little attention as possible.

Their selection and use for utilitarian reasons is a greater concern. Walks may be classified as primary or secondary in their roles. A *primary walk* is one that is used often and by large numbers of people. The foremost example in a residential design is the front walk that connects the entry with the street or the driveway or both. It should be a minimum of five feet in width to permit two people to approach side by side. It also needs to be constructed of a solid, durable material. A *secondary walk* connects areas used less often and by fewer people. An example is the walk around the side of a house, used to bring trash to curbside or to permit the mower to move from one area to the next. Secondary walks need only be about three feet wide. Some may be constructed of less durable

Figure 11-12 This pergola styled overhead won't make the food taste better, but it does create a sense of enclosure for this outdoor kitchen.

materials, such as loose aggregates (gravel, marble chips, and others).

Stepping stones usually function as secondary walkways or lesser used pathways. Their width needs to be sufficient to accommodate the anticipated traffic. Also, the individual stones need to be large enough to accept an entire foot and avoid creating a tiptoe effect. They should be placed to permit a natural stride pattern, Figure 11-24.

Driveways are equally troublesome as design elements. They often attract too much attention due to their expansive display of hardscape. Like the entry walk, they need to be minimized as much as possible by screening, use of conservative colors, and incorporation of materials used elsewhere in the curb view. Their width is usually determined by the size of the garage they adjoin. For a single-car garage, a drive width of 10 feet is minimal. It doubles for a two-car garage or anywhere that two cars must pass or be parked side by side.

A Comparison of Types of Surfacing					
Surfacing Type	Installation Cost	Maintenance Cost	Walking Comfort	Use Intensity	Seasonal or Constant Appearance
hard paving	highest	low	lowest	highest	constant
soft paving	moderate	moderate (1)	moderate	moderate (1)	constant
turf grass	lowest	high (2)	high	moderate (6)	constant (7)
groundcover	moderate	moderate (3)	N/A (5)	low	constant to seasonal (8)
flowers	moderate	high (4)	N/A (5)	lowest	seasonal

(1) Some replacement is required each year.
(2) Fertilization, weed control, watering, and mowing are necessary.
(3) Initial cost is high due to the hand weeding that is required. Once established, costs are moderate.
(4) Much hand weeding, watering, and fertilizing are required.
(5) Should not be used in areas with pedestrians.
(6) Use intensity is greatest where people do not continuously follow the same path.
(7) Many grasses are dormant in certain seasons and may change color.
(8) Appearance of these materials depends upon whether plants are evergreen or deciduous.

Water Features

Obviously water is a natural material and is not hard, so it is their manner of use and containment that qualify water features as hardscape. The uses of water features include fishponds, reflecting pools, fountains, waterfalls, streams, swimming pools and spas, and combinations of these and other features. In short, the uses of water are limited only by imagination. Water features are increasingly part of designed landscapes, both exterior and interior, Figures 11-25 and 26. The controlled and contrived use of water in landscapes has been with us for centuries. The only significant changes over time have been in the technology of the plumbing and in the sources of the water. Fountains of the Renaissance depended largely upon the energy of gravity, created when water dropped between levels, to propel the water high into the air. That effect can now be created with pumps and electricity. Historic water features also needed a natural water source to supply them, often resulting in the damming of small streams to create lakes and even the diversion of rivers from their natural course to new routes through the gardens of wealthy aristocrats. That impractical solution has been replaced with pumps and filters that conserve and recirculate a fixed water reservoir.

Design Concerns

Since there are so many types of water features, the treatment here must be very general. The landscape designer and client must ask and answer several questions as planning for the inclusion of water begins.

■ *Is the primary purpose of the water feature aesthetic or utilitarian?* A water feature used to raise the humidity in an arid region may not be as valued for its appearance as for its effectiveness.

■ *If the purpose is aesthetic, what is the intended appeal?* Some water features are most successful visually. The movement and sparkle of the water, often coupled with the artistry of a fountain or piece of sculpture, is the desired aesthetic. Other places it is simply the sound of moving water that is sought and most valued. The appearance is of secondary importance.

Figure 11-13 In regions where plant growth is slow and sometimes difficult to sustain, the use of hardscape materials is extensive. In this photo, the plants can be seen as providing visual accents, while the constructed materials bear responsibility for making the landscape usable.

Figure 11-14 Privacy to this conversation area of the patio is provided by the lattice fencing that encircles it.

Figure 11-15 Privacy is possible even where the landscape meets the street if the fencing style is carefully selected.

Figure 11-16 To reduce the visual weight of privacy screening, the hardscape can be combined with plant materials.

- *Is the water feature to look natural or crafted?* A fountain or formal swimming pool makes no pretense of being natural, but creating a natural appearing stream where no natural water source exists can be a challenge.

- *Is the client fully aware of the maintenance requirements of the water feature and prepared to provide them?* No one wants to see the clear water in his pond turn to a swampy green. No one wants to have to drain the spa after every use as though it was a bathtub. To avoid such occurrences, the water must be chemically treated to prevent algae and other microbial growth. Filters and skimmers must be cleaned to assure their continued effectiveness. Fountains and pumps must be drained at season's end in regions of the country where they could freeze and burst during the winter months. Underwater lights need to be replaced. Water plants and fish may have to be taken in at season's end as well.

- *How much does the client want to spend?* There are inexpensive recirculating fountains that can be set in place, filled with water, and plugged in for an instant effect. At the other end of the expense spectrum are custom-built

Figure 11-17 This white picket fence is a thematic match for the early American home that it adjoins.

Figure 11-18 The use of concrete pavers next to the concrete wall blocks creates a close integration of the hardscape materials.

Figure 11-19 Solid risers of wood are combined with treads of loose aggregates for coordination with the surrounding landscape.

water features requiring skilled laborers to install and the services of plumbers and electricians to make them operative. Discount stores, sculptors' studios, and many other alternatives can all provide water features depending upon the client and the project.

Constructing a Recirculating Water Feature

As an example of how a customized water feature can be installed, consider these general requirements and concerns.

Location. If the feature is to appear natural, there must be believability to the source of the water. A waterfall that seems to flow from the side of a building lacks credibility. Likewise, one that appears to have popped up in the center of a totally flat lot is equally incredible. A berm may need to be

Figure 11-20 To create more stylized and durable steps, brick treads are edged with granite slabs. The combination of smooth and textured materials as well as their color difference reduces the chance of someone falling on the stairs.

constructed and be of sufficient size to suggest that a small stream could actually originate within.

If the water feature is to support the growth of water plants, then it must be placed where it can receive sunlight during the day. If it is to contain fish, full sun for the entire day may be detrimental, so a location providing some shade may be preferred.

Containment. In one way or another the water must be contained to provide a closed system. Prefabricated forms are available from garden centers and other sources in a variety of shapes and sizes. Imaginative homeowners have been known

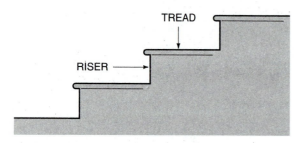

Figure 11-21 The parts of a step

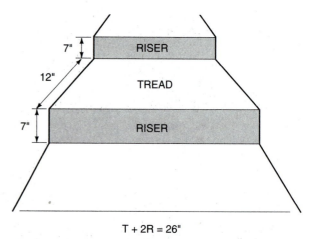

Figure 11-22 Calculating the dimensions of outdoor steps

to use buried bathtubs, wine casks, and an assortment of other containers for their basins. Larger installations may require that an area in the shape of the pond be excavated and lined with a heavy plastic membrane to retain the water.

The Apparatus. Water features need the water to move, even if only slightly. Otherwise they quickly become stagnant. A *pump* is essential. The size depends upon the quantity of water to be moved and the height to which it must be lifted. The *liner* was mentioned previously. It is usually black to aid the illusion of depth. It must be heavy enough to avoid tearing because it gets pressed against rocks beneath it and those added inside to give realism. A *skimmer* draws debris from the surface of the water and prevents it from clogging the pump. Usually the skimmer has a bag attached that collects the debris and holds it until it can be cleaned and dumped. *Flexible PVC (polyvinyl chloride) piping* allows the debris-free water to flow from its farthest or lowest point back to the pump. A *filter* may be used in lieu of or in addition to the skimmer to protect the pump's clogging. *Lighting fixtures,* either underwater or exterior to the feature may be used for special effects. An *aerator* in forms ranging from a simple bubbler to a spectacular fountain will keep the pond supportive of desired plants and fish and contribute to control of the microorganisms that can cloud the water. *Concealment devices,* such as gravel to hide the liner and rocks or plants to camouflage the skimmer and the pump, must also be used.

Keeping the Water Clean. The skimmer, filter, and aerator are essential to the maintenance of good water quality. In addition, it may be necessary to treat the water with chemicals or add beneficial bacteria to the water to attain optimum clarity. If chemical additives are used, their effects on plants, fish, or small animals that may drink from the pond should first be checked.

Powering Up and Maintaining the Water Level. A pump-driven recirculating water feature requires electricity to operate. Also, over time some of the water will evaporate and necessitate replacement. If the feature is not constructed near a building where these critical utilities exist, the cost of construction will increase because the utilities will have to be brought out to it.

Figure 11-27 offers a sectional look at a typical recirculating water feature. While the size of the feature may vary, the essential construction objectives are similar, whether building a small courtyard fountain or a large corporate park lake.

Outdoor Lighting

One of the fastest growing niches in the service offerings of American landscape contractors is outdoor lighting. Not too many years ago, there was only one nationally recognized manufacturer of outdoor lighting fixtures and lamps, Loran, Inc. of

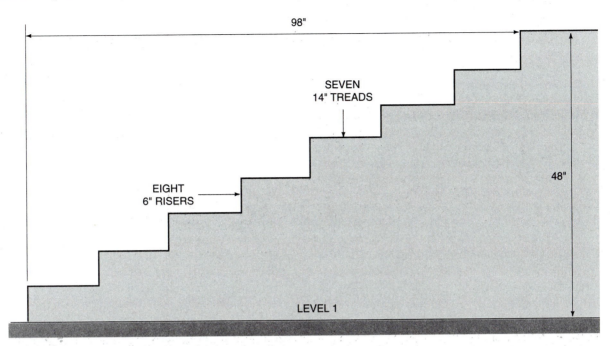

Figure 11-23 Planning space for outdoor steps

California, the manufacturer of *Nightscaping*. There are now many more in the marketplace. There are also a number of products that target the homeowner and amateur gardener market, such as solar walk lights and boxed sets of inexpensive lamps strung along a wire like holiday lights.

The lighting systems now being specified by landscape designers and contractors are 12-volt systems, and they are very different from the 120-volt systems used to illuminate outdoor roadways, parking lots, or recreational facilities. At first consideration, the difference between 12-volt and 120-volt systems would be thought related only to the size of the landscape being illuminated. While true in some cases, more often the difference between the systems is a difference of purposes. Security and safety concerns or the need to provide daytime brilliance at night is best assigned to the standard-voltage systems. Where the concern is for lighting effects and decoration, the low-voltage lighting systems are more appropriate. Not only are they less expensive to install and maintain, they offer

the designer much greater control over the impact that the lighting system will bring to the landscape. This brief introduction to outdoor lighting as hardscape will focus on the low-voltage systems.

Design Concerns

The first consideration should be whether the lighting plan will emphasize the *effect* or the *fixture*. There are many fixtures available that are definite eye catchers. Some are cloyingly cute; others are artsy, approaching sculpture. As such, they can assume the role usually assigned to lawn ornaments and can become either enriching or distracting. What they are lighting becomes of secondary importance to how they look while doing it. While there is a place for those types of fixtures, most landscape lighting designers choose and locate fixtures for the effects they create. Skillful lighting of the landscape mimics theatrical lighting by evoking reactions from the viewer. Whether done to extend the time of use in

Figure 11-24 Stepping stones must be large enough to fit the users' feet and be spaced so they permit a natural stride pattern.

the outdoor room, or to create moods, or to provide visual punctuation to the architecture of the landscape, low-voltage night lighting is one more tool used to enhance a residential or commercial property.

The System

Low-voltage outdoor lighting systems require 12-volt lamps and a step-down transformer to convert the 120-volt current to a 12-volt current. That reduction of the current is both necessary electrically and important to human safety, an important aspect of low-voltage lighting. The Underwriters Laboratory has affirmed that a shock of 12-volts or less will rarely if ever cause a breakdown in skin resistance sufficient to cause physical injury to a person.

The lamps for low-voltage systems are varied and should be selected to create the effects being sought by the designer on behalf of the clients.

Figure 11-25 Water features are increasingly popular, requiring landscape contractors to master the details of their construction.

Depending upon the lamp selected, effects can range from a thin pinpoint spotlight on a selected focal point to a soft broad illumination that duplicates moonlight. Unlike many of the lamps used in high-voltage systems that cast an orange color or one that is blue-green, the light quality of 12-volt lamps is true to the white light of sunlight. That purity of color coupled with the engineering of the lamps themselves makes 12-volt light easier to control than 120-volt systems. Some clients may also desire supplemental system features such as

Figure 11-26 Making the water feature look natural can be difficult. A clever blend of plant materials with the hardscape is necessary. If not done successfully, the water feature can stand apart from the rest of the design and never seem to belong.

timers or photocells to activate the system automatically. Others may want a dimmer switch to add additional control options to their enjoyment of the system.

Many landscape contractors are electing to offer night lighting services because low-voltage systems do not require the depth of knowledge needed by an electrician. Still, knowing how to determine the proper transformer size, what fixtures and lamps to select for the desired effects, and how to connect the low-voltage system to the 120-voltage of the property requires training. That has required a few states to require practioners to be licensed. Many of the lighting manufacturers and distributors offer instructional workshops for the purpose of providing contractors with the necessary training.

Creating the Lighting Plan

The best lighting plans will be developed concurrent with the total design of the landscape. However, many are added after the landscape has been in place for a while. Regardless of when the system is planned, certain questions should be addressed before beginning.

■ What were the objectives in the design of the building architecture and the landscape? How can the night lighting reinforce those objectives (it should never counter them)?

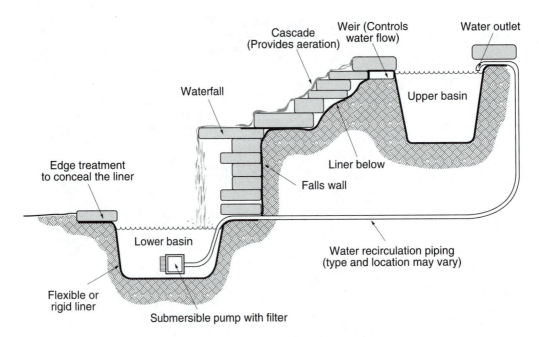

Figure 11-27   Cross-section of a water feature showing components

- What individual features of the building(s) and/or the landscape need to be enhanced?

- What individual features can and should be hidden or minimized at night?

- How will the clients use the landscape at night? How much light will be needed to support those activities without over-lighting the areas?

- Are different moods desired for different areas of the landscape?

- What areas of the landscape are potentially dangerous at night? Are there grade changes, steps, projections, pools of water, darkly shadowed areas near the garage or entry, or other features that might need safety or security lighting?

Once these questions are answered, the lighting designer can then decide the specifics of the plan, including what to light. Typical items that accept night lighting while retaining the design intent of the daytime landscape include specimen plants—light is used to emphasize the ongoing nature of their uniqueness, such as the branching habit or the bark texture, rather than short-term distinctiveness such as that found in flowers or fall color. Water features also embrace night lighting, which can bring their sparkle and movement into the nighttime, often enhancing their importance by the careful and clever use of colored lamps. Some of the most dramatic uses of night lighting can be created by illuminating selected areas or features of the building architecture. While outlining the roofline is best left to theme parks or the holiday season, a nicely lit entryway or soft uplighting of a building's columns can give rewarding emphasis to a structure. Even the casting of a plant's shadow against a building wall can provide a fresh appreciation of both components.

The best conclusion that can be offered for this brief presentation of night lighting is to say that the potential for enjoyment and enrichment is great, but so is the potential for waste. Lighting unimportant features or areas wastes money and insults the design of the landscape and the buildings. Avoid

the temptation to overuse it. Understatement is better and more tasteful than turning a lovely daytime setting into a nighttime theme park.

Achievement Review

A. Indicate if the following uses of hardscape are primarily (A) architectural, (B) engineering, or (C) aesthetic.

1. A timber retaining wall is used to create a level grade.
2. Concrete surfacing is used to support heavy foot traffic.
3. A fountain is placed into the design as a focal point.
4. A fence is placed at the edge of the outdoor room to provide privacy from the neighbors.
5. A canvas canopy is placed over a patio to create a more intimate ceiling effect for the outdoor room.

B. Explain how each of the following criteria may affect the choice of specific hardscape items and materials for a landscape.

1. cost
2. client preference
3. function
4. strength
5. weather effects
6. maintenance concerns
7. the other components of the landscape

C. Using the *Comparison of Surfacing Materials* chart at the end of the chapter, select the surfacing material that meets the following criteria.

1. It will be used as a raised patio surface. It must allow water to pass through it. It must be modular so that parts can be removed and replaced as needed.
2. It will be used as a surfacing for a picnic grove. It must be permeable to water and permit easy coverage of the stains of spills.

3. The surfacing will be used for multiple purposes: walkway, driveway, and as a patio. It will have a broom-swept finish applied.
4. The surfacing will be used for an ornately patterned walk. It is to be a modular surface so that different colors can be used to create the patterns.
5. The surfacing will be used as a patio. It is to be a hard modular surface but will be composed of irregularly shaped pieces.

D. Consult the *Comparison of Enclosure Styles and Materials* chart at the end of the chapter in selecting your answer to these questions.

1. Which of the following styles does not offer good security at a height of 6 feet?
   a. poured concrete wall
   b. post and rail fence
   c. vertical louvered fence
   d. rubble stone wall
2. Which style is most suitable as a security enclosure for a children's play area?
   a. lattice fencing
   b. split rail fencing
   c. chain link fencing
   d. grape stake fencing
3. Which style would look most appropriate when used in a rustic rural setting?
   a. split rail fencing
   b. picket fencing
   c. wrought iron grills
   d. chain link fencing
4. Which of the following provides the best support of grade changes?
   a. railroad ties
   b. solid board fencing
   c. stockade fencing
   d. slat fencing

5. Which enclosure material would best accept the silhouette of a plant projected against it
   a. louvered fencing
   b. railroad ties
   c. ashlar stone wall
   d. poured concrete wall

E. If the landscape necessitated a grade change of 56 inches, how many steps having 12-inch treads would be required?

F. Explain the function of each of the following items in the operation of a water feature.
   1. pump
   2. skimmer
   3. filter
   4. aerator

G. When interviewing a client who has requested a night lighting plan, what questions should be asked of the client by the lighting designer?

Suggested Activities

1. Using the two charts that compare surfacing and enclosure materials as a guide, collect color photo examples of the materials described in the charts. Prepare a personal comparison chart for your own reference and add to it over time. As sources, use mail order catalogues, garden center and lumber company advertisements, house and garden magazines, and trade journals.

2. Make a collection of photos of housing styles. Find examples of as many different styles as possible. House plan magazines, available at most newsstands, are a good source. Make a listing of the specific types of enclosure and surfacing materials that would look appropriate with each of the styles.

3. Invite a lumber dealer to talk with the class about wood. Which woods last longest? How can less expensive woods that rot quickly be made to last longer? What prefabricated fencing styles are most popular in your area?

4. Extend similar invitations to other hardscape specialists in the area, such as pool builders or lighting contractors.

5. If there is a manufacturer or major supplier of hardscape materials in your area, plan a visit to a brick-works, lumberyard, patio furniture shop, etc.

6. Demonstrate the different elevations of enclosure needed to provide total privacy in an area. Have two class members sit opposite each other (one on a staircase or ladder). Two other class members can then separate them by holding a large blanket between them. Measure the height needed to block the students' view of each other. Gradually elevate one student's vantage point, each time measuring the height of enclosure needed for view blockage. Alter the placement of the enclosure. Place it closer to the viewer first and then closer to the person being viewed.

7. Evaluate the comfort of nearby outdoor steps by walking up and down them to measure how closely they allow users to maintain a natural stride pattern. Then measure their riser and tread dimensions to determine how closely they correspond to the $T + 2R = 26"$ formula.

8. Demonstrate the requirements for a recirculating water feature in the classroom using small scale versions of the pump, filter, and tubing available where aquarium supplies are sold.

9. In a darkened room, demonstrate the differences in the light quality of incandescent, fluorescent, and 12-volt lamps by selecting a variety of landscape materials, such as green plants, flowers, different colors of pavers, and several different wood stains, and evaluate their color rendition under the different light sources.

A Comparison of Enclosure Styles and Materials

Style	Material	Security at 6-foot Height	Privacy at 6-foot Height	Noise Reduction at 6-foot Height	Wind Deflection at 6-foot Height	Grading Structure	Useful for Raised Beds	Comments
BASKETWEAVE FENCE	Wood	Yes	Yes	Moderate	Moderate	No	No	Available in prefabricated sections, attractive on both sides
BRICK WALL	Brick	Yes	Yes	Good	Good	Yes	Yes	An ideal material for free-standing retaining or seat walls; width will vary with height and function
CHAINLINK FENCE	Steel	Yes	No	None	None	No	No	Good for use around pet areas or to safeguard children's play areas
CONCRETE BLOCK WALL	Concrete	Yes	Yes	Good	Good	Yes	Yes	Less expensive than poured concrete, stone, or brick
GRAPE STAKE FENCING	Wood	No	Yes	Moderate	Moderate	No	No	A rustic style that weathers to an attractive gray; also comfortable in urban settings
LATTICE FENCE	Wood	No	Variable	No	Limited	No	No	Effectiveness as a screen depends upon how closely the lattice is spaced
LOUVERED FENCE	Wood	Yes	Yes	Moderate	Moderate	No	No	Louvers may be vertical or horizontal and are angled to provide privacy
PICKET FENCE	Wood or Iron	Depends on the height	No	No	No	No	No	High maintenance costs because of frequent painting needed
POST-AND-RAIL FENCE	Wood	No	No	No	No	No	No	Degree of formality varies with finish of lumber; a style valued for aesthetics more than for security or privacy
POURED CONCRETE WALL	Concrete	Yes	Yes	Good	Good	Yes	Yes	Can be smooth or textured, colored or inset with materials to add interest to the surface; requires reinforcing if it is to provide strength

A Comparison of Enclosure Styles and Materials *(continued)*

Style	Material	Security at 6-foot Height	Privacy at 6-foot Height	Noise Reduction at 6-foot Height	Wind Deflection at 6-foot Height	Grading Structure	Useful for Raised Beds	Comments
RAILROAD TIES	Wood	No	No	No	No	Yes	Yes	Ideal for rustic, natural enclosures; widely used for soil retention; care should be taken to assure that the ties have not been preserved with a phytotoxin
SLAT FENCE	Wood	Yes	Variable	Moderate	Moderate	No	No	Effectiveness as screen or wind deflector depends upon how closely slats are spaced
SOLID BOARD FENCE	Wood	Yes	Yes	Moderate	Good	No	No	Expensive but the best for security and privacy combined; maintenance easier than with other styles
SPLIT RAIL FENCE	Wood	No	No	No	No	No	No	A rustic style best used in rural settings; lumber is rough and unfinished
STOCKADE FENCE	Wood	Yes	Yes	Moderate	Good	No	No	A variation of solid board fencing
STONE WALL, ASHLAR	Stone	Yes	Yes	Good	Good	Yes	Yes	The stone is cut, usually at the quarry. Stones vary in their smoothness and finish. They are laid in a horizontal and continuous course with even joints.
STONE WALL, RUBBLE	Stone	Yes	Yes	Good	Good	Yes	Yes	The stone is not cut. No course is maintained. Small stones are avoided. Larger stones are used at the base of the wall.
WOOD RETAINING WALL	Wood	No	No	No	No	Yes	Yes	Wood must be preserved to avoid rapid decay; the preservation should not be phytotoxic; reinforcement necessary to assure strength
WROUGHT IRON GRILLS	Iron	Yes	No	No	No	No	No	Expensive and used mostly for aesthetics; grills may be continuous or used as baffles separately

A Comparison of Surfacing Materials

Material Description	Hard Paving	Soft Paving	Modular	Continuous and Solid
Asphalt: A petroleum product with adhesive and water-repellant qualities. It is applied in either heated or cold states and poured or spread into place.	Semihard; allows weeds to germinate and grow through it			•
Asphalt Pavers: Asphalt combined with loose aggregate and molded into square, rectangular, or hexagonal shapes. They are applied over a base of poured concrete, crushed stone, or a binder.	Semihard if not applied over concrete		•	
Brick: A material manufactured of either hard baked clay, cement, or adobe. While assorted sizes are made, the standard size of common brick is 2 1/4 × 3 3/4 × 8 inches.	•		•	
Brick Chips: A by-product of brick manufacturing. The chips are graded and sold in standardized size as aggregate material.		•	•	
Carpeting, Indoor/Outdoor: Waterproof, synthetic fabrics applied over a concrete base. They are declining in popularity. Their major contribution is to provide visual unity between indoor and outdoor living rooms.				•
Clay Tile Pavers: Similar to clay brick in comparison, but thinner and of varying dimensions (most commonly 3 × 3-inch, or 6 × 6-inch squares). They are installed over a poured concrete base and mortared into place.	•		•	
Concrete: A versatile surfacing that can be made glassy smooth or rough. It can also be patterned by insetting bricks, wood strips, or loose aggregates into it. Concrete is a mixture of sand or gravel, cement, and water. It pours into place, is held there by wood or steel forms, then hardens.	•			•

Slippery When Wet	Permeable to Water	Suitable for Vehicles	Suitable for Walks	Suitable for Patios
		•	•	Certain formulations are suitable. Others may become sticky in hot weather. *Note*: The application of a soil sterilant before applying the asphalt can eliminate the weed problem in walks, drives, and patios.
	If installed over crushed stone		•	•
	If installed in sand	•	•	•
	•		Edging needed to hold them in place	
	Provision must be made for surface water drainage or the carpeting becomes soggy			•
•			•	•
Only when smoothly finished		•	•	•

A Comparison of Surfacing Materials *(continued)*

Material Description	Hard Paving	Soft Paving	Modular	Continuous and Solid
Crushed Stone: Various types of stones are included in this umbrella term: limestone, sandstone, granite, and marble. Crushed stone is an aggregate material of assorted sizes, shapes, and durability.		•	•	
Flagstone: An expensive form of stone rather than a kind of stone. Flagstone can be any stone with horizontal layering that permits it to be split into flat slabs. It may be used as irregular shapes or cut into rectangular shapes for a more formal look. It is usually set into sand or mortared into place over a concrete slab.	•		•	
Granite Pavers: Granite is one of the most durable stones available to the landscaper. The pavers are quarried cubes of stone, 3 1/2 to 4 1/2 inches square, that are mortared into place. Various colors are available.	•		•	
Limestone: A quarried stone of gray coloration. Limestone can be cut to any size. It adapts to formal settings.	•		•	
Marble: An expensive quarried stone of varied and attractive colorations. It has a fine texture and a smooth surface that becomes slippery. Its use as surfacing is limited. It can be inset into more serviceable surfaces such as poured concrete.	•		•	
Marble Chips: A form of crushed stone, marble chips are more commonly used as a mulch than a surfacing. They are expensive compared to other loose aggregates; still they enjoy some use as pavings for secondary walks and areas that are seen more than walked upon.		•	•	
Patio Blocks: Precast concrete materials available in rectangular shapes of varied dimensions and colors. Limitless patterns can be created by combination of the sizes and colors. The blocks are set into sand or mortared over concrete.	•		•	

Slippery When Wet	Permeable to Water	Suitable for Vehicles	Suitable for Walks	Suitable for Patios
	•	•	Edging needed to hold the material in place	Limited use except beneath picnic tables where stains might spoil hard paving
Depends upon the rock used and how smooth the surface is			•	•
		•	•	Too rough
		•	•	•
•				Best used in dry climate where slipperiness will not be a frequent concern
	•		•	
			•	•

A Comparison of Surfacing Materials *(continued)*

Material Description	Hard Paving	Soft Paving	Modular	Continuous and Solid
Sandstone: A quarried stone composed of compacted sand and a natural cement such as silica, iron oxide, or calcium. Colors vary from reddish brown to gray and buff white. The stone may be irregular or cut to rectangular forms.	•		•	
Slate: A finely textured stone with horizontal layering that makes it a popular choice for flagstones. Black is the most common color, but others are available.	•		•	
Stone Dust: A by-product of stone quarrying. Stone dust is finely granulated stone, intermediate in size between coarse sand and pea gravel. It is spread, then packed down with a roller. The color is gray.		•		•
Tanbark: A by-product of leather tanning. The material is processed oak bark. It has a dark brown color and a spongy, soft consistency. It is ideal for children's play areas.		•	•	
Wood Chips: A by-product of saw mills, wood chips are available from both soft-woods and hardwoods. The latter decompose more slowly than the former. Wood chips have a spongy, soft consistency. They are often used as mulches.		•	•	
Wood Decking: Usually cut from soft-woods, the surfacing can be constructed at ground level or elevated. The deck is valuable as a means of creating level outdoor living spaces on uneven terrain. Space should be left between the boards to allow water to pass through and the wood to dry quickly. Use of a wood preservative will slow decay.	•		•	
Wood Rounds: Cross sections of wood cut from the trunks of trees resistant to decay, such as redwood, cypress, and cedar. The rounds are installed in sand. Individual rounds are replaced as they decay.	•		•	

Slippery When Wet	Permeable to Water	Suitable for Vehicles	Suitable for Walks	Suitable for Patios
		•	•	•
•			•	•
	•		•	
	•		Edging needed to hold the material in place	
	•		•	
	•		•	•
	•		•	•

Enrichment

Objectives:

Upon completion of this chapter, you should be able to

- define landscape enrichment
- distinguish between tangible and intangible enrichment
- distinguish between natural and fabricated enrichment items
- evaluate the value of selected enrichment items

Landscape Enrichment Is Something Extra

To understand the role that enrichment items play in the design and development of a landscape, picture again an indoor room. When the basic structure of an indoor room is completed, it has walls, ceiling, and a floor, but it is not yet fully usable. To make it functional and personal, it requires furniture, lighting, pictures, music, pets, and many other things that outfit the room for use by the people for whom it is intended. Although these items are not wall, ceiling, or floor elements, their importance as components of the indoor room is easily seen.

The outdoor room has a similar need to be made usable and personal. **Enrichment items** are com-
ponents of the outdoor room that are not essential to the formation of its walls, ceiling, or floors. While they may also serve as walls, ceiling, or floor materials, enrichment items are first selected to fulfill some additional purpose.

Tangible and Intangible/ Natural and Fabricated

Enrichment items are represented by a widely diverse collection of materials. Some are **tangible**, meaning that they can be touched; while others are **intangible**, meaning that they are sensed but cannot be touched. Some are naturally occurring, while others are fabricated, meaning manufactured. Consider this partial listing to aid in understanding the contribution that enrichment items make to the landscape.

Figure 12-1 Large stones provide an enriching backdrop for the rock garden plants growing among them.

Tangible, Natural Enrichment

Stones, such as rocks, boulders, and natural outcroppings, are valuable additions to any landscape. Where they are an existing site feature, such as rock outcroppings, the landscape designer is wise to incorporate them into the design. They can serve effectively as backdrops for border plantings or as the framework for the development of a rock garden, Figure 12-1. In other settings, naturally occurring or selected stones can serve as natural sculptures in the landscape, Figure 12-2. Where the design and size of the site permit, stones can also serve as alternatives to traditional outdoor furniture, permitting people to rest against them or children to climb safely over them, Figure 12-3.

Figure 12-3 Rock formations add natural enrichment to the landscape. (Photo courtesy of Colleen Pasquale)

Figure 12-2 Stone used as a natural sculpture

Figure 12-4 Stones play an important and traditional role in Japanese gardens.

Figure 12-5 Stones are an important natural enrichment feature in this southwestern landscape.

In Oriental garden design and in gardens of the American Southwest, stones have a long history of importance. The Chinese and Japanese peoples have traditionally assigned great spiritual and cultural significance to the selection and positioning of stones within their gardens, Figure 12-4. In the Southwest, stones have been one means of interfacing designed gardens with the wide open spaces that lie outside, Figure 12-5.

Specimen plants bring something extra to the landscape's design, which may last only for a few days or which may be effective year-round. Some plants have highly attractive and unusual growth habits, making them natural outdoor sculptures, Figure 12-6. Others are striking because of their blossom, foliage, fruit, or bark colors and textures. When pruned into unusual shapes, plants can become highly styled objects of art, Figure 12-7. Such pruning of plants into unusual shapes is termed *topiary pruning* and dates back several centuries. Because topiary pruning creates a high maintenance requirement for the garden, designers should use it sparingly as an enrichment technique.

Figure 12-6 The unusual silhouette of this spruce gives it strong eye appeal. It becomes an enriching focal feature.

Figure 12-7 The topiary pruned shrub as an enrichment item requires high maintenance and is used only in very formal settings.

Water is one of the most versatile enrichment materials because it can enhance the landscape's design in so many different ways. When occurring naturally on a site, as a stream, lake, or waterfall, water can enhance the recreational uses of the site. It also attracts wildlife that might otherwise not approach the site. Water mirrors the sunlight and ripples in the wind, Figures 12-8 and 12-9. It makes the landscape sparkle. It is the most popular of all recreational and aesthetic enrichments.

Animals run a close second to water as people's most favored form of tangible, natural enrichment. Like water, animals bring life and movement into the landscaped setting. Because they are not stationary, they add variation and surprise to the daily landscape. Whether it be the color of birds at a feeder, the scamper of a chipmunk across the patio, or a view of farm animals grazing on a distant hillside, people enjoy sharing their landscapes with animals of their choice. Designers can easily provide for the inclusion of wildlife in landscapes by designing outdoor rooms that serve as habitats for both humans and animals. Many plants produce berries that will attract birds and other small animals. A hedgerow can serve as an outdoor wall while simultaneously supplying nesting sites. Planning for the inclusion of feeding stations in the design will serve to assure the continued presence of wildlife through the winter as well, Figure 12-10.

Figure 12-8 The splash and sparkle of water make it appealing in the landscape.

Tangible, Fabricated Enrichment

Outdoor furniture has traditionally been one of two types: heavy and cumbersome or cheap and tacky. Neither type offers appropriate style or much comfort. Depending upon whether the landscape is residential, public, or commercial, furnishings for the landscape may include tables and chairs, lounges, play apparatus, and trash containers. In a long overdue break with tradition, modern landscapes are finally incorporating furnishings that enhance good designs rather than detracting from them, Figure 12-11. Outdoor furniture should be selected to match the theme and style of the outdoor room in which it is placed. Some furniture may be permanent, requiring sufficient durability to withstand the winter. Other furniture may be seasonal, requiring it to be taken in when the weather becomes inclement. While the inexpensive outdoor furniture available each spring in the supermarkets and discount stores will always find buyers, the landscape designer who would be of greatest service to his or her clients should encourage those clients to select their outdoor furniture as carefully as they have selected every other component of the outdoor room, being certain that it is comfortable, color coordinated, strong, and weather resistant.

Figure 12-9 Large, still bodies of water become the reflective mirrors of the landscape.

Figure 12-10 People and wildlife can coexist compatibly or in conflict. By careful selection of plants, preservation or elimination of habitat areas, and provision or lack of feeding stations, designers can encourage or discourage animal populations.

Outdoor art plays a role in the outdoor room that is similar but sometimes not identical to its role in the indoor room. In an indoor setting, pictures are selected for the walls and sculpture is placed on the coffee table for several reasons, not the least of which is because they fit the color scheme of the room or they are the right size to fill an empty space. In the outdoor setting the spaces are larger, and artwork must be larger as well. Whereas indoors, artwork may compliment the theme of the room, outdoors a piece of art may establish the theme for the room, Figure 12-12. In America's national landscape, public art is proving increasingly popular. The exterior walls of inner-city buildings are being brightened by the multi-story murals of local artists. Hotels, parks, campuses, and corporations are making art an important part of their landscapes, Figure 12-13. As a result, public appreciation for art and its importance to the quality of life is growing. It is predictable that most clients

will be receptive to the inclusion of some type of art enrichment in their landscapes, and designers and landscape architects should be sufficiently knowledgeable to offer proper professional guidance.

The selection and use of art is very subjective, but there are some guidelines worth noting:

a. Avoid cartoonish art, such as the plastic and plaster lawn ornaments that fill the discount stores every spring.

b. Avoid anything made of pressed plastic. It is lightweight and will likely fall over or blow away (and probably should).

c. Avoid using reproductions of historic art in contemporary settings. Cherubs and gargoyles were appropriate in the gardens of Europeans 300 years ago, but they are outdated today.

d. Be sensitive to the religious significance of some artwork, and don't use it in a way that would be offensive to certain viewers.

Figure 12-11   Outdoor furniture, such as this chair and table set, is weather resistant and can be made more colorful and/or more comfortable with the addition of cushions or a bright umbrella. It is reasonably priced and adaptive to numerous types of landscape themes and settings.

e. Use outdoor art selectively and conservatively. Don't turn the landscape into a gallery or a gift shop.

Pools and fountains use water again, but this time in fabricated forms. Whether as a swimming pool in the backyard or as a show-stopping display fountain in a shopping mall, water in fabricated forms appeals to people. The wise designer will consider the potential for using water whenever space, weather, and the client's budget permit, Figures 12-14 and 12–15.

The residential landscape can easily incorporate water's enriching qualities. Small prefabricated re-circulating fountains, available at a modest price, can bring new life to a patio area. Such fountains are self-contained, electrically powered units that operate with the simple flip of a switch. Also popular,

and affordable, are small fish ponds such as the one illustrated in Figure 12-16.

With a bit of imagination, landscape architects and designers can specify customized water features that will fit specific landscape sites and not rely solely upon catalogued products. Rock waterfalls are the specialty of numerous landscape contracting firms and stone masons around the country, so bringing the designers' ideas to reality is usually possible. Also, many items such as statues and tubs that were not initially designed as water features can be piped to spout water or lined to contain it.

Night lighting assures that the enjoyment of the landscape need not end when night falls. By including provision for the evening illumination of the garden, the landscape designer adds these and other values to the design:

Figure 12-12 In the coastal island setting of Hilton Head, South Carolina, this looming sculpture of Neptune strikes an appropriate theme.

Figure 12-13 This large sculpture and reflecting pool enrich the pedestrian mall of the New York State governmental center.

- Increased time of use

- Provision of greater safety and security for users

- Creation of special effects such as colored lighting, silhouette lighting, shadow effects, or patterns against buildings

- Maintenance of the same visual relationship between indoors and outdoors that exists during daylight hours

Not too long ago most backyard lighting consisted of floodlights mounted on the roof of the house or garage. The glare of white light was unattractive and more suggestive of a prison yard than a gracious garden. Such floodlights are still excellent security lights, but lighting design has advanced to softer, more attractive forms for the outdoor room, Figure 12-17. While fixtures and manufacturers vary,

there are five common techniques for lighting residential landscapes, Figure 12-18.

Walk lights offer both safety and decorative effects. They should be used wherever it is necessary to warn pedestrians that the walk is about to change direction or elevation, such as at steps or ramps.

Silhouette lighting outlines plants when placed behind them. The viewer sees a dark plant form against a background of light.

Shadow lighting places the light source in front of the plant and causes a shadow to be cast onto a wall or other flat surface behind the plant.

Down lighting creates patterns of light and leaf shadows on the ground. The light fixture is placed high in a tree and directed downward.

Figure 12-14 The crowd of people enjoying this public fountain is testimony to the power of water as a desired component of outdoor urban spaces.

Up lighting is the reverse of the above. The light fixture is placed at the base of the object being illuminated. It is directed upward.

Two precautions are worthy of note when planning the illumination of landscapes. One is to position the lights so that they do not shine into the eyes of the landscape's users. That necessitates that the designer anticipate the probable locations of the users at night. The other precaution is to be sure that the level of brightness outside the building equals that inside. If the brightness levels are not equal, the separating glass door or window reflects like a mirror. The value of the lighting effects is then lost to the indoor viewer looking out.

Intangible Enrichment

The inability to touch certain types of enrichment does not diminish their importance to the landscape. Intangible enrichment appeals to the senses, primarily to the client's senses of hearing, smell, and sometimes taste. Consider the following contributions to the enjoyment of a landscape, and judge their importance.

- The sound of splashing water from a fountain or cascade
- The fragrance of plants in bloom
- The rustle of wind through a planting of trees
- The sounds of birds, crickets, frogs, and other wildlife
- The ringing of distant church bells or carillons
- The lapping of waves against the shore

When such enrichment occurs naturally on a site, the designer has only to take advantage of it by encouraging its continuance. When it does not occur in significant amounts, intangible enrichment can be developed by the designer. Careful selection of plant species can encourage the nest-

Figure 12-15 The combination of water with classical sculpture is a familiar theme. Here the concept is updated with more contemporary art.

Figure 12-16 A backyard fishpond can bring enjoyment to the smallest family living area.

Figure 12-17 Lighting in the style of an earlier era adds quaintness to this seating area.

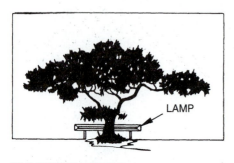

SILHOUETTE LIGHTING. . .
THE LAMP IS PLACED
BEHIND THE PLANT.

UP LIGHTING GIVES HIGH
LIGHT AND SHADOW
PATTERNS TO OBJECTS
ABOVE THE LAMP.

SHADOW LIGHTING. . .THE
LAMP IS IN FRONT AND
THE WALL IS BEHIND THE
PLANT.

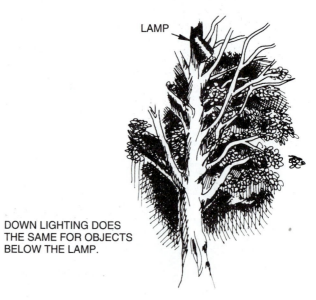

DOWN LIGHTING DOES
THE SAME FOR OBJECTS
BELOW THE LAMP.

WALK LIGHTS GIVE
SPECIAL EFFECTS, MARK
CHANGES IN DIRECTION,
AND OFFER SAFETY.

Figure 12-18 Five common outdoor lighting techniques

ing of songbirds or provide suitable habitats for frogs or other "musical" animals. Plants known to produce fragrant flowers can be placed beneath bedroom or kitchen windows or near conversation areas of a patio where they are certain to be appreciated. Music systems can be designed into the outdoor setting with speakers hidden discreetly out of sight. Even a simple wind chime can bring enriching sounds to a private area. Additionally, the sense of taste can be enlisted by placing plants with edible and tasty fruit in areas of the landscape where people can easily sample them at certain times of the year.

For clients who are or have family members who are in some way physically impaired, intangible enrichment can play a significant role in their enjoyment of the outdoor rooms of the landscape.

Is the Enrichment Item Necessary?

It is possible and not uncommon to design a landscape with too many enrichment items. When this occurs, the effect is one of clutter, with the design principles of simplicity and unity falling victim to excess. Amateur gardeners are common culprits in the over-enriching of their home landscapes, but beginning designers are also susceptible to the temptation.

A few simple guidelines can be helpful in planning the enrichment of a landscape. First, be sure that the item being considered is truly enriching to the landscape. It must fulfill a role other than being a wall, ceiling, or surfacing material. Second, most tangible enrichment items attract the viewer's eye strongly and should be used sparingly as focal points in the landscape. Finally, when in doubt about the value of or need for an enrichment item, remove it. Then envision standing back and observing where it was. If nothing would be missed and the design would work as well without it, the enrichment item is probably unnecessary. If a visual or sensory hole would be left in the design, then the enrichment item probably belongs there.

Achievement Review

A. Indicate which of the following characteristics apply to enrichment items (E), nonenrichment items (N), or both (B).

1. Their major function is to provide shelter.
2. Their major function is to shape the outdoor room.
3. Their major function is something other than serving as a wall, ceiling, or floor element in the outdoor room.
4. They may be tangible items.
5. They may be tangible or intangible.
6. They could be used as focal points of the design.
7. They could protect people from a rainstorm.
8. They may require periodic maintenance.
9. They are sometimes created by nature.

B. Indicate if the following natural enrichment items are tangible (T) or intangible (I).

1. a boulder
2. a lake
3. the sound of a bird
4. a distorted old pine tree
5. wind whistling through trees
6. a waterfall
7. the sparkle of a waterfall
8. the sound of a waterfall
9. berries on a shrub
10. the taste of berries

C. Give four examples of natural, tangible enrichment items. Do not use any of those listed in question B.

D. Give four examples of natural, intangible enrichment items. Do not use any of those listed in question B.

E. Indicate whether the following enrichment items are natural (N) or man-made (MM).

1. birdbath
2. sundial
3. lake
4. statue
5. large boulder
6. chaise lounge
7. picnic table
8. chipmunk
9. outdoor lights
10. wind in pine trees

F. Select the best answer from the choices offered.

1. Where is the light for silhouette lighting located?
   a. behind the plant
   b. above the plant
   c. in front of the plant

2. Where is the light for shadow lighting located?
   a. behind the plant
   b. above the plant
   c. in front of the plant

3. What is a good way to test the value of an enrichment item in a landscape?
   a. Sell it and determine its worth.
   b. Remove it temporarily.
   c. Add another one like it.

4. Permanent outdoor furniture should be attractive and _____.
   a. lightweight
   b. weather resistant
   c. upholstered

Suggested Activities

1. Demonstrate different lighting techniques in the classroom. Use a small flashlight for the light source. Use a box to represent the building or wall background. Select a small leafed branch from a shrub to represent the plant. Darken the classroom and try up lighting, down lighting, silhouette lighting, and shadow lighting the plant and building.

2. This activity demonstrates the necessity of designing outdoor lighting so that the level of brightness is close to that of the indoors. Go outside on a bright, sunny day and try to see into the school building from the road or sidewalk. The windows, acting as mirrors, will cause the inside of the building to appear dark.

3. Study the sidewalk system around the school building. Where are walk lights needed for safety?

4. Invite a sculptor to visit the class to discuss original outdoor art: its costs, common locales, time needed for production, and modern preferences. If possible, arrange a class visit to the artist's studio or to a museum of modern art.

5. Make a bulletin board display of outdoor furniture, including pictures and information about materials and costs. As sources, use newspapers, department store catalogues, landscape trade journals, and gardening magazines.

6. List all of the natural enrichment items in your home landscape. Indicate if they are tangible, intangible, or have characteristics of both.

7. Do a similar study of a park near the school. Which landscape has more natural enrichment, the park or the home?

CHAPTER 13

Landscaping in the Age of Technology

Objectives:

Upon completion of this chapter, you should be able to

- discuss the current uses of modern technology in the landscape industry
- explain the major types of computer programs currently popular in the industry
- project future benefits of technology to the landscape industry

Technology Is Dynamic

Technology is an umbrella term that is used to explain how things get done. Manufacturers have specific technologies that get their products to market. Farmers have specific technologies that result in the production of their plant and animal products. While some products and services have been available for many years, the technology used to create them has evolved, becoming increasingly more sophisticated. Thus, to understand fully the technology of an industry, it must be viewed as specific to the particular tasks common to that industry, and it must also be accepted as a moving target, changing through time. Today's cutting edge technology is tomorrow's display in the Smithsonian.

Landscaping is mostly a service industry. Landscapers do not manufacture things. Rather they use products produced by others to fabricate new environments or to sustain or improve existing environments. It is a highly labor intensive industry, and historically it has been slower than many other industries to embrace new ideas and techniques. However, that has changed in recent years. While there will likely always be a hand-crafted quality to the work of landscapers, much has happened in the last decade to accelerate changes in the way landscapers do business and accomplish their jobs. Still, even as we now discuss it, the discussion must be regarded as a temporary stopping point along a moving timeline that assures ongoing change.

The Necessary and the Novel

Landscape trade publications as well as magazines directed to homeowners and hobbyists are filled with articles and advertisements for new products that will outperform their antiquated or bourgeois predecessors. Industry trade exhibitions promote new tools, machines, materials, software programs, and management aids. Motivational books and high-priced consultants promise fresh insights and formulas for success to landscape leaders who feel the need for outside assistance. It is easy to get swept up in new things and new ideas, so the wise landscape practitioner will carefully evaluate new materials and new technology before investing. Not all technology is truly new—it may just be different. Also, not all new technology is necessary just because it is possible.

Sources of New Technology

Technological advances in both materials and methods come to the country from sources both within the country and from abroad. *Product and equipment manufacturers* in all of the world's developed countries continually bring new items to the landscape industry, which is itself an international industry. A new paving material produced in Canada may find eager buyers in the United States. A tool or vehicle redesign in Europe may find a receptive market here as well. As the world's strongest economy, the United States is the marketplace of choice for every other nation that has something to sell, so if a new product is available overseas, it quickly finds its way to this country. Our own manufacturers are also eager to develop new products for American consumers, so the landscape industry is continuously presented with new items. Some are highly specialized and may do only one thing. Others may have multiple capabilities. Most are initially presented to the industry at regional and national trade shows, Figure 13-1. Many are also marketed to smaller groups and to individual companies using videos, direct mailings, and the Internet.

Other technological advances result from *research at universities*. Often given financial support by industry organizations, the scientists and engineers at these schools are able to test, evaluate, and make recommendations about new materials or techniques that can improve the way landscapers practice. Trade magazines willingly publish the results of these university studies, giving them rapid and widespread acceptance.

At the heart of the current technology surge is the *computer*. Once a mystery to many people, the computer has moved from the shadows to center stage in our lives. In the landscape industry, many of the tasks in both field and office are being done differently today than a few years ago due to the seemingly omnipresent computer.

The Forms and Types of Technology in Landscaping

Collectively, over the past 10-plus years, the landscape industry has applied some type of new technology to these practices:

- correspondence
- tracking client accounts
- accounting
- payroll
- mailing lists
- inventory control
- billing
- project scheduling and monitoring
- specification writing
- take-offs
- cost estimating
- plant selection
- landscape design
- capture, storage, and display of geographic data
- graphic imaging
- irrigation design

Figure 13-1 Trade shows are one of the best ways for landscape professionals to see what is new in equipment, software, and other products created for the industry. (Courtesy of the Green Industry Expo)

- promotion of company services

- communication between the company and vendors, subcontractors, and other professionals

- communication between workers within the company

- communication between the company and current or potential customers

Landscape companies still acquire customers in the traditional ways, through media advertising and word-of-mouth referrals. However, many companies now use the Internet to present a Web-page description of their services that is more descriptive than what can be done in a brochure or other form of advertisement.

Many of the traditional practices of the office workplace are now commonly done using software that expands the capabilities of office administrators, reduces the time needed for repetitive tasks, minimizes errors, and improves the level of support services provided to customers and to company personnel outside the office. Not only has the appearance of the office workplace changed in recent years, but so have the skills required of staff who work in the office. Even the location of the office workplace may have changed, since technology now permits some work to be accomplished at home as efficiently as at an office desk. For small companies, some of their office work, such as payroll preparation, may be contracted out to specialized firms that can do the work more efficiently and cost effectively than the company could do it in-house.

In the field, workers now have more immediate and direct access to their supervisors and each other by means of cell phones and two-way radios. With less down time resulting from miscommunications, missing tools and materials, confused travel directions, and a host of other delays that occur when people cannot speak immediately to each other, the landscape company is able to operate more efficiently. Cell phones, pagers, and call forwarding also assure that salespeople and others with direct customer links do not miss calls from potential or current clients while they are away from their desks or at project sites. With personal data assistants (PDAs), field supervisors can keep and retain all job site records in a convenient and compact log.

Perhaps nothing is more indicative of how technology responds to time and trends than the newly engineered way that heavy field equipment is steered and controlled. Traditional controls are being replaced with new devices that are suggestive of the joysticks that young people become familiar with as they grow up playing video games. By combining the ease and familiarity of control with hydraulic operation, the technology of today permits workers of every size, regardless of their strength, to operate large and powerful pieces of equipment, Figure 13-2.

Landscape designers and landscape architects now commonly use computer-aided drafting (CAD)

systems to produce all or part of the drawings required for a project. While the computer has not yet learned how to initiate a new design concept, CAD systems do enable the designer to accelerate the processes that have traditionally been done with T-squares and pencils. Computer drafting has increased mainly because of a PC-based CAD software program named *AutoCAD®* (a registered trademark of AutoDesk, Inc.). Because of the necessity for specialized hardware, software, and user training, AutoCAD is an expensive investment for a company. It is not ready to use on delivery to the office, especially if no one on the staff has received prior training in its use. In the hands of a skilled operator, AutoCAD can be used to create landscape designs and irrigation designs. However, the time involved to produce the design can be lengthy, particularly if a presentation-quality drawing is desired. As a result, there are now a number of discipline-specific software interfaces that cus-

Figure 13-2 Technology has replaced muscle power in many areas of landscaping, enabling people of all strengths and sizes to accomplish in hours tasks that formerly required days. (Courtesy of Chapel Valley Landscape Company)

tomize the general AutoCAD system for a specific type of design and drafting. For the landscape profession, a number of landscape and irrigation design-specific programs have been developed. Several target the infrequent designer, and some are even promoted to homeowners. Many landscape designers and most landscape architects are currently using a software program called LANDCADD®. Developed by a group of practicing professionals and released commercially in 1984, LANDCADD is currently marketed by Eagle Point Software of Dubuque, Iowa. Since its inception, the program has advanced through many generations of revisions and currently offers modules that permit a company to purchase only those that it needs, avoiding the requirement of paying for ones it cannot use. At the heart of the software system is the Site Planning and Landscape Design Module. It permits creation of two- and three-dimensional (2-D and 3-D) plan views, elevations, and perspectives of landscape features and entire sites. Textures, colors, enlargements of special areas, lettering, and a variety of other graphic features are possible. As the design develops, the plant list is automatically created and tabulated. Designers can select from a palette of symbols or develop and add those of their own design—meaning the program can be customized to a particular company or individual. Figure 13-3 illustrates a landscape plan developed using LANDCADD. Three-dimensional views of buildings, terrain, and grading plans are also possible using the system's Planning and Design Module. The Irrigation Design Module also enjoys widespread use within the industry, Figure 13-4. Other modules permit cut and fill calculations, material quantification, cost estimating, drainage calculations, plant selection, road alignment, and other calculations vital to landscape architecture and landscape construction.

A wealth of very specific software is now available to improve the accuracy, comprehension, or visual presentation of designs. Both 2-D and 3-D modeling software is available, as is video imaging, which superimposes a landscape concept over a photographic image of the undeveloped landscape, Figure 13-5. Currently somewhat cumbersome as high-tech gadgetry goes, the technique requires one or more photographs captured by a conventional camera, a digital camera, or a video camera. The photo of choice is then entered into the system's memory and later brought up on a monitor, and the computer is used to superimpose color photo images of plants, pavements, pools, lawn, walls, fences, and other landscape elements over and in front of the building. The system also permits the designer to modify the photo to illustrate the impact of removing existing features, changing colors, or even remodeling the architecture. The resulting image gives customers a full color print suggesting how their landscape will look before the project even begins. It is an excellent sales tool, and the developing programs promise to make it even easier to use, assuring its use for a wider variety of projects in the future.

In addition to the data that a designer is able to obtain and enter into the CAD system personally, the current technology permits the acquisition of data from other sources as well as the transmission of data to others involved in the project. Digital technologies now enable a landscape designer to orchestrate the movement of large amounts of information from distant sources or other software programs into any project being developed. Geographic information systems (GIS) provide support to design problems requiring knowledge of an area's geography. Several are now available and provide databases that supply planetary maps that can be integrated in digital form with CAD systems. Unlike the flat maps that predate them, GIS maps recognize the curvature of the earth and can provide precise locations of any point on a site and any site or study area on the surface of the planet. Once the location of a site or other area is identified, the GIS can provide the designer with a bounty of information that can impact and influence decisions on how the site can or cannot be developed. Information as vital as the site's existing vegetation, soil type(s), elevation, ownership, zoning, flood potential, and suitable uses can be obtained from a GIS map. Its value to landscape architects and others involved in large project development is readily apparent.

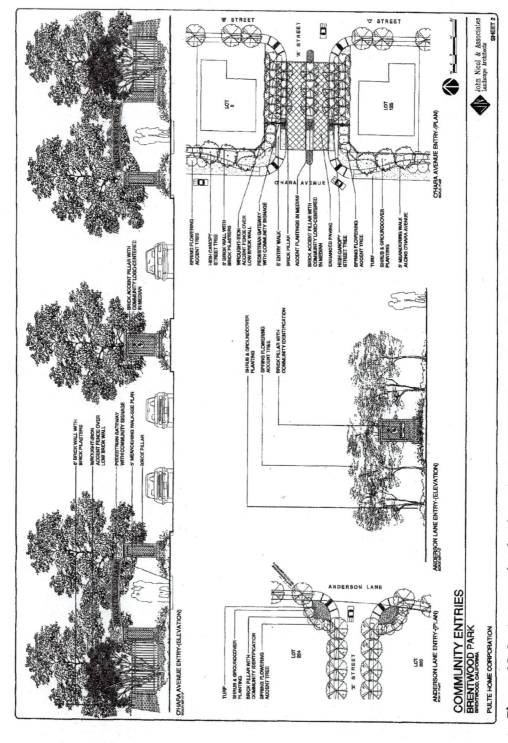

Figure 13-3 An example of a plan using LANDCADD to produce the design and lettering. Trees were added using a paste-up technique. (Courtesy of John Nicol and Associates, Lafayette, CA)

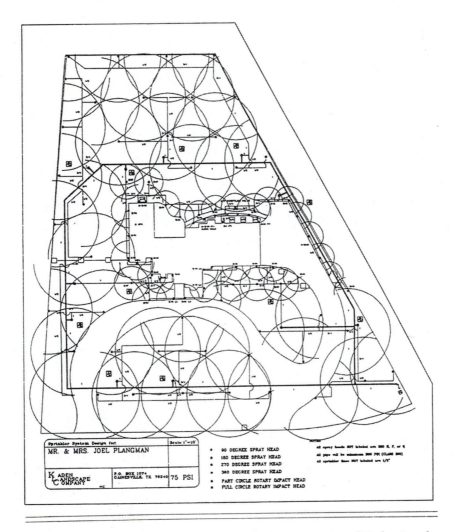

Figure 13-4 A LANDCADD irrigation design (Courtesy of Kaden Landscape)

Other computer software is revolutionizing the presentation appearance of landscape designs. LANDCADD continues to expand its font styles to provide new ways of lettering. Scanners can capture and incorporate virtually any image and bring it into a presentation. Photo-manipulation software, such as Adobe Photoshop, and graphic design software, such as Adobe Illustrator, enable a designer to prepare exciting 2-D and 3-D presentations of their ideas for client review. Programs like Microsoft PowerPoint permit the presentation of

design proposals to large audiences, even permitting the addition of sound, music, video clips, and other effects.

Spreadsheet software has almost completely replaced the ledger books and estimating worksheets of years past. LANDCADD offers an estimating module, and there are dozens of others available in the marketplace. Some are generic and applicable to many industries. Others have been designed to meet the specific needs of the green industry and landscape firms. Some are

Figure 13-5 Before and after example of computer imaging (Courtesy of Design Imaging Group)

construction focused, while others are more easily adapted by maintenance firms. Some are specific to lawn care companies. Still other companies have chosen to develop their own software programs to assure that their spreadsheet design and information output is exactly what they need.

Landscapers are now able to quantify materials from designs and convert the data into precise bids in a fraction of the time required by conventional measuring and counting methods. Last-minute changes to design or maintenance specifications can be quickly entered and the bid updated. Less time is required for calculations that in the past were time consuming, tedious, and subject to human error. At the same time, the accuracy of the calculations has probably improved, since the computer's insistence on certain data input has compelled companies to keep close track of labor hours, material costs, and equipment usage.

Input devices for numerical calculations from drawings include a keyboard, a digitizer, and some type of scanner, such as the stylus shown in Figure 13-6. The digitizer enables the scanner to make direct counts of items, such as plants, and to make direct measurements of lengths, areas, volumes, and perimeters from drawings. It thereby enables the drawn data to be converted to numerical equiv-

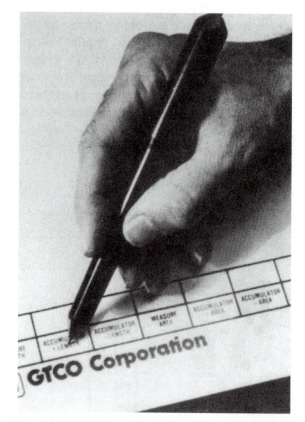

Figure 13-6 The stylus is one of several types of input devices used for computerized calculating.

alents—a necessary step for the preparation of an estimate or bid.

Where drawings are not involved, estimating programs can be carried in hand-held computerized devices that enable a landscape salesperson or estimator to work directly at a site and provide a client with an immediate estimate of project costs. Some companies are even using graphic imaging on-site to provide customers with a variety of suggestions for ways to improve their landscapes. It is simply a technological updating of the on-site sketching that landscapers have done in the past to help clients visualize the potential for their property.

The Future

The landscape industry has climbed aboard the technology express train. At the helm are young men and women who not only accept technological change but expect and welcome it. Although rooted in a past that was part hobbyist and richly gardenesque, the industry has now moved forward and upward, aware of its role as a multibillion dollar national industry of the new century. There will be no turning away from the technological advances that are already nearing the marketplace.

Design and estimating operations that currently depend on scanners and keyboards will soon respond to voice commands once the bugs in the technology are resolved. Landscape architects will soon have software that will create models of their proposals that will enable them to predict and evaluate the performance of systems (such as water movement) as they are designed but before they are built. Virtual reality software will enable designers and clients to visually walk through projects, designing as they go, before a single spade of earth has been turned.

Customers may soon be able to prepare their own estimates for maintenance projects by entering their own dimensions and other data into a form posted on a company's Web page.

Field operations are going to change as well. Mowers are being developed that operate without human assistance. They graze like mechanical sheep over the lawn, cutting only where programmed. New manufacturing technology will continue to produce synthetic variations of classical materials. Most will offer an attribute of greater strength or aesthetics, easier application, resource conservation, or cost reduction.

Even the ways that companies recruit personnel will be further affected by technology. Companies are already incorporating job announcements in their Web pages and contacting potential employees through e-mail. The near future is likely to see applicants preparing electronic resumes that add photos and sound, thereby permitting them to display their past projects, portfolios, and accomplishments more fully.

We live in an age of information and technology. The speed of their expansion is unprecedented. It has been and will continue to be difficult for educational programs and textbooks to keep current with that expansion. Students preparing for entry into the industry must be familiar and comfortable with technology. You need to have mastery of that which is current, be it computer centered, research based, or mechanically engineered. Whether through formal instruction, short courses, reading, or apprenticing, it is important to update your knowledge and competencies regularly.

Achievement Review

1. What are the three major forces behind technological advance?

2. Name three ways that landscape designing has been impacted by recently developed computer software.

3. Identify at least five tasks done by members of the office staff that can now be accelerated using a computer-driven program.

4. What three things are needed to enter data into computer programs used for preparing take-offs and estimates?

5. What uses are companies making of their Web pages on the Internet?

6. Recalling the *advantages* of technology as described in this chapter, can you think of any *disadvantages*?

7. If you were trying to convince a reluctant landscape company owner to modernize his or her company by investing in computer programs and other types of high-tech equipment, how would you counter the perceived disadvantages given in your answer to the above question?

Suggested Activities

1. Using a personal computer and a spreadsheet program such as Microsoft Excel, do the practice exercise at the end of Chapter 14 (Pricing the Proposed Design).

2. Contact local dealers for a demonstration of a CAD design program and video imaging.

3. Visit a landscape equipment dealership to see the most recently introduced innovations in landscape field equipment.

4. Invite a representative from the local telephone company to give a presentation on the latest communication devices being promoted to small businesses.

5. Have members of the class go an entire week without using their own high-tech devices, such as cordless phones, cell phones, pagers, calculators, personal computers, e-mail, the Internet, etc. Ask them to record such things as additional time required to do specific tasks and their degree of frustration resulting from delayed communications or misunderstandings.

6. Have students prepare an electronic resume that describes themselves as in a printed resume, but that also incorporates a personal photo and images of landscape projects they have worked on (real or pretend). If possible, incorporate the student's voice and/or a video clip in the resume. Create the resume as a file that can be sent to a prospective employer.

CHAPTER 14

Pricing the Proposed Design

Objectives:

Upon completion of this chapter, you should be able to

- explain the difference between cost and price
- explain the difference between an estimate and a bid
- describe landscape specifications
- prepare a design cost estimate

One of the questions that every client can be expected to ask is "What is this landscape proposal going to cost me?" The landscape contractor will ask a similar question, "What will it cost my company to build this landscape?" Although both have used the word *cost*, they do not seek the same information. The client's question really means, "What will be the total outlay of money necessary to get this landscape done from start to finish? When all is done, how much will I have spent?" The landscape contractor's question really means, "What expenses of constructing this project must I recover to not lose money?" The landscape architect or designer must know the answer to both questions. He or she must design the project to stay within the client's budget for the project. In so doing, the landscape architect or designer must have a realistic understanding of the contractor's costs in order to anticipate the actual cost of building the project.

Cost and Price

Although the terms **cost** and **price** are often used interchangeably, they do not always mean the same thing. *Cost* refers to the recovery of expenditures. *Price* refers to an outlay of funds.

The cost of landscape services is the sum of material costs, labor costs, equipment usage costs, and overhead costs.

Cost = Materials + Labor
+ Equipment use + Overhead

These are all expenses assumed by the landscape contractor during the construction of a landscape.

If they are not recovered through payment by the client, then the landscaper will lose money.

Price is what the client pays to have the landscape built. It includes the price of the landscape architect or designer's services plus the price of the landscape contractor.

Price = Landscape design services
+ Landscape contracting services

The price of the landscape design services will include the designer or landscape architect's fee for designing the landscape, modifying it as requested by the client, and representing the client whenever necessary. Representation of the client may include obtaining necessary permits, attending zoning meetings, conferring with individuals or groups involved in the project, selecting the landscape contractor, supervising construction of the project, and providing final inspection of the project to certify completion. The landscape architect's services may be calculated as per hour fees or as a percentage of the total value of the project. They will include cost recovery for all the time and any materials expended by the landscape architect. They will also include his or her profit on the project. The profit will be added to the costs to determine the price for landscape design services.

The landscape contractor will use a similar approach to determine the price for landscape contracting services. To the costs of materials, labor, equipment use, and overhead will be added two things; contingency and profit. The **contingency** allowance is something like insurance for the landscape contractor. Due to the unpredictable things that can disrupt a construction schedule, it is not possible to calculate precisely the costs of labor and materials. Delays caused by weather, labor problems, miscalculations of material quantities, or builders not finishing on time can wipe out the expected profit. The inclusion of a contingency fee in the price given the client can safeguard the landscape contractor's profit. However, if it is too large it may make the price too high and unacceptable to the client.

Price = Landscape design services + Labor
+ Materials + Equipment use
+ Overhead + Contingency + Profit

NOTE: **Overhead** costs include administrative salaries, advertising costs, rent or mortgage payments, office expenses, accountant and attorney fees, subscriptions, memberships, insurance premiums, and equipment maintenance. They are operational costs of the firm. Some are general overhead costs and cannot be assigned just to a specific project. A portion of general overhead costs is charged to each project and included in the price. Other overhead costs can be assigned totally to a particular project. Costs such as portable toilet rental or temporary utility hookups can be fully assigned to the project for which they are needed. These are termed project overhead costs.

Estimates and Bids

An **estimate** is an approximation of the price that a customer will be charged for a landscape project. Estimates may be prepared by landscape architects or by landscape contractors. The time of preparation may range from minutes to hours, depending upon the size of the project and the level of accuracy needed. To keep a project within the client's budget limits, estimates are done repeatedly during the design stage. Price overruns only result in client disappointment and a waste of the designer's time.

When a price estimate is presented to a client, it is important for the client to understand that the price is not a firm quotation. Rather, it is a close approximation, subject to some change as actual costs are later determined.

A **bid** is a definite offer to provide the services and materials specified in the contract in return for the price agreed upon. Once confirmed by the signature of all parties, the price is binding upon the landscape company to do exactly what it has said it will do. If changes are made by the client or landscaper, the changes must be agreed to by everyone involved. As a result of the changes, the price may be changed, since the changes affect the original bid. Should unforeseen conditions delay the project or otherwise increase costs, the bid price remains fixed. In such cases, the landscaper's profit declines. Therefore, bids must be calculated carefully to assure that all costs are included.

When the landscape project is small, such as a residential property, there may be only one landscape firm working with the client. In such a case, the bidding process is usually informal and negotiable between the landscaper and the client. With larger projects, contracts are certain to require an exact bid, and negotiation may not be possible. On large projects, it is likely that several contractors will be competing for the same job. The firm with the lowest bid has a good chance of being selected, so a bid cannot be greatly overpriced or underpriced. These extremes will lose either jobs or profits for the landscape firm.

Specifications

Large landscape projects usually need more than drawings to explain fully what is required in the design and what quality standard is acceptable. **Specifications** are a listing of the materials, quality standards, and time schedules required to build a particular landscape. They may be prepared by the landscape architect or a professional specification writer. Copies of the specifications and the design are provided to each landscape contractor who wants to submit a bid for a project. The specifications must be carefully written to assure that the designer and client get what they are expecting from the landscape contractor. For example, specifications that describe a required species as "eight sugar maples" could result in bids ranging from a few dollars to several thousand dollars. This is because there is neither an indication of the size of the plants at the time of installation nor a specific requirement for bare-rooted or balled and burlapped plants. A client expecting a near-mature appearance in the finished landscape

would be greatly disappointed by the sparseness of sapling plants. Precisely written specifications prevent client disappointment and discourage deceptive bidding.

Writing specifications can be very tedious work. Every plant in the design must be described by size at planting, root form, soil preparation, and pre- and post-transplant care. Each surfacing installation technique must be explained fully. Mulch depths, lighting fixtures, seed blends, and hundreds of other details must be fully described.

The landscape contractor uses the specifications and the drawings to prepare the bid. Each plant, every brick, every cubic yard of concrete or mulch must be counted. Then the labor and the supervisory time required for their installation are calculated. After that, charges for contingency, overhead and profit are added, and the bid is totaled. Specifications are discussed in greater detail in Chapter 18.

Preparing the Estimate

Landscape estimates and bids are prepared as spreadsheets. As the data is assembled, it is presented in columns that itemize quantities, descriptions of materials or services, unit costs of materials, unit costs of installation, and total costs for each item. The data is also grouped into categories that organize the estimate for easy reading and referral. A typical design cost estimate includes the following:

- Cost of site clearing and other preparation
- Cost of plant materials
- Cost of construction materials
- Cost of turfgrass
- Allowance for overhead
- Allowance for contingencies
- Fee for landscape designing
- Allowance for profit
- Name of the estimator and date of the estimate

Table 14-1

Estimate for Plant Materials

Quantity (1)	Unit (2)	Description (3)	Material Cost (4)	Installation Cost (5)	Total + 35 percent (6)
2	EA	*Acer platanoides*, 5', B & B	$ 35.00	$16.30	$138.51
1	EA	*Betula pendula*, 6', B & B	31.90	23.48	74.76
4	EA	*Cornus florida*, 4', B & B	45.00	13.80	317.52
15	EA	*Euonymus alatus*, 18", BR	8.40	11.60	405.00
200	EA	*Vinca minor*, 2" pots	1.05	0.50	418.50
			Total for plant materials (7)		$ 1,354.29

Estimate for Construction Materials

Quantity (1)	Unit (2)	Description (3)	Material Cost (4)	Installation Cost (5)	Total + 25 percent (6)
208	SY	Crushed limestone paving, 4" thick, compacted on a compacted subgrade using a motor grader	$ 7.00	$ 0.32	$1,903.20
90	SF	Timber retaining wall with deadmen, inc. excavation and backfilling	6.00	7.00	1,462.50
1	EA	Fountain, including jet, pump, drain, overflow, water linkage, underwater lighting, electric hookup, and basin	3,500.00	700.00	5,250.00
600	SF	Slate pavers, 1¼" thick, irregular fitted in a bed of sand with dry joints	6.40	4.50	8,175.00
			Total for construction materials (7)		$16,790.70

NOTES:
1. *Quantity* is the number of each separate item used in the design.
2. *Unit* is the form of measurement used to count the item. Examples include square yard (SY), square feet (SF), cubic yard (CY), cubic feet (CF), each (EA).
3. *Description* gives an explanation of what will be done or supplied. For plants it includes the species name, size at installation, and root form (BR for bare root, CT for container, and B & B for balled and burlapped). For construction materials, it includes briefly the materials to be used, means of installation, and other details needed to identify what is to be done.
4. *Material Cost* is the cost of one unit meeting the descriptions given.
5. *Installation Cost* is the charge for labor and any equipment needed to accomplish one unit of the task or service described.
6. *Total + Percent* is the sum of the material and installation cost for one unit multiplied by the quantity of units. To this is added a percentage allowance for overhead and profit. In the examples, 25 percent is used for the O & P of construction, and 35 percent is used for O & P of plant materials. The latter allows for the greater loss of plant items due to their perishability.
7. *Category Subtotals* are underlined and set off to the right for easy totaling later.

Table 14-1 consists of two partial estimates, including an explanation of each column.

The fee for landscape design services is handled in several ways, depending upon the size of the project, the credentials of the designer, the policy of the firm, and the laws of the state. In some states, only accredited landscape architects are permitted to charge for their design services. Design-build

Price Estimate for the Design and Development of the Property of Mr. and Mrs. John Doe, 1234 Main Street, Cleveland, Ohio

I. PLANT MATERIALS

Quantity	Unit	Description	Material Cost	Installation Cost	Total + 35 percent
6	EA	*Celtis occidentalis*, 1½" cal. B & B	$ 125.00	$31.00	$1,263.60
5	EA	*Cornus florida*, 7', B & B	101.00	45.00	985.50
20	EA	*Euonymus alatus*, 15", BR	7.00	10.55	473.85
15	EA	*Philadelphus coronarius*, 3', BR	5.40	15.63	425.86
50	EA	*Vinca minor*, 2" pots	1.05	0.50	104.63
			Total for plant materials		$ 3,253.44

II. CONSTRUCTION MATERIALS

Quantity	Unit	Description	Material Cost	Installation Cost	Total + 25 percent
500	SF	Concrete pavers, 3⅛" thick, interlocking dry joints on 2" sand base, with 4" gravel subbase	$ 2.48	$ 1.53	$2,506.25
300	SF	Pressure treated timber wall, 6" × 6" timbers, gravity type, inc. excavation and backfill	6.00	7.00	4,875.00
1	EA	Flagpole, aluminum, 25' ht, cone tapered with hardware halyard and ground sleeve	1,060.00	59.02	1,398.78
			Total for construction materials		$ 8,780.03

III. TURFGRASS

Quantity	Unit	Description	Material Cost	Installation Cost	Total + 35 percent
900	SY	Kentucky bluegrass sod, on level prepared ground, rolled and watered	$ 1.00	$ 1.48	$3,013.20
			Total for turfgrass		$ 3,013.20

IV. TOTAL COST OF ALL MATERIALS AND INSTALLATION (See Note 1)	15,046.67
V. CONTINGENCY ALLOWANCE (See Note 2)	1,504.67
VI. FEE FOR LANDSCAPE DESIGN SERVICES (See Note 3)	1,805.60
VII. TOTAL COST FOR COMPLETE LANDSCAPE DEVELOPMENT (See Note 4)	**$18,356.94**

NOTES:
1. The total cost is obtained by adding the subcategory totals.
2. The contingency is taken as a percentage of total costs in IV. In the example, 10 percent is used.
3. The design fee is taken as a percentage of IV also. In the example, 12 percent is used.
4. The final figure should be the most distinctive on the spreadsheet. It should be the bottom line figure, with no others below it to create confusion.

firms often have salaried designers on the payroll. In such companies, the design costs are usually treated as overhead costs assigned to specific projects. In either case, the fee for designing, drafting, and overseeing the installation of a project to meet full client satisfaction can be estimated at 8 to 15 percent of the total cost of materials and installation. Usually a large project will use the lower percentage of the range, while a small project will use the higher percentage.

The contingency allowance may be calculated in a similar manner. Using the total cost of materials and installation as a base, a percentage may be added to cover contingencies.

The completed estimate or bid includes a client's name and address, the name of the estimator, and

the date of preparation. It may also include an expiration date, after which the figures will no longer be honored because of their unreliability.

A designer will usually present the price estimate at the time the design is shown to the client. A landscape contractor will present the estimate or bid during the negotiation stage of the project. In both cases, the explicit figures shown in the examples are deleted from the proposal given to the client. Only the total figure is presented. A complete price estimate is included in this chapter. In this example, no special site preparation was required.

Practice Exercise

Prepare a complete price estimate for the property of Mr. and Mrs. Byron Lord, 1238 N. Grand Street, Harwich, Massachusetts. Use the following data and follow the format outlines in this chapter. No site preparation is required.

PLANT MATERIALS REQUIRED

4 Chinese redbud (*Cercis chinensis*), 5' ht., B & B, cost $51.00 each and $16.25 to install

20 Golden forsythia (*Forsythia spectabilis*), 18" ht., BR, cost $3.00 each and $11.63 to install

2 Sugar maple (*Acer saccharum*), 1 1/4' cal, B & B, cost $58.00 each and $30.85 to install

12 Carolina rhododendron (*Rhododendron carolinianum*), 2' ht., B & B, cost $39.00 each and $12.50 to install

10 Japanese barberry (*Berberis thunbergi*), 15" ht., CT, cost $12.50 each and $10.55 to install

20 Vanhoutte spirea (*Spiraea vanhouttei*), 1 gal. CT, cost $4.35 each and $10.55 to install

300 Japanese pachysandra (*Pachysandra terminalis*), 2" pot, cost $0.83 each and $0.50 to install

1 Babylon willow (*Salix babylonica*), 8' ht., B & B, cost $44.70 and $45.00 to install

12 Floribunda roses (*Rosa floribunda*), 18" ht., BR, cost $6.60 each and $11.63 to install

2 Red oak (*Quercus rubra*), 6' ht., BR, cost $21.50 each and $23.48 to install

NOTE: Profit and overhead allowance: 35 percent of the cost of materials and installation.

CONSTRUCTION REQUIRED

90 linear feet (LF) of chain link fencing, 4' high, with galvanized posts and top rail and bonded vinyl 9 gauge fabric @ $4.37 per LF. Installation cost is $1.56 per LF.

1 chain link gate, 4' high, 3' wide, with 2 galvanized gate posts, bonded vinyl 9 gauge fabric @ $218.00 each. Installation cost is $8.49 each.

110 square feet of brick pavers, 2 1/4" thick, set over a finished subgrade, dry joints, 2" sand base @ $2.31 per SF. Installation cost is $2.71 per SF.

4 walk lights, 10' high, plain steel pole set into a cubic foot of concrete, with distribution system and control switch @ $2,000.00 each. Installation cost is $500.00 each.

NOTE: Profit and overhead allowance: 25 percent of the cost of materials and installation.

LAWN REQUIRED

1,000 square yards of Kentucky bluegrass sod, rolled and watered @ $1.00 per SY. Installation cost is $0.72 per SY.

NOTE: Profit and overhead allowance: 35 percent of the cost of materials and installation.

Fee for landscape design services is to be 12 percent of the total cost of all materials and installation. Contingencies will be included as 10 percent of the total cost of all materials and installation. Use your name and the current date as the name of the estimator and the date of the estimate.

Achievement Review

A. Complete the sentences below by inserting the most suitable word.

1. The price of the landscape is of concern to the client. The _drawings_ of building it is of concern to the landscape contractor.
2. All _labor_ must be determined before a landscape contractor can add on his profit correctly.
3. The _cost_ of the landscape includes allowances for overhead, contingencies, and profit.
4. The cost of doing business is assigned as a percentage to every client's price estimate. It represents a general _estimate_ cost.

5. The cost of temporary water and electrical hookups at a remote work site while a landscape is being built represent _8 – 15%_ overhead costs.
6. Cost = Materials + _____ + Equipment Use + _____
7. Price = _____ + Materials + Labor + Equipment Use + Overhead + _____ + Profit

B. Define the following terms.

1. estimate
2. bid
3. specifications

C. List the components of a typical landscape price estimate.

SECTION 2

Landscape Contracting

CHAPTER 15

Landscape Calculations

Objectives:

Upon completion of this chapter, you should be able to

- describe the calculations needed for landscape take-offs
- explain the importance of organization and standardization when doing take-offs
- perform typical calculations required for take-offs

The Landscape Take-Off

As a landscape project progresses from the design stage to the point where it has received client acceptance and is ready to be built, it widens in scope and begins to involve additional landscape professionals. Creating substance from concept by turning symbols on the plan into numbers on an order form is the responsibility of the landscape contractor. The calculation of quantities from plans and specifications is termed the **take-off**.

Take-offs are done at different stages during the development of a landscape project, and each is done with different levels of refinement. Partial take-offs may be done by the landscape designer or landscape architect while the design is being developed as a means of checking to assure that the project is staying within the allotted budget. Approximate take-offs may be done by contractors

who are asked to give an estimate of the cost or time required to construct a project. A final take-off is an exact calculation of quantities and time required to bring the design off the drawing board and into the ground.

Take-offs are also done by landscape management firms seeking to determine the quantities of time and materials required to maintain a landscape for a specified length of time to the standards prescribed by a client.

The calculation of material quantities and time requirements needed to install a design or maintain an existing landscape is necessary so the landscape company can (1) determine its costs and (2) set a price for the client. Only after material quantities are determined can the company seek the best prices and sources of required materials. Only after installation or maintenance techniques are known can the time requirements be determined. Only after time requirements are calculated can the labor needs be measured and labor costs ascertained.

The Necessary Calculations

The individual usually assigned to do the landscape take-off is the **estimator**. Working with tools such as the scale, protractor, triangle, planimeter, calculator, graph paper, and/or the computer, the estimator counts, measures, and calculates every item needed for the landscape and every minute of time required as labor. Typical calculations made during a take-off include:

a. Unit counts — Examples: the number of each plant species, the number of bricks, the number of walk lights

b. Surface area calculations — Examples: the lawn area to be sodded, the size of the patio to be paved, the amount of flagstone pathway required

c. Volume calculations — Examples: the amount of soil displaced by the soil balls of the plants, the amount of concrete needed to pour the patio to desired thickness

d. Time calculations — Examples: the number of hours required to clear a woodlot, the number of hours required to build a stone wall

Sources of the Data

The information needed for the take-off comes to the landscape estimator from three sources: the drawings, the written specifications, and the landscape firm's records of crew performance capability. The drawings provide the estimator with much of the quantity data, including most of the dimensions. The written specifications may provide additional quantities as well as an explanation of desired qualities of materials, sources of supply, and information about special methods of installation or maintenance or frequency of performance.

Anything that eventually translates into costs for the landscape company must be examined by the estimator and turned into a quantity that is measurable. The measurability of the data is important if cost and price figures are to be derived from it. Certain units of measurement are less precise than others and should be avoided whenever there is a more precise means of expressing the quantities required to complete a project. For example, it is less precise to list 100 bales of mulch than to list the need for 650 cubic yards. Bales vary in size and volume, so the actual quantity of material provided is imprecise, and the time required to spread it will be equally imprecise. However, if the plan calls for the mulch to be spread four inches deep, then the amount of coverage provided by 650 cubic yards is calculable.

From the calculation of quantities, the estimator must then determine the number of hours of labor required to install the materials or perform the task required. For tasks that the company's crews perform regularly, the estimator can make an accurate calculation of the number of hours required to accomplish the task. For tasks that are unfamiliar or unusual, the estimator will have no historical figures to draw upon and must calculate more cautiously.

Standardized Technique Is Important

A large landscape project, whether it is a construction project or a management project, can necessitate hundreds of calculations. To avoid mistakes, it is essential that the estimator establish a method for doing the take-off and then follow that method every time a take-off is done. In large companies, the method of doing take-offs becomes part of the corporate culture. Calculations are made the same way and in the same order every time, regardless of who does the take-off.

The reason for standardization is simple. Doing it the same way every time assures that nothing is overlooked, which could happen if the order or the method of counting and calculating was optional from project to project. Following a standard procedure also allows the estimator to establish stopping points along the way and then start again at a later time without making a mistake.

No one system for take-offs has industry-wide acceptance, nor does there need to be. Most companies need only to develop a system that works for them. Here is a typical sequence for making counts and calculations that might be used for the take-off of a landscape construction project.

Typical Sequence of Calculations for Landscape Take-Offs

I. Quantity Calculations
    A. Plant materials
        1. Make a list of all plants required
            a. List each separate species
            b. List the sizes required for each species
            c. List the quantities of each size required
        2. Make a list of all support materials required
            a. List the number of stakes
            b. List the specific guying materials
            c. List the quantities of guying materials
            d. List the quantities of tree coils and tree wraps
        3. Make a list of turf needs
            a. Describe each species or blend if more than one and whether it is to be applied as seed, sod, sprigs, or plugs
            b. Calculate the surface area(s) to be installed or renovated
    B. Materials associated with the plantings and turf
        1. Make a list of all mulches to be applied
        2. Make a list of all landscape fabrics to be used
        3. Calculate the surface area of the planting beds
        4. Calculate the volume(s) of mulch required
        5. Make a list of all soil amendments to be used
        6. Calculate the volumes of soil amendments required
        7. Calculate the volume of soil requiring disposal
        8. Make a list of all edging materials to be used
        9. Calculate the quantity of edging(s) required
    C. Hardscape materials
        1. Make a list of all hardscape materials required
            a. List and describe each separate material
            b. List and describe all associated support materials such as molds and forms, hardware, stains, etc.
        2. Calculate the quantities of each item required
        3. List and describe all prefabricated hardscape features to be provided, such as fountains, benches, lights
    D. Preliminary services
        1. List all services required to prepare the site for plant or hardscape installations
            a. List and describe each separate service, such as brush removal, excavation, or grading
            b. Calculate the areas and/or volumes affected

II. Time Calculations (These will be based upon specific crew compositions and equipment availabilities known to the estimator.)
    A. Layout times
        Calculate the time required to mark out the planting beds and other landscape features and place the plants

B. Preparation times

Calculate the time requirements for each preliminary service required

C. Installation times

1. Estimate the time required to cut out the planting beds
2. Estimate the time required to prepare the soil for planting
3. Estimate the time required to install the plants
4. Estimate the time required for staking/guying
5. Estimate the time required for applying tree coils and wraps
6. Estimate the time required to spread landscape fabrics
7. Estimate the time required to install edgings
8. Estimate the time required to spread mulches
9. Estimate the time required for finishing details, such as pruning damaged wood and creating soil rings
10. Estimate the time required to install and/or repair lawn areas
11. Estimate the time required to clean up the site, including the time required for hauling away rubbish, excess soil

III. Maintenance Calculations

A. Short-term requirements

1. List all follow-up maintenance requirements such as watering, weeding, removing staking materials
2. Estimate the materials and time required to meet the obligation

B. Long-term requirements

If the company obtains the contract for continuing maintenance of the property, a separate take-off will probably be done at this point

In this example, calculations of materials are grouped together, as are calculations of time.

In other systems the materials, quantities, and labor times for each separate task, area, or phase of the project may be grouped together. Companies that use computerized estimating systems may have their time requirements already tied to certain tasks, so that when the task is described and measured, the calculation of labor hours required is instantaneous.

Once the take-off is completed, the estimator applies cost figures for materials, labor, and overhead to the numbers to arrive at a cost to the company. After that, the desired profit is added and a price to the client is finalized.

Typical Calculations Required for Take-Offs

Most of the calculations required for landscape take-offs use simple algebra, geometry, and old fashioned arithmetic. What follows are some examples and sample problems that illustrate how take-off calculations are made.

Linear and Surface Area Measurements

Table 15-1 contains the linear equivalents and formulas most commonly used to express lengths and measure surface areas.

Sample Problem 1:

In this example, what is the surface area of the patio? Of the lawn?

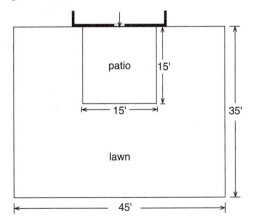

Table 15-1 Linear and Area Measurement

Units of Linear Measurement

12 inches..1 linear foot (LF)
3 linear feet ....1 linear yard
1,760 yards...1 statute mile
5,280 feet ....1 statute mile

Units of Area Measurement

144 sq. inches...1 sq. ft. (SF)
9 sq. ft..1 sq. yd. (SY)
4,840 sq. yds. ....1 acre
43,560 sq. ft. ....1 acre
640 acres ....1 sq. mile

Formulas for Areas

Code

π = 3.14
A = area
r = radius

d = diameter
c = circumference
b = length of arc

Θ = angle in degrees
S = short radius of ellipse
L = long radius of ellipse

Circumference and Area of a Circle

$c = 2\pi r$

$A = \pi r^2$

Area of a Sector

$A = \dfrac{br}{2}$

$A = \pi r^2 \times \dfrac{\Theta}{360}$

Circumference and Area of an Ellipse

$c = 2\pi \sqrt{\dfrac{S^2 + L^2}{2}}$

$A = \pi SL$

Area of a Square

$A = side \times side$

$A = side^2$

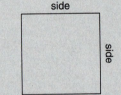

Area of a Rectangle

$A = length \times width$
$A = base \times height$

width or height

length or base

Area of a Triangle

$A = \frac{1}{2}(base \times height)$
$A = \frac{1}{2}(length \times width)$

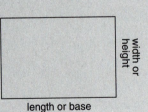

width or height

length or base

Solution:

I. The patio, since it has sides of equal length, is a square. Use the formula for the area of a square to find the area of the patio.

$$A = side \times side$$
$$A = 15\ LF \times 15\ LF$$
$$= 225\ SF$$

[Note that area is always expressed in square units and that multiplication of any units is only possible when the units match. In this example, it would not be possible to multiply linear feet by linear yards or any other dissimilar unit without first converting one unit to match the other.]

II. The lawn and patio complex, since it has unequal sides, is a rectangle. (1) Use the formula for the area of a rectangle to find the area. (2) Then subtract the area of the patio to determine the remaining lawn area.

(1) A = 45 LF × 35 LF

= 1,575 SF [the lawn/patio complex]

(2) 1,575 SF
− 225 SF [the patio]
1,350 SF [the area of the lawn]

Sample Problem 2:

What is the surface area of this patio?

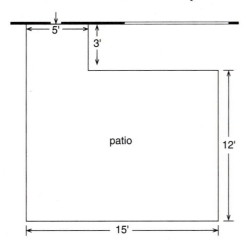

Solution:

The patio is two connecting rectangles. One rectangle is 15 LF × 12 LF; the other is 5 LF × 3 LF. (1) Use the formula for the area of a rectangle to determine the two areas, then (2) add the areas together.

(1) 15 LF × 12 LF = 180 SF

5 LF × 3 LF = 15 SF

(2) 180 SF + 15 SF = 195 SF [the area of the patio]

Sample Problem 3:

In this example, what is the surface area of the pool? Of the planting bed?

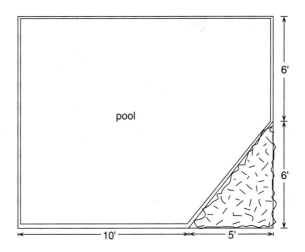

Solution:

I. If the planting bed did not exist, the pool would be a rectangle. Use the formula for the area of a rectangle to determine the area of the space before the planting bed is considered.

A = 15 LF × 12 LF

= 180 SF

II. The planting bed is a triangle. Use the formula for the area of a triangle to determine the area.

A = 1/2 (5 LF × 6 LF)

= 1/2 (30 SF)

= 15 SF [the area of the planting bed]

III. Subtract the area of the planting bed from the larger area to determine the area of the pool.

180 SF − 15 SF = 165 SF [the area of the pool]

Sample Problem 4:

What are the areas of flower beds A and B? How many feet of edging would be required to encircle flower beds A and B?

Flower Bed A

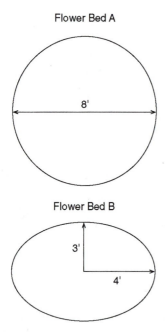

Flower Bed B

Solution:

I. Flower bed A is a circle. Use the formula for the area of a circle to determine the area. If the diameter is 8 LF, then the radius is half of that, 4 LF.

$$A = \pi r^2$$
$$= 3.14 \,(4 \text{ LF} \times 4 \text{ LF})$$
$$= 3.14 \,(16 \text{ SF})$$
$$= 50.24 \text{ SF [area of flower bed A]}$$

II. Flower bed B is an ellipse. Use the formula for the area of an ellipse to determine the area. If the short radius (S) is 3 LF and the long radius (L) is 4 LF, the formula is applied this way:

$$A = \pi SL$$
$$= 3.14 \,(3 \text{ LF} \times 4 \text{ LF})$$
$$= 3.14 \,(12 \text{ SF})$$
$$= 37.68 \text{ SF [area of flower bed B]}$$

III. Use the formula for the circumference of a circle to determine the length of edging required for flower bed A.

$$C = 2\pi r$$
$$= 2 \,(3.14 \times 4 \text{ LF})$$
$$= 2 \,(12.56 \text{ LF})$$
$$= 25.12 \text{ LF [edging required for flower bed A]}$$

IV. Use the formula for the circumference of an ellipse to determine the amount of edging required for flower bed B.

$$C = 2\pi \sqrt{\frac{S^2 + L^2}{2}}$$
$$= (2 \times 3.14) \sqrt{\frac{S^2 + L^2}{2}}$$
$$= 6.28 \sqrt{\frac{9 + 16}{2}}$$
$$= 6.28 \sqrt{\frac{25}{2}}$$
$$= 6.28 \sqrt{12.5}$$
$$= (6.28)\,(3.54 \text{ LF})$$
$$= 22.23 \text{ LF [edging required for flower bed B]}$$

Sample Problem 5:

What is the area of the deck surrounding the spa?

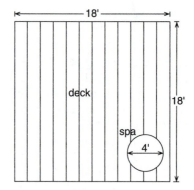

Solution:

I. The deck is a square. Use the formula for the area of a square to determine the area of the deck if the spa was not there.

A = side × side

= 18 LF × 18 LF

= 324 SF [area of the deck/spa complex]

II. The spa is a circle. Use the formula for the area of a circle to determine the area of the spa.

A = πr^2

= 3.14 (2 LF × 2 LF)

= 3.14 (4 SF)

= 12.56 SF [area of the spa]

III. Subtract the area of the spa from the area of the deck/spa complex to determine the surface area of the deck surrounding the spa.

  324.00 SF
− 12.56 SF
  311.44 SF [area of the deck around the spa]

Sample Problem 6:

In this example, what are the surface areas of the brick patio, the deck, the planter, the red flowers, and the white flowers?

Answers:

For this problem, the answers are provided, but not the explanation of the solutions. See if you can attain these answers using the formulas from Table 15-1 and the methods explained in Sample Problems 1–5.

Area of the brick, including that beneath the deck:	116.86 SF
Area of the deck:	96.00 SF
Area of the planter:	25.00 SF
Area of the red flowers:	24.00 SF
Area of the white flowers:	24.00 SF

Volume Measurement

Table 15-2 contains the units of measurement and conversion equivalents most commonly used to calculate volumes of landscape materials.

Table 15-2   Volume Measurement
1,728 cubic inches ....1 cubic foot (CF)
46,656 cubic inches ....1 cubic yard (CY)
27 cubic feet....................................1 cubic yard (CY)
1,000 cubic yards..1 MCY*

*NOTE: M is an abbreviation for 1,000. Thus 3,000 CY could be expressed as 3 MCY.

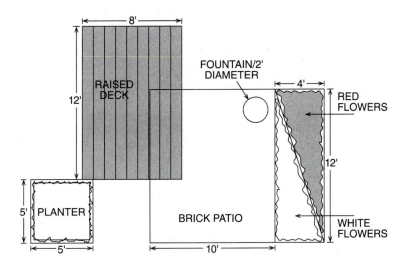

The calculation of volume requires three dimensions, two horizontal and one vertical, such as length, width, and thickness (or depth). A formula for volume calculation could be

$$Volume = Length \times Width \times Depth$$

Another way of expressing the same calculation would be

$$Volume = Surface\ Area \times Depth$$

This is because surface area has been previously calculated by multiplying the two sides or horizontal dimensions.

It has been noted before and is appropriate to restate here that these formulas will only work when all dimensions or measurements are expressed in the same units. Unlike units cannot be multiplied. It is not possible to multiply feet and inches or inches and yards until one has been converted to the equivalent of the other. In the sample problems that follow, some conversion equivalencies are illustrated. These are followed by several sample problems that require the calculation of typical landscape volumes.

Sample Problem 1:

Convert 4 CY to the equivalent number of CF.

Solution:

$$4 \times 27 = 108\ CF$$

Sample Problem 2:

Convert 405 CF to the equivalent number of CY.

Solution:

$$405 \div 27 = 15\ CY$$

Sample Problem 3:

Convert 8,550 CY to the equivalent number of MCY.

Solution:

$$8,550 \div 1,000 = 8.55\ MCY$$

Sample Problem 4:

A patio is 20 LF by 18 LF. It is to be paved with concrete 4 inches thick. What volume of concrete is required?

Solution:

I. The first calculation must be the conversion of unlike units to a common unit. Since two of the dimensions are already in linear feet and since the volume of concrete is more commonly expressed in cubic feet than cubic inches, it is logical to convert the 4-inch thickness of the concrete to an equivalent number of linear feet (See Table 15-1).

$$4\ inches \div 12\ inches = 0.33\ LF$$

II. With all three dimensions now in the same units of measurement, the volume can be calculated.

$$Volume = 20\ LF \times 18\ LF \times 0.33\ LF$$
$$= 118.8\ CF\ of\ concrete$$

Sample Problem 5:

A lawn requires a layer of topsoil to be spread over it prior to a new seeding. The lawn is 80 LF by 74 LF. The topsoil is to be 3 inches thick. How many CY of topsoil are required?

Solution:

I. Converting from one unit to another introduces the probability of whole numbers becoming fractions. In turn, when partial numbers are multiplied or divided, the resulting answer is usually rounded off, and that introduces inaccuracies. Therefore, when making conversions during calculations that allow several options, as in this problem, it is best to select the option that results in the fewest partial numbers being entered into the calculations.

Begin by converting the three inches to linear feet so that all three dimensions have the same units.

$$3\ inches \div 12\ inches = 0.25\ LF$$

II. Next, determine the volume of topsoil in cubic feet.

$$Volume = 80\ LF \times 74\ LF \times 0.25\ LF$$
$$= 1,480\ CF$$

III. Finally, while the volume of topsoil is correctly calculated, it is not expressed in units of cubic yards, so the 1,480 CF must be converted to CY.

$$1,480 \text{ CF} \div 27 \text{ CF} = 54.81 \text{ CY of topsoil}$$

Sample Problem 6:

A swimming pool is 30 LF × 20 LF × 5 LF. It rests on a bed of gravel, which is 4 LF deep, and extends outside the area of the pool by 3 LF in all directions. How many cubic yards of gravel will be required?

Answer:

The answer is 138.67 CY of gravel required. Using Table 15-2 and the methods illustrated in the previous sample problems, work through this problem until you arrive at the same answer.

Time Calculations

The unit of measurement for time is hours. Labor costs are calculated on the basis of how many units of material or service can be accomplished per hour by a person or crew of known composition and capability, using prescribed equipment and techniques. In the sample problems that follow, various methods of calculating time are illustrated.

Sample Problem 1:

If a landscape crew can install 400 SY of sod in an 8-hour day, how many hours will be required to install 280 SY?

Solution:

The solution requires an algebraic comparison, with X representing the number of hours required to install the sod.

$$\text{I.} \quad \frac{400 \text{ SY}}{8 \text{ hrs.}} = \frac{280 \text{ SY}}{X \text{ hrs.}}$$

$$\text{II.} \quad 400 X = (280)(8)$$
$$= 2,240$$

$$\text{III.} \quad X = 2,240 \div 400$$
$$= 5.6 \text{ hrs. required}$$
$$\text{to install the sod}$$

Sample Problem 2:

If the crew can clear by hand 620 SY of light density brush in an 8-hour day, how many hours will be required to clear 1 acre of such brush?

Solution:

Although the numbers are larger, the solution is the same as in Problem 1. Establish a comparison between what is known and what is sought. Consult Table 15-1 for the equivalency between SY and acres.

$$\text{I.} \quad \frac{620 \text{ SY}}{8 \text{ hrs.}} = \frac{4,840 \text{ SY}}{X \text{ hrs.}}$$

$$\text{II.} \quad 620 X = (8)(4,840)$$
$$= 38,720$$

$$\text{III.} \quad X = 38,720 \div 620$$
$$= 62.45 \text{ hrs. required}$$
$$\text{to clear the brush}$$

Sample Problem 3:

Flower plantings of 6 MSF can be installed at the rate of 75 plants per hour when spaced 1 foot apart. If the contract requires the flower plantings to be rotated three times per season, how many total hours will be spent installing flowers?

Solution:

I. If spaced 1 foot apart, one installation of flowers will require that 6,000 plants be installed.

$$6 \text{ MSF} = 6,000 \text{ SF}$$
$$6,000 \text{ SF @ 1 plant/SF} = 6,000 \text{ plants}$$

II. If there are three rotations per season, then the total number of flowers to be installed for the year needs to be calculated.

$$6,000 \text{ plants per rotation} \times 3 \text{ rotations} =$$
$$18,000 \text{ plants}$$

III. Finally, establish the algebraic comparison of what is known to what is sought. As before, X represents the unknown.

A. $\dfrac{75 \text{ plants}}{1 \text{ hr.}} = \dfrac{18,000 \text{ plants}}{X \text{ hrs.}}$

B. $75\,X = (1)\,(18,000)$

$= 18,000$

C. $X = 18,000 \div 75$

$= 240$ hrs. required
to install flowers

Sample Problem 4:

A lawn of 30 MSF is mowed with two different pieces of equipment. 26 MSF of the lawn is cut using a 60-inch wide riding mower that travels at 8 mph. The remainder of the lawn is cut with a 36-inch wide walking mower that averages 2 mph. How much time is required for one mowing of the lawn?

Solution:

I. To begin, it is necessary to envision the lawn being reshaped into two pieces. One piece will be 5 LF (60 inches) wide, and the other will be 3 LF (36 inches) wide. Both pieces are of indeterminate length, which must be calculated.

II. To calculate the length of the first piece, divide the known area [26,000 SF] by the known width [5 LF] to determine the unknown length.

26,000 SF ÷ 5 LF = 5,200 LF
to be cut by the riding mower

III. Calculate the length of the second piece by dividing the known area [4,000 SF] by the known width [3 LF] to determine the unknown length.

4,000 SF ÷ 3 LF = 1,333 LF to be cut by
the walking mower

IV. Next, convert each length to miles. See Table 15-1.

5,200 LF ÷ 5,280 LF = .98 miles to be cut
by the riding mower

1,333 LF ÷ 5,280 LF = .25 miles to be cut
by the walking
mower

V. Next, set up the algebraic comparisons of what the mowers can cut per hour to what must be cut in each area. That permits a determination of the number of hours required by each machine to cut its section of the lawn.

A. The riding mower

$\dfrac{8 \text{ miles}}{1 \text{ hr.}} = \dfrac{.98 \text{ miles}}{X \text{ hrs.}}$

$8\,X = (1)\,(.98)$

$= .98$

$X = .98 \div 8$

$= .12$ hr. to cut 26 MSF once

B. The walking mower

$\dfrac{2 \text{ miles}}{1 \text{ hr.}} = \dfrac{.25 \text{ miles}}{X \text{ hrs.}}$

$2\,X = (1)\,(.25)$

$= .25$

$X = .25 \div 2$

$= .13$ hr. to cut 4 MSF once

VI. Sum the time requirements for the two sections of the lawn to determine the total time requirements for mowing the lawn once.

.12 hr. + .13 hr. = .25 hr. required
per mowing

Sample Problem 5:

If in the previous problem the lawn was so cluttered with trees, poles, and other obstacles that the mowers have only 70 percent efficiency, how much time is required for a single cutting?

Answer:

The reduced efficiency applies to the rate of speed of the mowers. It does not affect the amount of lawn requiring mowing. Although capable of operating at 8 mph on perfectly flat, open land, the riding mower will instead average 5.6 mph (8 mph × 0.70), and the walking mower will average 1.4 mph (2 mph × .70).

Using these figures, redo problem 4 and see if you can derive these answers:

> The riding mower will require 0.18 hr. to cut 26 MSF of lawn.

> The walking mower will require 0.18 hr. to cut 4 MSF of lawn.

> The total time required for one complete mowing is 0.36 hr.

Quantity Calculations and Conversions

Other calculations often required during landscape take-offs include the determination of quantities needed to meet the specifications of a contract. The sample problems and solutions that follow illustrate how quantity calculations are made.

Sample Problem 1:

> If the recommended rate of application for grass seed is 2 lbs./MSF, what quantity of seed is required to establish a new lawn of 15,000 SF?

Solution:

> I. Convert 15,000 SF to its equivalent in MSF units. (See Table 15-1.)

$$15,000 \text{ SF} \div 1,000 \text{ SF} = 15 \text{ MSF}$$

> II. Multiply the rate of application for 1 MSF by the total number of units (15).

$$2 \text{ lbs./MSF} \times 15 = 30 \text{ pounds of seed required}$$

Sample Problem 2:

> If the rate of application for lawn fertilizer is 4 lbs. of nitrogen per thousand square feet, how many pounds of fertilizer with a 10-6-4 analysis are required to fertilize 20,000 SF?

Solution:

> I. Calculate the number of MSF units needing fertilization.

$$20,000 \text{ SF} \div 1,000 \text{ SF} = 20 \text{ MSF of lawn to be fertilized}$$

> II. Calculate the amount of actual nitrogen (N) in one pound of 10-6-4 fertilizer.

$$1 \text{ lb. of fertilizer} \times .10 = 0.10 \text{ lb. of actual N}$$

> III. Calculate the number of pounds of 10-6-4 fertilizer required to provide 4 lbs. of actual nitrogen/MSF.

$$4 \text{ lbs. (needed)} \div 0.10 \text{ lb.} = 40 \text{ lbs. of } 10\text{-}6\text{-}4/\text{MSF needed}$$

> IV. Calculate the total number of pounds of fertilizer required to apply 4 lbs. of N/MSF to 20 MSF.

$$40 \text{ lbs. of fertilizer/MSF} \times 20 = 800 \text{ lbs. of } 10\text{-}6\text{-}4 \text{ fertilizer}$$

Sample Problem 3:

> In the above problem, how many pounds of 10-6-4 are required to fertilize a 5-acre lawn?

Solution:

> I. Changing the size of the area to be fertilized does not alter the amount of elemental nitrogen in the fertilizer. It still requires 40 lbs. of 10-6-4 to deliver 4 lbs. of elemental N to 1,000 SF of turf. What does have to be determined are the number of MSF units in 5 acres. (See Table 15-1.)

$$\text{A. } 43,560 \text{ SF} = 1 \text{ acre}$$

$$\text{B. } 43,560 \text{ SF} \times 5 = \text{The number of SF in 5 acres}$$
$$= 217,800 \text{ SF}$$

$$\text{C. } \frac{217,800 \text{ SF}}{1,000 \text{ SF}} = \text{The number of MSF units in 5 acres}$$
$$= 217.8 \text{ MSF to be fertilized}$$

> II. Multiply the total number of MSF units to be fertilized by the amount of fertilizer required per MSF to determine the total requirement for the 5 acres.

$$217.8 \times 40 \text{ lbs./MSF} = 8,712 \text{ lbs. of } 10\text{-}6\text{-}4 \text{ fertilizer}$$

Sample Problem 4:

Concrete dye is mixed into concrete at the rate of 4 ozs. per 100 CF. How many ounces are required to color 50 CY?

Solution:

I. Determine the number of cubic feet in 50 cubic yards. (See Table 15-2.)

27 CF/CY × 50 CY = 1,350 CF

II. Make an algebraic comparison between the quantity known and the quantity being sought. As before, X represents the unknown amount.

A. $\dfrac{4 \text{ ozs.}}{100 \text{ CF}} = \dfrac{X \text{ ozs.}}{1{,}350 \text{ CF}}$

B. $100 X = 5{,}400$

C. $X = 54$ ozs. of dye required

Sample Problem 5:

An 800 SF concrete patio, 4 inches thick, is to be poured over 6 inches of crushed stone subsurface. How many CY of concrete are required and how many CY of crushed stone are required?

Solution:

I. The amounts of concrete and crushed stone must be calculated from the known surface area and depth dimensions. To do that requires that the units of measurements be converted to match. The simplest conversion is to turn the 4-inch and 6-inch measurements into equivalent LF units.

A. 4 inches ÷ 12 inches = 0.33 LF of concrete

B. 6 inches ÷ 12 inches = 0.50 LF of crushed stone

II. Multiplying the surface area by the thicknesses of each material will produce the quantities required. The unit of measurement will be CF.

A. 800 SF × 0.33 LF = 264 CF of concrete required

B. 800 SF × 0.50 LF = 400 CF of crushed stone required

III. Convert the volumes from cubic feet to cubic yards.

A. 264 CF ÷ 27 CF = 9.78 CY of concrete required

B. 400 CF ÷ 27 CF = 14.81 CY of crushed stone required

Sample Problem 6:

A planter is 5 LF in length, 3 LF wide, and 4 LF deep. It is to be filled with a soil mix that is 60 percent actual soil and 40 percent conditioners. The planter will be filled to a depth of 3 1/2 LF with the conditioned soil; then 6 inches of mulch will be added. How many cubic yards of actual soil will be required to fill the planter? How many cubic yards of mulch?

Answers:

The answers are provided below, but not the method of deriving them. Based upon the calculations shown and explained in the above sample problems, solve this problem and see if your answers match these.

The amount of actual soil is 1.17 CY.

The amount of mulch is 0.28 CY.

Soil Compaction Calculations

Soil occupies different volumes of space depending upon whether it is in a natural, undisturbed state or has been loosened by digging, thereby adding more air to the soil's composition. It also changes volume when air is driven out by compaction, as when compressed by heavy traffic. The degree of expansion or compaction in comparison to the natural, undisturbed state is dependent upon the type of soil. The more clay particles there are in the soil, the greater will be the expansion and compaction potential of the soil. The more sand particles the soil contains, the less will be the difference in the soil's volume as it changes from the undisturbed state to the expanded or contracted state.

While the differences between a soil's volume as it changes from one state to another is of little concern when installing only a few plants, it becomes a greater concern on large projects. When large quantities of soil are disturbed through grading or extensive construction, the landscape contractor may need to provide for the acquisition or the removal of large amounts of soil. As with every other task, handling material translates into time and money. Table 15-3 compares the volumes of different soil types in three different states. Undisturbed field soil is said to be *banked*. The unit of measurement is BCY (Bank Cubic Yard). When disturbed and mixed with additional air, the volume is expressed as LCY (Loose Cubic Yard). If compressed and the amount of air in the soil reduced, the volume measurement is termed CCY (Compact Cubic Yard).

Sample Problem 1:

To what volume would 5 BCY of loam expand after digging?

Table 15-3 A Comparison of Soil Expansion and Compaction

Soil Type	As a BCY	As a CCY	As a LCY
Clay	1	0.85	1.25
Clay loam	1	0.91	1.24
Clay/gravel mix	1	0.95	1.18
Loam	1	0.95	1.23
Sand	1	0.98	1.12
Sandy loam	1	0.97	1.20
Sandy clay	1	0.87	1.27
Sand and gravel	1	0.99	1.11
Gravel	1	0.98	1.12

Solution:

Multiply the number of BCY of loam soil by the numerical equivalent for 1 LCY of comparable soil.

$$5 \text{ BCY} \times 1.23 = 6.15 \text{ LCY of loam}$$

Sample Problem 2:

To what volume would 8 BCY of sandy clay expand after digging?

Solution:

$$8 \text{ BCY} \times 1.27 = 10.16 \text{ LCY of sandy clay}$$

Sample Problem 3:

To what volume would 15 BCY of clay and gravel compress after compaction?

Solution:

$$15 \text{ BCY} \times 0.95 = 14.25 \text{ CCY of clay/gravel mix}$$

Sample Problem 4:

To what volume would 6 LCY of loam compress after compaction?

Solution:

I. Convert 6 LCY to the equivalent number of BCYs by dividing by the expansion equivalent for 1 LCY of loam.

$$6 \text{ LCY of loam} \div 1.23 = 4.88 \text{ BCY of loam}$$

II. Convert 4.88 BCY of loam to the equivalent number of CCYs.

$$4.88 \text{ BCY} \times 0.95 = 4.64 \text{ CCY of loam soil}$$

Sample Problem 5:

To what volume could 7 CCY of sandy loam expand?

Answer:

Using the methods shown in the previous problems, see if you can arrive at the correct answer.

The answer is 8.66 LCY of sandy loam soil.

Pit and Ball Calculations

Estimators must further deal with soil volume calculations when large numbers of plant materials are installed. A certain volume of soil is removed when the planting pit is dug. A different volume of soil, contained within the new plant's root ball, is placed into the pit. That creates two additional volumes: the volume of material required to backfill the pit around the soil ball, and the volume of soil that cannot fit back into the pit and becomes waste. Table 15-4 contains the formulas that explain how these various volumes are determined.

Sample Problem 1:

A planting pit has a volume of 10.6 CF. The plant being installed has a soil ball volume of 5.2 CF. What is the volume of soil in the backfill, and what volume of waste must be disposed of?

Solution:

I. Determine the volume of backfill using the formula from Table 15-4.

$$PV - BV = BF$$
$$10.6 \text{ CF} - 5.2 \text{ CF} = 5.4 \text{ CF of backfill}$$

II. Determine the volume of waste using the formula from Table 15-4.

$$BV + SA = WV$$

Table 15-4 Pit and Ball Volume Calculations

Key	Formulas
PV = Pit volume	$PV - BV = BF$
BV = Ball volume	
BF = Backfill volume	$BV + SA = WV$
SA = Solid soil additives	
WV = Waste volume	

Since there are no solid soil additives in the backfill, SA has a value of 0. The calculation becomes

$$5.2 \text{ CF} + 0 = 5.2 \text{ CF of waste}$$

When there is no replacement of the original pit soil with solid additives such as peat moss or sand, then the waste volume is equivalent to the volume of the soil ball.

Sample Problem 2:

Four shrubs are to be installed. Each plant has a ball volume of 8.9 CF. Each planting pit has a volume of 16.0 CF. What is the total volume of soil to be excavated? The total ball volume? The total backfill volume? The total waste volume?

Solution:

I. Multiplying each known volume by 4 provides necessary quantity information.

A. Total ball volume = $8.9 \times 4 = 35.6$ CF

B. Total pit volume = $16.0 \times 4 = 64.0$ CF

II. Applying the formulas from Table 15-4 generates the remaining volumes.

C. Total backfill volume =
$64.0 \text{ CF} - 35.6 \text{ CF} = 28.4 \text{ CF}$

B. Total waste volume =
$35.6 \text{ CF} + 0 \text{ CF} = 35.6 \text{ CF}$

Sample Problem 3:

Six trees, each with a soil ball volume of 14.0 CF, are being installed in pits, each with a volume of 23.8 CF. The backfill mix is to be 50 percent original soil and 50 percent sand/peat mix. What is the total volume of soil to be excavated? Total ball volume? Total backfill volume? Total volume of soil in the backfill? Total volume of solid soil additives? Total volume of waste requiring disposal?

Solution:

A. Total pit volume = 23.8 CF × 6 = 142.8 CF

B. Total ball volume = 14.0 CF × 6 = 84.0 CF

C. Total backfill volume =
142.8 CF − 84.0 CF = 58.8 CF

D. Total volume of soil
in the backfill = 50 percent of total BF
58.8 CF × .50 = 29.4 CF

E. Total volume of soil additives =
50 percent of total BF
58.8 CF × .50 = 29.4 CF

F. Total volume of waste =
84.0 CF + 29.4 CF = 113.4 CF

There are two means of determining the volume of soil removed from planting pits and contained within plant root balls. Both methods are based upon the diameter and depth of individual pits and balls. One method uses formulas, while the other method is a chart derived from the formulas, Table 15-5. When the pit and ball dimensions fall within the chart, it is timesaving to use it. When the dimensions of either the pit or the ball are not on the chart, then the formulas can be used to calculate the volumes.

Volume of the soil ball in CF =

$$\left(\text{ball diameter in LF}\right)^2 \times \left(\text{ball depth in LF} \times \frac{2}{3}\right)$$

Volume of the pit ball in CF =

$$\frac{\pi \times \left(\text{pit diameter in LF}\right)^2}{4} \times \text{pit depth in LF}$$

Sample Problem 4:

Using Table 15-5, determine the total volume of seven planting pits, each one 5 LF in diam-

eter and 3 LF deep. Determine the total ball volume for seven plants, each with root balls with diameters of 4 1/2 LF and depths of 2 3/4 LF. Finally, calculate the total backfill volume and the total soil waste.

Solution:

I. Consulting Table 15-5 reveals that the volume of one 5' × 3' pit is 58.88 CF. The volume of one 4 1/2' × 2 3/4' soil ball is 37.12 CF.

II. Multiplying each volume by the total number of pits and balls (seven each) gives the total pit volume and total ball volume for the planting.

A. Total pit volume = 58.88 CF × 7 pits
= 412.16 CF

B. Total ball volume = 37.12 CF × 7 balls
= 259.84 CF

C. Total backfill = 412.16 CF − 259.84 CF
= 152.32 CF

D. Total soil waste = 259.84 CF + 0 CF
= 259.84 CF

Sample Problem 5:

Use the formulas to calculate the volume of a planting pit that is 8 LF in diameter and 4 LF in depth. Then calculate the volume of a soil ball that is 7 3/4 LF in diameter and 3 3/4 LF deep.

Answer:

The pit volume is 200.96 CF. The ball volume is 148.95.

You can test your understanding of the formulas by creating sample problems of your own using dimensions that are in the chart and seeing if you calculate the same volumes as are shown in the chart.

Table 15-5   Planting Pit and Root Ball Volumes (Cubic feet per plant)

Diameters	1'	1 1/4'	1 1/2'	1 3/4'	2'	2 1/4'	2 1/2'	2 3/4'	3'	3 1/4'	3 1/2'
Depths											
Pit 1'	0.79	1.23	1.77	2.40	3.14	3.97	4.91	5.94	7.07	8.29	9.62
Ball	0.67	1.04	1.50	2.04	2.67	3.37	4.17	5.04	6.00	7.04	8.17
Pit 1 1/4'	0.98	1.53	2.21	3.01	3.93	4.97	6.13	7.42	8.83	10.36	12.02
Ball	0.83	1.30	1.87	2.55	3.33	4.22	5.21	6.30	7.50	8.80	10.21
Pit 1 1/2'	1.18	1.84	2.65	3.61	4.71	5.96	7.36	8.90	10.60	12.44	14.42
Ball	1.00	1.56	2.25	3.06	4.00	5.06	6.25	7.56	9.00	10.56	12.25
Pit 1 3/4'	1.37	2.15	3.09	4.21	5.50	6.95	8.59	10.39	12.36	14.51	16.83
Ball	1.17	1.82	2.62	3.57	4.67	5.91	7.29	8.82	10.50	12.32	14.29
Pit 2'	1.57	2.45	3.53	4.81	6.28	7.95	9.81	11.87	14.13	16.58	19.23
Ball	1.33	2.08	3.00	4.08	5.33	6.75	8.33	10.08	12.00	14.08	16.33
Pit 2 1/4'	1.77	2.76	3.97	5.41	7.07	8.94	11.04	13.36	15.90	18.66	21.64
Ball	1.50	2.34	3.37	4.59	6.00	7.59	9.37	11.34	13.50	15.84	18.37
Pit 2 1/2'	1.96	3.07	4.42	6.01	7.85	9.94	12.27	14.84	17.66	20.73	24.04
Ball	1.67	2.60	3.75	5.10	6.67	8.44	10.42	12.60	15.00	17.60	20.42
Pit 2 3/4'	2.16	3.37	4.86	6.61	8.64	10.93	13.49	16.33	19.43	22.80	26.44
Ball	1.83	2.86	4.12	5.61	7.33	9.28	11.46	13.86	16.50	19.36	22.46
Pit 3'	2.36	3.68	5.30	7.21	9.42	11.92	14.72	17.81	21.20	24.87	28.85
Ball	2.00	3.12	4.50	6.12	8.00	10.12	12.50	15.12	18.00	21.12	24.50
Pit 3 1/4'	2.55	3.99	5.74	7.81	10.21	12.92	15.95	19.29	22.96	26.95	31.25
Ball	2.17	3.39	4.87	6.64	8.67	10.97	13.54	16.39	19.50	22.89	26.54
Pit 3 1/2'	2.75	4.29	6.18	8.41	10.99	13.91	17.17	20.78	24.73	29.02	33.66
Ball	2.33	3.65	5.25	7.15	9.33	11.81	14.58	17.65	21.00	24.65	28.58
Pit 3 3/4'	2.94	4.60	6.62	9.02	11.78	14.90	18.40	22.26	26.49	31.09	36.06
Ball	2.50	3.91	5.62	7.66	10.00	12.66	15.62	18.91	22.50	26.41	30.62
Pit 4'	3.14	4.91	7.07	9.62	12.56	15.90	19.63	23.75	28.26	33.17	38.47
Ball	2.67	4.17	6.00	8.17	10.67	13.50	16.67	20.17	24.00	28.17	32.67
Pit 4 1/4'	3.34	5.21	7.51	10.22	13.35	16.89	20.85	25.23	30.03	35.24	40.87
Ball	2.83	4.43	6.37	8.68	11.33	14.34	17.71	21.43	25.50	29.93	34.71
Pit 4 1/2'	3.53	5.52	7.95	10.82	14.13	17.88	22.08	26.71	31.79	37.31	43.27
Ball	3.00	4.69	6.75	9.19	12.00	15.19	18.75	22.69	27.00	31.69	36.75
Pit 4 3/4'	3.73	5.83	8.39	11.42	14.92	18.88	23.30	28.20	33.56	39.38	45.68
Ball	3.17	4.95	7.12	9.70	12.67	16.03	19.79	23.95	28.50	33.45	38.79
Pit 5'	3.93	6.13	8.83	12.02	15.70	19.87	24.53	29.68	35.33	41.46	48.08
Ball	3.33	5.21	7.50	10.21	13.33	16.87	20.83	25.21	30.00	35.21	40.83
Pit 5 1/2'	4.32	6.75	9.71	13.22	17.27	21.86	26.98	32.65	38.86	45.60	52.89
Ball	3.67	5.73	8.25	11.23	14.67	18.56	22.92	27.73	33.00	38.73	44.92
Pit 6'	4.71	7.36	10.60	14.42	18.84	23.84	29.44	35.62	42.39	49.75	57.70
Ball	4.00	6.25	9.00	12.25	16.00	20.25	25.00	30.25	36.00	42.25	49.00

Diameters	3 3/4'	4'	4 1/4'	4 1/2'	4 3/4'	5'	5 1/4'	6'	6 1/2'	7'	7 1/2'
Depths											
Pit 1'	11.04	12.56	14.18	15.90	17.71	19.63	21.64	28.26	33.17	38.47	44.16
Ball	9.37	10.67	12.04	13.50	15.04	16.67	18.37	24.00	28.17	32.67	37.50
Pit 1 1/4'	13.80	15.70	17.72	19.87	22.14	24.53	27.05	35.33	41.46	48.08	55.20
Ball	11.72	13.33	15.05	16.87	18.80	20.83	22.97	30.00	35.21	40.83	46.87
Pit 1 1/2'	16.56	18.84	21.27	23.84	26.57	29.44	32.45	42.39	49.75	57.70	66.23
Ball	14.06	16.00	18.06	20.25	22.56	25.00	27.56	36.00	42.25	49.00	56.25
Pit 1 3/4'	19.32	21.98	24.81	27.82	31.00	34.34	37.86	49.46	58.04	67.31	77.27
Ball	16.41	18.67	21.07	23.62	26.32	29.17	32.16	42.00	49.29	57.17	65.62
Pit 2'	22.08	25.12	28.36	31.79	35.42	39.25	43.27	56.52	66.33	76.93	88.31
Ball	18.75	21.33	24.08	27.00	30.08	33.33	36.75	48.00	56.33	65.33	75.00
Pit 2 1/4'	24.84	28.26	31.90	35.77	39.85	44.16	48.68	63.59	74.62	86.55	99.35
Ball	21.09	24.00	27.09	30.37	33.84	37.50	41.34	54.00	63.37	73.50	84.37
Pit 2 1/2'	27.60	31.40	35.45	39.74	44.28	49.06	54.09	70.65	82.92	96.16	110.39
Ball	23.44	26.67	30.10	33.75	37.60	41.67	45.94	60.00	70.42	81.67	93.75
Pit 2 3/4'	30.36	34.54	38.99	43.71	48.71	53.97	59.50	77.72	91.21	105.78	121.43
Ball	25.78	29.33	33.11	37.12	41.36	45.83	50.53	66.00	77.46	89.83	103.12
Pit 3'	33.12	37.68	42.54	47.69	53.13	58.88	64.91	84.78	99.50	115.40	132.47
Ball	28.12	32.00	36.12	40.50	45.12	50.00	55.12	72.00	84.50	98.00	112.50
Pit 3 1/4'	35.88	40.82	46.08	51.66	57.56	63.78	70.32	91.85	107.79	125.01	143.51
Ball	30.47	34.67	39.14	43.87	48.88	54.17	59.72	78.00	91.54	106.17	121.87
Pit 3 1/2'	38.64	43.96	49.63	55.64	61.99	68.69	75.73	98.91	116.08	134.63	154.55
Ball	32.81	37.33	42.15	47.25	52.65	58.33	64.31	84.00	98.58	114.33	131.25
Pit 3 3/4'	41.04	47.10	53.17	59.61	66.42	73.59	81.14	105.98	124.37	144.24	165.59
Ball	35.16	40.00	45.16	50.62	56.41	62.50	68.91	90.00	105.62	122.50	140.62
Pit 4'	44.16	50.24	56.72	63.59	70.85	78.50	86.55	113.04	132.67	153.86	176.63
Ball	37.50	42.67	48.17	54.00	60.17	66.67	73.50	96.00	112.67	130.67	150.00
Pit 4 1/4'	46.92	53.38	60.26	67.56	75.27	83.41	91.96	120.11	140.96	163.48	187.66
Ball	39.84	45.33	51.18	57.37	63.93	70.83	78.09	102.00	119.71	138.83	159.37
Pit 4 1/2'	49.68	56.52	63.81	71.53	79.70	88.31	97.36	127.17	149.25	173.09	198.70
Ball	42.19	48.00	54.19	60.75	67.69	75.00	82.69	108.00	126.75	147.00	168.75
Pit 4 3/4'	52.44	59.66	67.35	75.51	84.13	93.22	102.77	134.24	157.54	182.71	209.74
Ball	44.53	50.67	57.20	64.12	71.45	79.17	87.28	114.00	133.79	155.17	178.12
Pit 5'	55.20	62.80	70.90	79.48	88.56	98.13	108.18	141.30	165.83	192.33	220.78
Ball	46.87	53.33	60.21	67.50	75.21	83.33	91.87	120.00	140.83	163.33	187.50
Pit 5 1/2'	60.71	69.08	77.98	87.43	97.41	107.94	119.00	155.43	182.41	211.56	242.86
Ball	51.56	58.67	66.23	74.25	82.72	91.67	101.06	132.00	154.92	179.66	206.25
Pit 6'	66.23	75.36	85.07	95.38	106.27	117.75	129.82	169.56	199.00	230.79	264.94
Ball	56.25	64.00	72.25	81.00	90.25	100.00	110.25	144.00	169.00	196.00	225.00

Achievement Review

A. List the four different types of measurements and calculations usually required for landscape take-offs.

B. Indicate which of the following units are measurements of

(A) Unit counts   (C) Volume
(B) Surface area   (D) Time

1. square yards
2. square feet
3. hours
4. each
5. cubic yards
6. miles
7. cubic inches
8. mins/MSF

C. Why is it important that the take-off be done the same way each time?

D. Do the following calculations using appropriate tables from the chapter.

1. A 12 LF × 15 LF concrete patio is to be built at grade level. One 15' side is against the house. The remaining three sides are to be edged with standard brick. Each brick is 8 inches in length. How many bricks will be required to edge the patio?

2. What is the surface area of the patio?

3. If the concrete for the patio is 5 inches thick and it is poured over a subsurface of crushed stone that is 7 inches thick, how many cubic yards of concrete and crushed stone are required?

4. If a crew can excavate soil at the rate of 26 BCY/hr., dump and spread crushed stone at the rate of 12.5 SY/hr., then pour and finish concrete at the rate of 125 SF/hr., how much time will be required to accomplish all of these tasks for this patio?

5. Twenty trees, each with a root ball 3 1/2 LF in diameter and 3 LF deep, are to be installed into pits that are 5 LF in diameter and 3 LF deep. The pits are to be backfilled with original soil. How much soil will need to be disposed of? What is the quantity of backfill?

Suggested Activities

1. Obtain a set of plans and specifications from a local landscape contractor and prepare a take-off. Follow the sequence for calculations outlined in this chapter.

2. If the use of Microsoft Excel is available, there are templates available that will permit the preparation of scenarios of various business situations. These can enable students or others seeking practice to write typical purchase orders and/or invoices for landscape projects and then do the calculations necessary to arrive at the required quantities. To access the templates, once into Excel, select File, then New, then Spreadsheet Solutions.

3. Once the information and techniques in Chapters 14 and 29 (Pricing the Design and Pricing Landscape Maintenance) have been learned, use the templates described above to create practice scenarios that will allow students to prepare spreadsheets and arrive at final project costs and prices.

CHAPTER 16
Understanding Contracts and Contractors

Objectives:

Upon completion of this chapter, you should be able to

- list and describe the components of a contract
- describe the different contractual associations that occur in the landscape industry
- explain the different ways that landscape contractors are selected for jobs

The Importance of Contracts

A **contract** is an agreement, usually between two parties, that describes certain services and/or materials that will be provided in return for an established payment or other compensation. There is nothing new or unusual about the concept of contracts or the definition given here. People have been making contracts for centuries. When a parent tells a child that he can go out and play after he eats everything on his plate, the conditions of a contract are specified. When two people wed, they do so in the presence of witnesses and simultaneously agree to love, honor, and cherish . . . for as long as they both shall live . . . until death parts them. Thus, the conditions of the partnership are described: what each will do, for what length of time, and under what conditions the agreement will be ended. Perhaps the simplest and most traditional of all con-

tracts has been the handshake, often referred to as a "gentlemen's agreement." By looking each other in the eye and agreeing to honor conditions expressed orally, the handshake contract has been accepted and honored by many societies for a long time. It is an important part of business today.

Still other types of contracts that are entered into by most people at one or more times during their lives are loans, credit card agreements, labor agreements, buying a car or home, renting, and hundreds of other arrangements wherein some item or service is provided in return for payment. Take a minute to analyze the last major purchase that you made that required borrowing money. How many different parties were involved? You and a bank, parent, or friend who provided the money, and the supplier of the item purchased were all parties in the transaction. What were the conditions for payment to the supplier and for repayment to the lender? How many separate agreements were made before the transaction was complete?

Contracts are part of the way we live, play, and do business today. Our challenge is not merely to begin the study of contractual relationships, but rather to understand more fully what we are already involved with. In particular, it is important to understand the importance and uses of contracts within the landscape industry. Although the handshake and verbal commitment are still part of the landscape profession, it is most common and most wise to put the agreement between parties into written form. Thus, most modern contracts are written to assure mutual understanding of the terms of the contract. Written contracts are also less likely to result in litigation if they are carefully prepared and understood in advance by all parties.

The Components of a Contract

Contracts range in size and complexity from a single page to voluminous documents. Some are so simple they can be prepared on a standardized, pre-printed form, Figure 16-1. Others are so complex that a team of lawyers is needed to prepare and interpret them. Yet all contain certain features that are common to every contract.

a. Names and addresses of the parties involved

b. Date of contract preparation

c. Description of the work to be accomplished and the materials or services to be provided, including the location where it is to be accomplished

d. The terms of completion, including such things as inspection, the completion date, and the penalty for not completing on time

e. The terms of payment

f. The signatures of both parties or their legal representatives

g. The date of signing

Certain of the features can be extensive and complex, causing the contract to stretch across many pages and through numerous clauses. In addition, there may be drawings or other attachments to the signed documents that are also part of the contract. All of these will be described and discussed in later chapters.

Types of Contractual Associations

A landscape project may involve clients and/or their representatives, architects, engineers, landscape architects or designers, landscape construction firms, specialty construction firms, maintenance firms, and/or material suppliers. Numerous contracts must be negotiated to bind all parties together in a manner that will accomplish the project properly.

The **client** is the person(s) or organization that owns and provides financing for a project. Funding may be from the client's personal resources, or there may be another source. The **contractor** is a party in contract with the client or the client's representatives. The *prime contractor* or general contractor is the firm directly responsible to the client for construction of the project. Depending upon the project and the company's capabilities, the prime contractor may build the entire project. More often, the prime contractor retains full responsibility for management control and supervision of the project, but hires one or more specialty firms to construct portions of the project. *Subcontractors* are firms in contract with the prime contractor. They may do the electrical work, or the plumbing, or pool construction, or masonry, or countless other specialized construction tasks that the prime contractor is unable to do or could not do as economically as the subcontractor can do it. The subcontractor and prime contractor enter into a relationship known as a *subcontract*. The subcontractor is usually accountable to the general contractor, not the project owner. Responsibility for the satisfactory performance of the subcontractor on the owner's project is assumed by the general contractor. Figure 16-2 illustrates

CONTRACT

At Your Service Landscape Company
Your Full Service Landscape Professionals
1515 West Roosevelt Parkway   Lincoln, Nebraska 12043
518 234-5646

Client Name   Telephone   Date

Address   Project Title

                                                Project Location

We agree to provide the following services and materials on behalf of the client and project identified above:

WE AGREE to furnish the material and labor, complete and in accordance with the above specifications and with the attached guarantees for the sum of

_____ dollars ($_____)

Payment is to be made as follows: _____

Authorized Signature for the Company _____ Date _____

ACCEPTANCE: The above prices, descriptions of work, conditions, and terms of payment are satisfactory. You are authorized to do the work as specified.

Authorized Signature for the Client _____ Date _____

Figure 16-1   A simple one-page contract

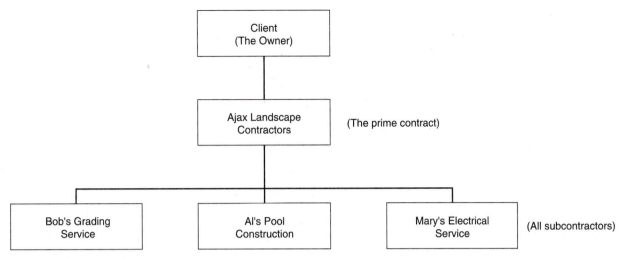

Figure 16-2 A typical prime contract and subcontractor relationship

a typical prime contract and subcontract relationship for a residential landscape project.

In the example, Ajax Landscape Contractors could be a design-build firm. They would have prepared a landscape plan, received approval from the owners of the property, and entered into a prime contractual relationship to build the landscape. They may use their crew to do the finish grading and installation of the lawn and all plantings. All other construction work would be subcontracted to three specialty firms, who will be under contract to Ajax Landscape Contractors, not the homeowners. Therefore it will be the responsibility of Ajax, as the prime contractor, to assure that all work done by the subcontractors is coordinated to begin and end on schedule and without conflict among the various subcontractors. Ajax must supervise the work of the subcontractors to assure that the desired quality is obtained and that the client is satisfied. Ajax also bears responsibility for keeping the subcontractors on schedule so that the completion date that the client has specified is honored.

Figures 16-3 and 16-4 illustrate variations on the contractual relationships that are introduced by the presence of a landscape architect in the project.

The situation shown in Figure 16-3 establishes the landscape architect as the prime contract holder on the project and usually empowers the landscape architect as the owner's representative in all matters that involve the project and the subcontractors. As the principal professional on the project, the landscape architect has the responsibility for selecting the landscape contracting firm that will actually build the landscape. In addition to the drawings and renderings prepared by the landscape architect to help the client visualize the proposed design, the landscape architect must prepare detailed construction drawings and written specifications that cover every aspect of the project. Only in that way can subcontractors know exactly what is expected of them. Should questions or problems arise at any stage of the project, the subcontractor's direct line of appeal is to the landscape contractor or to the landscape architect, depending upon with which party they share their contract.

Figure 16-4 illustrates a situation typical of a larger landscape project in which the client contracts directly with both a landscape architect and a landscape contracting firm. The client may or may not have sought the assistance or advice of the landscape architect in selecting the landscape contractor. Often in this type of contractual relationship, the landscape architect may still be designated as the client's representative in dealing with

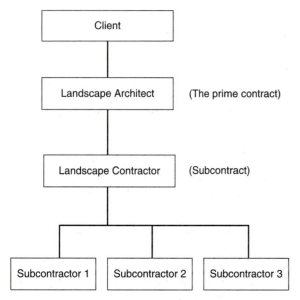

Figure 16-3 Contractual associations with a landscape architect holding the prime contract

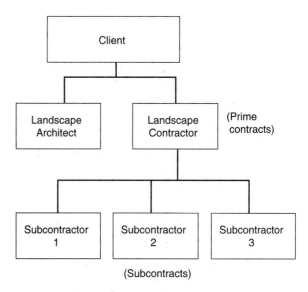

Figure 16-4 Contractual associations with a landscape architect and landscape contractor, each holding contracts directly with the client

the landscape contractor when the client lacks the time or expertise to deal with technical matters. However, the direct responsibility of both professionals is to the client.

On a very large construction project, the landscaping may be only one part of the work being designed and implemented. There may be an interdisciplinary team of professionals responsible for the project. Figure 16-5 diagrams such a project and the contractual relationships that they contain.

As complex as the example appears, an actual construction project can be even more so. Yet as long as every participant fulfills the commitment of each contract and subcontract, most projects get built to the satisfaction of all parties.

The contracts that bind the client to the interdisciplinary team are all separate. Still, one professional is usually designated as the overall coordinator of the project to assure that the work progresses in a logical sequence with minimal confusion and congestion at the site. The general contractor may or may not be a landscape firm. On a large project, the general contractor is likely to be

the coordinating overseer of the subcontractors, maintaining the schedule and quality standards established by the client's professional team. It is the responsibility of the various construction specialists and their subcontracted firms and suppliers to move on and off the site, do their work, and depart in time for the next subcontractor to begin work. The need for communication and frequent consultation among all participants in the construction project is obvious.

How Landscape Contractors Are Selected

There are three ways that landscape contractors can obtain work:
1. Direct solicitation of clients
2. Selection by the client for noncompetitive reasons
3. Selection by the client through the competitive bidding process

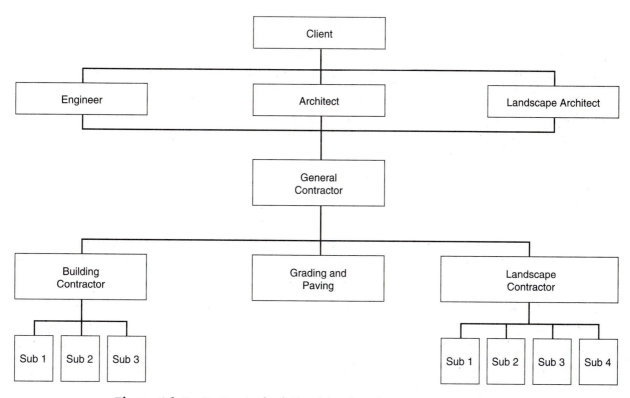

Figure 16-5 Contractual relationships for a large commercial project

Direct Solicitation

Rather than waiting for the telephone to ring or for a new customer to walk through the door, a landscape contracting firm has numerous options to exercise in attracting new clients. Small firms in thin market areas, new firms, and expanding firms are the most frequent users of this means of finding new jobs. Direct solicitation is used more commonly to seek residential contracting than commercial work; however, for an expanding company that desires to become known to architects and landscape architects, the direct approach also has value.

The direct solicitation of work takes different forms. They include such techniques as:

a. Media advertising, principally the newspapers, and the telephone yellow pages; to a lesser extent radio and television.

b. Blanket mailings to all property owners within targeted neighborhoods or developments. Such mailings will emphasize the past projects of the landscape firm and their experience and quality of work.

c. Participation in community flower shows and competitions that put the company's name before a public group with a common interest in gardens.

Figures 16-6 and 16-7 show examples of media advertising and direct mail advertising. Figure 16-8 illustrates the type of garden exhibit promotion that some firms use to display their names and talents before the public.

Noncompetitive Selection

Every landscape contracting firm aspires to attain a status within its market area that will assure it a full season of production without the need to advertise or compete with other firms for the same jobs.

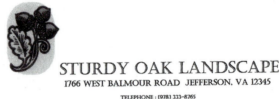

Ready or Not, Here Comes Spring!

Jefferson's largest and most respected landscape company is currently quoting for summer delivery of landscape construction, installation and maintenance services.

We will be pleased to discuss your requirements for decks, fencing, terraces, or other specialized construction features. We can also provide you with landscape lighting, automated irrigation systems, and imaginative water features. A full line of plant services, including complete installations, large tree transplanting, and total maintenance care is available.

Please give us a call to begin planning your summer landscape. It will be our privilege to serve you.

STURDY OAK LANDSCAPE
1766 WEST BALMOUR ROAD JEFFERSON, VA 12345

TELEPHONE : (978) 333-8765

Figure 16-6 An example of attitude advertising. It seeks to plant the seed of an idea into the minds of the readers and motivate them to contact the landscape company.

Advertising costs money, and competitive bidding requires time and money, while not always resulting in new jobs.

Noncompetitive selection of a landscape contractor by the client or the client's representative means that the contract specifications and the price for doing the job will be negotiated between all parties until a satisfactory agreement is reached. The negotiated contract is the preferred method of doing business because it eliminates or clarifies most areas of uncertainty before construction begins.

Much private landscape construction is done by firms that were personally chosen by the owner or landscape architect. Reasons for the selection of a particular firm vary. Frequently a firm is chosen because of the professional reputation it has established within its market area or within the industry of a region. Being known as a firm that does quality work, that begins and completes a project on schedule, that charges fairly and per-

forms ethically, is the essence of a good reputation. This is an excellent basis for noncompetitive selection by a client.

Another reason for direct selection by a client is the proven ability to perform the work, as evidenced by past projects. The client and/or his representative is able to anticipate the performance of the landscape contractor, and that gives his confidence an important boost by removing one uncertainty from the development process. As projects are worked on, the landscape contracting firm should be clearly identified with tasteful signs and efficient workers wearing uniforms that identify the firm. Projects should be professionally photographed throughout the construction period for later use in brochures, award competitions, and for working with prospective clients.

Being familiar to the landscape architect who will actually select the contractor is another way that many firms obtain contracts. Because the

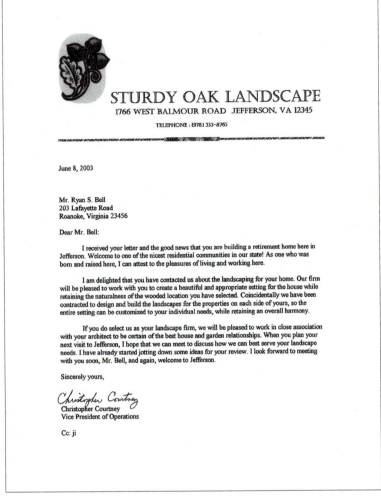

Figure 16-7 An example of direct mail advertising

landscape architect wants to do the best job possible, he or she may select the landscape contractor from a brief list of dependable firms who have done similar projects before. It is important for the landscape contractor to maintain good relations and frequent communications with such landscape architects and similar architectural and engineering firms. Keeping them aware of past and present projects that the landscape contractor is building

can cause them to call that contractor when they are involved in comparable projects.

Yet another reason why a landscape contractor may be selected to build a landscape without having to engage in competitive bidding is its proximity to the project site. Reduced travel time between the project and the contractor's base of operation can mean faster construction at less cost, at least as perceived by the client. Thus the landscape

Figure 16-8 A typical flower show display garden. Participation in such shows often generates new customers for landscape firms.

contractor who is nearby may have a decided advantage over one farther away.

These four reasons, reputation of the firm, past projects, familiarity to the landscape architect, and proximity to the project site, are all common reasons why landscape contractors receive work without engaging in the bidding process.

Competitive Bidding

In the building trades, most general contracts are awarded on the basis of competitive bidding. In landscape construction, many contracts are awarded through this process. When the project is in the public sector, competitive bidding is nearly always required by law. Large private projects are also commonly awarded from among an assortment of proposals.

Competitive bidding requires that each company seeking selection as the contractor on a project submit its proposal or bid in a specified form, at a specified time and place, and usually sealed in an envelope. The sealed bids may or may not be opened in public, and then the client or clients' representative evaluates the bids.

If the lowest bid does not exceed the amount budgeted by the client, then it will usually be accepted by the client. A contract will be offered to the successful bidder, and that firm will then become the contractor for the work specified.

Bids are solicited by clients in several ways. Owners of private projects may seek bids by extending "invitations to bid" to contractors who seem to have the capability of doing the work. Public projects are usually advertised in newspapers and through other media to assure that no one is denied the opportunity to bid.

The Advantages and Disadvantages

The advantages of competitive bidding are mostly on the side of the clients since they seek construction of projects at the lowest possible cost. However, bidders also benefit since partiality and other forms of unfairness are eliminated, particularly in public work. Competitive bidding easily identifies those firms that are not financially astute, since their bids are likely to be significantly higher or lower than the majority. Thus an additional advantage of competitive bidding is that it elicits a true market value for the project.

The disadvantages of competitive bidding are also borne disproportionately by the client. The requirements of the project may limit the number of firms that can qualify to do the work and submit a bid. Thus the client's options are reduced. Also, despite safeguards such as the pre-qualification of bidders, public clients cannot always be certain that the lowest bidder possesses all of the expertise or financial security needed to complete the project after the contract is signed. An additional disadvantage falls to every contracting firm that devotes time to the preparation of a bid proposal and does not get the contract. However, despite the disadvantages, competitive bidding is likely to continue as the major means of obtaining contracts for public project construction.

Achievement Review

A. List the components of a basic landscape contract.

B. Define the following terms:
    1. client
    2. contractor
    3. prime contractor
    4. subcontractor

C. Consider the following situation. Robert Jones is a homeowner desiring a sizable residential property development. He hires an architect to build the house and a landscape architect to design the exterior areas. Both professionals have contracts directly with the client. The architect engages the services of a surveyor and an engineer. The landscape architect and the architect agree on a general contractor to build the house and prepare the site for the landscaping. The landscape contracting firm engaged to install the landscape on the prepared site will report directly to the landscape architect, not the general contractor.

Diagram the above relationships and label each relationship as a *prime* contract or a *subcontract*.

D. List the three ways that landscape contractors are selected for projects.

E. What four reasons may cause a landscape firm to be selected without competitive bidding?

F. Compare the advantages and disadvantages of competitive bidding to the client. Make the same comparison to the landscape contractor.

CHAPTER 17

Contract Documents

Objectives:

Upon completion of this chapter, you should be able to

- explain the need for documents as part of a contract
- describe and distinguish between the numerous documents that may be part of construction contracts

The Need for Documents

A construction project typically involves many different firms. A landscape firm is normally involved in numerous projects simultaneously. It would be impossible for the work of one major project or one active company to be accomplished efficiently without all performance requirements, scheduling, and other terms of work written down for repeated referral throughout the project or season of operation. Without the commitment of the terms of the contract to writing, most projects would come to a chaotic standstill, with all parties on the telephone to their attorneys.

Some of the documents that are part of a contract are initially used during the bidding phase of a project. Once a contractor or subcontractor is selected, many of the bidding documents become part of the contract.

The Documents

While the building industry can generate scores of documents to accomplish a total construction project, the emphasis here will be on those documents that may affect a landscape contractor.

Standard Form of Agreement

Comprehension of this document necessitates a two-part understanding. First, as a form of agreement the document concisely states the relationship and obligations of the client and the major party or parties responsible for accomplishment of the project. The agreement is usually between the client and his architect, engineer, and/or landscape architect. It may also be used when direct contractual relationships are established between the client and the general contractor. It is a formal statement of the work to be accomplished, any supplementary conditions that are required, and the price agreed to. Second, it is standardized in that it follows a format recognizable to everyone working within the construction industry. That means that the document has proven fair and workable for all parties using it in the past. It further means that loopholes, weaknesses, and vague areas have been eliminated by prior users of the form, thus reducing the possibility of misunderstanding between the current parties. Standardized forms that have been used before by the parties of a contract are familiar in their content and arrangement. That offers the users the assurance that all components and implications of the agreement are understood.

There are several sources of standard forms of agreement and other standardized contract documents. The federal government and many state and local governments have prepared standard documents in an effort to treat all bidders equitably and to assure that various aspects of their laws are applied as required and without bias. Other sources include several important professional and business organizations. Foremost among them are the American Institute of Architects (AIA), the Associated General Contractors of America, the Engineers' Joint Contract Documents Committee, the American Society of Landscape Architects (ASLA), and the National Landscape Association (NLA). The standardized documents of the AIA enjoy the widest acceptance for private building construction projects. Those of the ASLA are most frequently used when a landscape architect holds the prime contract on a project. Small projects of landscape contractors commonly use one of the NLA forms.

The very nature of standardized forms suggests that there will frequently be provisions or requirements that are unique to a particular project and will not be addressed within the general wording of the form. In such cases, the standard form may be modified to include these special features, or it may refer to attached supplementary conditions that will detail them. Because the clauses that constitute these special features are not standardized, they require careful reading by all parties before the contract is signed.

Invitation to Bid

This document is an advertisement to all interested members of the construction industry that a project is seeking a builder. More common to public agency projects than to privately owned projects, the invitation to bid usually appears for a required period of time in newspapers, trade journals, or other media and is the government's way of offering equal access to the project by all qualified bidders.

Some variation in form exists for this document. Even the title varies, being called a "Presolicitation Notice" by the federal government, and a "Notice to Bidders" or "Advertisement for Bids" by other agencies. Regardless of the title, the advertisement usually gives a brief description of the project; states the deadline for bidding; the place where bids will be received and (perhaps) opened publicly; the location and cost of plans, specifications, and other contract documents; requirements for bonding; and other guidelines for bidding.

Less formal invitations to bid may be used by private owners who do not choose to advertise their projects publicly. Instead, the client may screen and select a few prime contractors deemed reputable and qualified to do the work and invite them to prepare bids for the project. This method offers the client the advantage of competitive bidding while minimizing the risk of an unqualified contractor submitting a low bid and then being unable to accomplish the project. Figures 17-1 and 17-2 are examples of public and private invitations to bid.

Drawings and Specifications

At the heart of every construction contract are the illustrations and written descriptions that describe the project in finite detail. The drawings will be prepared by the architects, engineers, and/or landscape architects working on the project. For landscape contractors the drawings of concern may include planting plans; construction drawings for all landscape items such as outdoor furniture, walks,

NOTICE TO BIDDERS

The Municipal Sewage Treatment Plant of the Town of Cobleskill will receive Bids for the landscape installation at the new entry on Draker Road until 7:00 PM on Monday, February 24, 2003, at the Town Hall. At that time all Bids will be opened and read aloud to all interested parties.

Contract documents in the form of written specifications and a 12-page set of drawings are available for review at the Town Hall. Individual sets of the documents can be purchased for $75.00. That is refundable if all documents are returned undamaged within one week of Bid openings.

A Bid Bond is required from the Bidder or an acceptable surety agent in the amount equal to 5 percent of the total Bid. This will serve as a guarantee that the Bidder will accept and execute the Contract if selected. Further, it will serve as a guarantee that the Bidder will post both Performance and Labor & Material Bonds within seven days of being awarded the Contract.

The Municipal Sewage Treatment Plant of the Town of Cobleskill reserves the right to accept none of the submitted proposals or to reject specific technicalities of a proposal that are not in the best interest of the Plant.

It is agreed that the officials of the Treatment Plant may not delay longer than 30 days from the date of the Bids for the purpose of reviewing the submitted Bids and investigating the qualifications of the Bidders before awarding the Contract.

Figure 17-1 A typical public invitation to bid, usually published in newspapers

fencing, lighting, and pavings; irrigation designs; and grading plans. The specifications provide written explanations of requirements for the project. They describe such things as the materials to be used in the construction, dates of beginning and completion, quality standards, and other factors that are not part of the drawings, but which are essential to the contract. The specifications are prepared by or under the supervision of the project's designers and are usually prepared in accordance with the format established by the Construction Specifications Institute (CSI).

Initially made available to all interested contractors during the bidding period, the drawings and specifications become part of the contract after the successful bidder is selected.

Bid Form

Both the AIA and ASLA have standardized forms for submission of bids. An example of a bid form is shown in Figure 17-3.

The document is most common to public bids where all bidders must be treated equally, and the lowest qualified bid is likely to be accepted.

Noncollusion Affidavit

This document may be required by owners of large projects as a means of assuring that bidders have not discussed the project prior to bidding and established a minimum bid among themselves.

THE IMOGENE SMALLWOOD HOSPITAL
LINCOLN, NEBRASKA

June 19, 2003

Mr. Rodney Ryan, VP Sales
At Your Service Landscaping, Inc.
1515 West Roosevelt Parkway
Lincoln, Nebraska 12043

Dear Mr. Ryan:

The Imogene Smallwood Association is requesting bids for the maintenance and upkeep of all surrounding grounds as specified in the attached Request for Proposal (RFP). The hospital desires to change from operating our own in-house grounds maintenance operation to having the work done by contract.

While the specifications attached represent the hospital's anticipated needs and quality standard, you (and the other vendors) are invited to take exception to any specific item and propose a better and/or more cost effective way of meeting our needs.

It is the hospital's intent to solicit meaningful proposals in order to give us a range of alternatives from which to choose. To facilitate this, we will be pleased to work with your organization in answering any questions concerning the requirements outlined in the RFP.

Certain key days should be noted. A bidders' conference is scheduled for 9:00 AM on Tuesday, June 24, 2003 in the hospital's board room. Vendor proposals must be submitted to the hospital on or before July 9, 2003. There will be no acceptance or consideration of bids submitted after that date.

Your response to this RFP, including cost data and documentation of your company's qualifications, will be our primary basis for final selection. Therefore we suggest submitting as complete a response as possible.

I look forward to the receipt of your proposal and to the development of a cordial professional relationship in the future.

Sincerely yours,

Frances A. Dickinson

IMOGENE SMALLWOOD HOSPITAL ASSOC.
Frances A. Dickinson
Administrator

Figure 17-2 A private invitation to bid, which is sent directly to specific landscape firms

BID SUBMISSION FORM

NAME OF BIDDER _____

NAME OF AUTHORIZED REPRESENTATIVE _____

PROPOSAL FOR SITEWORK PLANTING AT THE
STATE UNIVERSITY OF NEW YORK AT
COBLESKILL, NY

A. The work proposed herein will be completed on or before September 1, 2003. In the event that the undersigned fails to complete the specified work on or before that time or by the time to which such completion may have been extended in accordance with the Contract Documents, the undersigned agrees to pay to The University liquidated damages in the amount of five hundred dollars for each calendar day of delay in completing the work.

B. The undersigned hereby declares that all Bidding and Contract Documents have been carefully examined and that he/she has personally inspected the actual site of the work and accepts all quantities and conditions, and understands that in signing this Bid Submission Form he/she waives all future right to contest the Contract for reasons of misunderstanding.

C. The undersigned further understands and agrees to do, perform, and complete all work in full accordance with the Contract Documents and to accept as full compensation the amount of the Total Bid.

D. **TOTAL BID**

    a. Bid for supplies and materials $ _____

    b. Bid for work and labor, excluding
       the aforesaid supplies and materials $ _____

    c. Total Bid (Total of a + b) $ _____

Authorized Signature for the Bidder _____

Date of Signing _____

Figure 17-3   A typical bid form

Such consultation is termed *collusion* and is illegal because it denies the owner the truly lowest bid, which is the purpose of competitive bidding. By signing the **noncollusion affidavit,** each bidder certifies that the bid submitted was calculated without consultation with other bidders. An affidavit is as legally binding upon the signer as ver-

bally swearing before a judge in court. Figure 17-4 exemplifies a noncollusion affidavit.

Bid Bond

Consider this scenario. A client and his landscape architect have spent several months preparing their

Affidavit of Noncollusion

By submission of this Bid, the person signing on behalf of the Bidder swears under penalty of perjury that to the best of knowledge and belief:

A. The prices quoted in this Bid were derived independently and without collusion, consultation, communication, or agreement with any other bidder or competitor with intent to restrict competition.

B. The prices quoted in this Bid have not been knowingly disclosed by the Bidder nor will they knowingly be disclosed by the Bidder to any other bidder or competitor prior to the time of the announced Bid Opening.

C. No attempt has been or will be made by the Bidder to discourage any other person or company from submitting a bid and thereby restrict competition.

AUTHORIZED SIGNATURE FOR THE CONTRACTOR:

In compliance with and pursuant to Section 107-A of the General Municipal Law, I hereby affirm that the above statements are true.

Signed _____

Title _____

Company Name _____

Company Address _____

_____

Date of Signing _____

Figure 17-4   A noncollusion affidavit

project for bidding and waiting through the bidding period, eager to get the contractor selected and construction started. They have met with dozens of would-be bidders, walked the site repeatedly, and answered hundreds of questions about the project. Finally, the bids are opened, a bid selected, the winner is notified, and the losers turn to other projects. The owner and landscape architect urge the selected contractor to sign and return the contract; but the contractor hesitates and delays. Finally, the

contractor announces that he has changed his mind. He never really wanted the contract. He just wanted to gain some experience at bidding. Now the client is back at square one. The bidding process must begin again; the project is delayed; the client is both angry and inconvenienced.

A **bid bond** is a required security deposit that is submitted by a contractor at the time a bid is submitted. It serves as a guarantee that the selected bidder will honor the bid if selected. Required on nearly all large public projects and many private ones, the bid bond requirement is frequently stated as a percentage of the maximum bid price. The most common requirement is 5 to 10 percent, although on construction projects a larger bid bond may be required.

The bid bond is the most common form of bid security. Contractors who submit numerous bids requiring bonding will obtain the bond from a financial organization in business for that purpose. Known as a surety, the bonding organization charges the contractor a fee for the temporary loan of the money, which is like interest on a short-term loan. Companies that post bonds frequently do not want their capital tied up in security bonds, so they may pay an annual service charge to a surety bondsman to fund their bonds as needed.

Other forms of bid security may include negotiable securities, such as certified checks or cashier's checks. When a successful bidder is selected, the client retains the security until the contract is signed and required performance bonds are posted. The bid bonds of the unsuccessful bidders are returned soon after all bids are opened. Occasionally the bonds of the next lowest bidder(s) are retained as a safeguard until the contract is signed.

In the event that the chosen contractor declines to sign the contract or to meet all of the contractual requirements, the owner can recover some of the resulting damages from the defaulting contractor's bond security. Two types of bid bonds are used in the construction industry. One is known as a liquidated damages bond. With this bond, the owner claims the entire amount of the bond as the contractor's penalty for default. The other type is a difference-in-price bond. It allows the owner to withhold enough from the bond of the defaulting contractor to pay the difference between the defaulted low bid and the price of the next highest bid, up to the full value of the bond.

Performance Bond

Understanding a **performance bond** may be easier after considering another example. Consider the predicament of the client who has a signed contract for a project under construction, then is told that the project will not be completed because the contractor or a major subcontractor has gone bankrupt; or has accepted a larger, more profitable project; or has lost key, uniquely skilled personnel; or dozens of other reasons for default that may be the responsibility of the contractor. The damages to the client can be imagined. There may be a hole in the backyard where the swimming pool was supposed to be. The family reunion that was supposed to be held in the new backyard had to be moved to a park because the new landscape was unfinished. The grand opening of the business was less than grand with construction equipment and piles of mud greeting the guests.

The purpose of a performance bond is to provide security that can be drawn upon by the owner to recover damages resulting from contractor default after the project has started. The bond often covers any warranty period that is required by the contract, as well.

The performance bond must provide enough money to the owner from the surety to complete the contract if default occurs. Most performance bonds are for 100 percent of the contract price, and some require security of more than 100 percent to allow for increased prices should default occur. As with bid bonding, the actual cost to the contractor for providing the bond is the charge assessed by the surety source. Unlike the bid bond, which is usually held for only a short time by the client, the performance bond may be held in part or in total for a long period of time. The greater value of the bond and the length of the loan it represents makes the cost of the performance bond more expensive for the contractor.

Not all companies qualify for the same levels of bonding, any more than all individuals represent the same loan risk to their lending institutions. Only larger landscape construction firms, with reputations for stability and dependability, are likely to find surety support when large performance bonds are required. Requiring proof of bonding is one way of assuring the owner that only qualified bidders and contractors are working on the project.

Subcontract and Material Bid Invitations

These forms are similar to the invitation to bid forms described previously, except they are used by a contractor to solicit bids from subcontractors and material vendors whose services or products are needed for the project. An additional difference from the invitation to bid used by the owner or owner's representative is the less formal nature of these invitations. They may be as brief as a post-card announcement or a letter generated on an office word processor. The invitation will usually identify the project and the material or service for which a quotation is needed. It will also state the place where proposals should be submitted, and perhaps the person to whom they should be submitted. If drawings or specs are involved, there will be information about where they can be obtained. Information regarding other bidding documents may be included, and the deadline for submission of the price quote is always included.

Subcontract

This document confirms the contractual relationship between the subcontractor and the prime contractor. It may be a standard formal document or, with small projects, it may be written but less formal. Formal or informal, the subcontract must detail the agreements and obligations of both the prime contractor and the subcontractor. Since the subcontractor's agreement is with the prime contractor, not the owner, the completeness of the subcontract is as important as that of the prime contract, because it alone protects the subcontractor. The subcontractor has no legal line of appeal or obligation to the owner.

There is strong support among many subcontractors who regularly work on large projects for the use of a standard subcontract form. A form that is familiar because it has been used before and checked carefully by attorneys for both parties has less chance of presenting difficulties in the business relationship. A standard subcontract form is shown in Figure 17-5.

Notice to Proceed

This document is common to large projects. It is a written notice sent by the owner to the contractor, directing the contractor to begin work on the project. The contract will frequently specify that the contractor must begin work on the project within a specified number of days after receipt of the notice to proceed. The benefit of this document is that it relieves the contractor of liability in cases where the owner may not yet hold clear title to the property where the project is located.

Labor and Material Payment Bond

Contractors and subcontractors rely on other businesses to furnish materials and/or labor to support the work they are obligated to perform on a project. For example, a local electric company may be required to bring power to an undeveloped site to permit lighting and the use of power tools; a company that provides portable toilets or trash removal may be engaged in support of the project. While having nothing to do with general or landscape construction, their services to the companies involved are essential. In the event of default by the contractor or subcontractor to whom they were bound by contract, the supplier might not get paid. The owners could not be expected to pay because they were not a party to the contract. The labor and material payment bond provides third-party suppliers of utilities, fuels, telephones, equipment rentals, and/or labor with a surety-backed security that payment will be forthcoming soon after the supplier has furnished the last of the materials or labor that it was obligated to provide. Usually written notice of the fulfillment of the obligation is required.

Change Order

It is predictable that on many projects alterations in the original contract will be made. The owner or architect may have a new idea while the project is still under construction. The contractor may find it impossible to acquire an item believed available at

SUBCONTRACT AGREEMENT

This agreement is established this 25th day of October, 2003, by and between AT YOUR SERVICE LANDSCAPING, hereinafter called the **General Contractor** and GREAT LAKES POOL CONSTRUCTION, hereinafter referred to as the **Subcontractor**.

The General Contractor and the Subcontractor agree to the following:

SECTION 1   JOB SITE

The job site is the rear of the residence at 1234 Main Street in Lincoln, NE.

SECTION 2   DESCRIPTION OF SERVICES

a. The work to be performed is outlined in the attached specifications.

b. Subcontractor shall provide all of the labor, materials, tools, equipment, and supervisory personnel needed to perform properly the work described in the specifications. All work will be performed in a professional and safe manner in accordance with accepted industry quality standards as set forth by the American Pool and Spa Institute.

c. Subcontractor shall at all times keep the premises free of accumulated litter, waste materials, and other debris resulting from the work. Removal of such material shall be at the subcontractor's expense.

d. The timely completion of the work is essential and is an important element for full compensation. The work must be started on or before April 20, 2004, and completely finished no later than May 20, 2004. The Subcontractor may request in writing for an extension of times for delays attributable to weather or other uncontrollable events. Approval is at the discretion of the General Contractor.

SECTION 3   INSURANCE COVERAGES

a. Subcontractor agrees to carry and maintain at its expense these minimum insurance requirements:

- Workers' compensation and Employer's Liability
- Comprehensive general liability
- Comprehensive automobile liability
- Umbrella excess liability coverage of at least $1,000,000.

b. Subcontractor agrees to indemnify and save harmless the General Contractor against any and all liabilities, damages, losses, claims, demands, expenses, and actions including personal injury, death, or property damage that arise from any act of negligence by the Subcontractor, its owners, employees, or agents while engaged in the performance of the services described in this subcontract.

SECTION 4   STANDARD CONDITIONS AND PROVISIONS

a. Subcontractor shall act as an independent contractor, employing and directing all personnel required to perform the specified services, securing all required permits or licenses, and complying with all federal, state, and local ordinances, laws, rules, and regulations.

b. Subcontractor shall be liable for any damages to existing property on the site, property that is in its care, or property that is being installed until the General Contractor has fully accepted and paid for such work.

Figure 17-5   A subcontract agreement

c. Subcontractor has thoroughly evaluated the site, the work to be done, and the general work conditions. No additional payment will be made for alleged or unknown contingencies or difficulties. Subcontractor also agrees to all terms and conditions of the Contract that the General Contractor has with the Owner and the Owner's Agents.

d. No extra work or changes in the work or terms of the subcontract will be recognized or compensated unless agreed to in writing before the work is done or the changes made.

e. Subcontractor shall pay and discharge all costs and expenses including reasonable attorney fees that may be incurred or expended by the General Contractor to enforce the provisions of this Subcontract.

f. Should the Subcontractor fail to provide the contracted services herein described, the General Contractor may contract with another company to complete the work and the additional cost to the General Contractor shall be deducted from any sums due the Subcontractor.

g. Subcontractor shall guarantee all work performed for a period of one year.

h. Subcontractor shall not contact or deal directly with the owner of the project where the work is being done or with any of the owner's family or agents. All communication and correspondence shall be directed to the General Contractor who will solely deal with the Owner.

i. Subcontractor shall immediately, within 24 hours, give written notice to the General Contractor of any accidents, injuries, damages, or additional costs or charges incurred during and in performance of this Agreement.

j. Nothing appearing herein shall represent a waiver of any right that the General Contractor may have.

SECTION 5   COMPENSATION

Payment is to be made net 30 days after completion and acceptance by AT YOUR SERVICE LANDSCAPING and after receipt by same of a completed full waiver of lien and a detailed invoice.

SIGNATURES TO THE AGREEMENT

For the Subcontractor: For the General Contractor:

_____   _____
(Authorized Representative)   (Authorized Representative)

Print Name _____   Print Name _____

Title _____   Title _____

Telephone _____   Telephone _____

Date _____   Date _____

Figure 17-5   A subcontract agreement *(continued)*

the time of contract signing. Unexpected developments at the site may complicate the schedule. The possible reasons necessitating a change in some part of a contract are endless. The change order document provides for those changes after all parties involved in the contract have agreed upon the change. The **change order** describes the amendment of materials, methods, price, and/or time. It requires the signatures of all parties and the dates of their signing. It then becomes a part of the contract that it has changed.

In construction contracts for large projects, there is usually a clause that describes how changes will be handled. Most changes are initiated by the owner, and the contract clause describing change orders will typically allow the owner that prerogative, providing a fair price or time allowance is given to the contractor. When the owner initiates the change, he or his architect, engineer, or other professional representative must present a description of the change, including revised drawings and specifications to the contractor for review. This must be done before the change order is signed. If the contractor requires a contract change, he or she must present the change in writing to the owner and include revised prices and time schedules. This should be done before the change is initiated at the project site. Should the contractor carry out a change before all parties have signed the change order, the additional work may not be paid for. In Figure 17-6 a sample change order is illustrated.

Certificate for Payment

Large construction projects that require many months or even years to complete must provide for periodic payment to the contractor as the project proceeds. This permits the contractor to meet his payroll and pay his subcontractors and suppliers. The contract agreement will contain a payment clause allowing for partial payments as substantial portions of the project are completed. A request for payment will be submitted by the contractor to either the owner or his architect, engineer, or landscape architect. Following approval of the request,

the owner's representative will issue the certificate for payment to the owner, who will then send the amount due to the contractor within an agreed-upon time period. Should the owner not pay by the time required, the contractor has the right to stop work on the project. Neither the issuance of a certificate for payment nor payment by owner represent acceptance of any part of the contractor's work by the owner. The owner is still free to claim defective work or default by the contractor at a later date.

Certificate of Final Completion

Two things are accomplished when a certificate of final completion is issued by the architect or other prime professional. It certifies that the project is ready for occupancy by the owner and ready to be used for its intended purpose. It also usually means that the contractor is entitled to most of the money remaining to be paid him. Usually 95 percent of the contracted amount is paid. The remainder is held by the owner until all guarantees, claims against the project, and other contingencies are settled. This latter withholding is called *retainage*. Final payment will return the retainage to the contractor.

The Complete Contract

The complete construction contract will always include the agreement form and may include some or all of the documents described previously. There is an unofficial correlation between the size of the project and the number of documents that comprise the contract. Large projects commonly involve more complex contracts than smaller projects. With the complexity comes the need for more documents to accomplish the project smoothly and on schedule. Some small landscape contractors never need more than the basic agreement form to establish a legal business relationship with their clients.

Where there are numerous documents involved in the contract, the agreement form will contain clauses that refer to each of the documents. The supplementary documents then serve to amplify the clauses.

CHANGE ORDER

At Your Service Landscape Company
Your Full Service Landscape Professionals
1515 West Roosevelt Parkway
Lincoln, Nebraska 12043

Client Name   Change Order Number

Address   Date of Change

Telephone Number

Project Name and Address if Different   Job Number

Date of Original Contract

CHANGES TO BE MADE
All changes become part of and in compliance with the original contract which these changes amend.

It is agreed that the above changes will be made at this revised price $

Previous Contract Price $

Revised Contract Total $

Client signature_____ Date _____

Company authorized signature _____ Date _____

Figure 17-6   A typical change order

Achievement Review

A. Explain why there may be numerous documents comprising the complete contract for a construction project.

B. Indicate which contract document is described by the characteristics below. Select your answers by number from these choices.

1. Standard form of agreement
2. Invitation to bid
3. Drawings
4. Specifications
5. Bid form
6. Noncollusion affidavit
7. Bid bond
8. Performance bond
9. Subcontract and materials bid invitations
10. Subcontract
11. Notice to proceed
12. Labor and material payment bond
13. Change order
14. Certificate for payment
15. Certificate of final completion

_____ A standardized form used for the submission of bids, permitting all bidders to be treated equally

_____ Provides monetary security to the owner of a project if a contractor defaults after the project has started

_____ A document that binds the principal parties of a contract. The form is acceptable to all parties because it has been used for past contracts and the possibilities of misunderstandings are reduced

_____ Documents that may include planting plans and construction schematics for landscape items

_____ Confirms the contractual relationship between the prime contractor and a specialty firm doing work on behalf of the contractor

_____ Surety-backed security for third-party suppliers

_____ An advertisement to all interested members of the construction industry that a project is seeking a builder

_____ Advertisements used by contractors to solicit bids from subcontractors and material vendors

_____ A written notice sent by the project owner to the contractor authorizing the commencement of work on a project

_____ Written explanations of requirements for the project

_____ A means of assuring the owner that bidders have not discussed the project prior to bidding and established a minimum bid among themselves

_____ Certifies that the project is ready for owner occupancy and that the contractor is entitled to most of the remaining money due

_____ A required security deposit to guarantee that the selected bidder will honor his/her bid

_____ A document that allows for amendments to the original contract

_____ Issued to the owner by the architect, engineer, or other principal, the document authorizes partial, periodic payments to the contractor

Specifications

Objectives:

Upon completion of this chapter, you should be able to

- explain the concepts of project manual and Masterformat
- distinguish between standard and customized specifications
- explain the value and importance of uniformity
- identify the originators of specifications
- describe different methods of specifying
- list sources of specification data
- distinguish between closed and open specifications

As described earlier, specifications are the written instructions that describe how a project is to be built. Whereas the drawings for the project provide visualization and quantitative measurements, the specifications explain quality standards, describe materials, and establish the technical requirements of the project. Comparatively new to the construction industry, specifications have made an especially slow entry into landscape contracting. Until recently, landscape specifications were notes jotted down on the corners and edges of site plans and construction drawings. Even building construction relied upon this risky, but accepted, means of conveying the intent of the designers and engineers. However, recent years have seen growth in the use of synthetic building materials and in uninhibited litigation. The entry of so many new construction materials and products into the marketplace requires carefully prepared descriptions of materials that are acceptable to the client and the means of using or installing them. Equally significant in explaining why specifications have taken on such importance in contracting is the unparalleled increase in recent years of lawsuits brought by owners against contractors, contractors against owners, and subcontractors against contractors. Fault, liability, and damages are at the center of most contract lawsuits, and the courts frequently refer to the specifications for guidance in attempting to determine who did what to whom.

Today, specifications rival drawings in their importance to construction contracting. As computer technology develops, drawings may lose their importance almost entirely, even as quantitative measures. With the computer able to calculate quantities from material descriptions and dimensions entered into it by architects and engineers, the drawing may serve little purpose except to aid the client's visualization of the project. Even then, the drawing may be a computer-generated monitor image.

Progressing from the scribbled notations of 30 years ago to bound volumes of several hundred pages today, specifications require careful preparation by writers who are experienced in construction, who understand the materials and methods they are describing, and who are accomplished technical writers. Specification writing is no job for amateurs. The specification writer is an important member of the architect-engineer-landscape architect team. While one of the design professionals may prepare the specifications for a project, it is more common for him or her to work with a professional specification writer who either works as part of the office staff or is hired as an independent affiliate for the project.

Project Manuals

To aid in the organization of the paperwork and the dates, and comprehension of the chain of authority and responsibility, the documents pertaining to large projects are bound together in a **project manual**. The specifications are included in the manual, along with the bidding and contract forms. Frequently the manual is referred to as "the specs," which is a clue to the importance of the specifications to construction projects today, Figure 18-1.

A typical project manual is divided into four parts:
1. Bidding information and bidding documents
2. Contract forms (agreements, bonds, certificates)
3. General and supplementary conditions of the contract
4. The specifications

Masterformat

In an attempt to create order from potential chaos, professional specification writers have endorsed and initiated efforts to structure and standardize the mass of materials within each project manual. Although there is still no one system that is used by all branches of government and the private construction industry, one system enjoys widespread acceptance. **Masterformat** was developed by the Construction Specifications Institute (CSI) in counsel with Construction Specifications Canada (CSC), so the system is used throughout North America.

Project manuals that use the Masterformat system are first divided into two parts:
1. Bidding requirements, contract forms, and conditions
2. Specifications

All of the non-specification documents are included in the first part. The second part is the categorization of the specs into 16 divisions. Each division is further subdivided into sections called "broadscope," "mediumscope," and "narrowscope." Using key words and a simple system of numbering, each division outlines the tasks and materials common to that area of construction. The numbering system is standardized, meaning that the same five-digit number is used for the same item whenever the Masterformat system is used for preparation of the specifications in broadscope and mediumscope formats. Further detailing of the specifications into narrowscope format is not as structured. Masterformat permits the individual specification writer to choose specific narrowscope sections, usually either by adding more digits or by adding a decimal with more numbers after the decimal. Other narrowscope identifica-

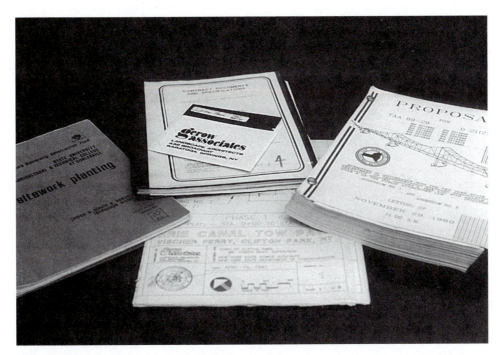

Figure 18-1 Project manuals include all of the contract documents, a set of plans, and frequently a computer disc of the details.

tion techniques include use of a letter after the five-digit number or some combination of decimal, letters, and numbering.

Standard and Customized Specifications

While Masterformat is widely accepted as the industry standard for project manual organization, it does not provide text material from which actual specifications may be written. It is the skeletal form upon which the specification writer builds the body of qualitative descriptions that are the specifications.

A number of commercially prepared guide specifications are available for use by professional specification writers. These guide specifications offer extensive pre-written descriptions of the materials and techniques used in the construction industry. The guide specifications range in their quality and completeness from those that are fully detailed and totally comprehensive (thereby allowing their use with little or no editing), to those that are general fill-in-the-blank types. The two most widely used systems of prepared guide specifications are Masterspec and Spectext. Masterspec is a product of the AIA, and Spectext is produced and distributed by the Construction Sciences Research Foundation. Of the two, at present Masterspec enjoys a plurality of use and acceptance among the many commercially available guide specifications on the market. Both Masterspec and Spectext are available in both paper version and in computer usable form. Both systems generally follow the naming and numbering system of Masterformat.

While most landscape contractors do not write specifications, it is still worth noting that the federal and state governments, through their many agencies, are the sources of many additional guide specifications. Few are equal to Masterspec or Spectext in quality or completeness, being written primarily for the type of work done specifically by each agency. Landscape contractors who bid on government jobs may encounter these specifications.

The use of commercially prepared guide specifications permits something approaching standardization in the writing of specifications. Specifiers are able to copy and edit entire sections or selected paragraphs from the prepared guides.

Some specification writers prepare their own guide specifications or use those developed by and for the offices in which they work. For firms that specialize in certain types of construction projects, office guide specifications have usability. They permit faster and easier editing in preparation of new specifications since the specifier is able to reuse large portions of the specs prepared for similar projects in the past. The greatest problem in the use of office guide specifications is keeping them current.

Most office guides are not written from scratch by the specifiers. Instead, they are started from one of the commercially prepared guides, then edited to exclude those sections that are not relevant to the work of the office. As the commercial guides are updated, the office guide specifications must be updated too.

Uniformity in Preparation of Project Manuals

The construction industry at large is moving rapidly toward standardized methods and materials. Manufacturers of construction materials are uniting in associations that publish specification standards for the best use of their products. Even growers of plant materials present their products in conformance with standards established by the American Nursery and Landscape Association. See Figure 18-2.

Masterformat has been an important contributor to standardized naming and numbering of specifications sections. Using the Masterformat system in association with accepted industry methods and quality standards, specification writers are able to prepare project manuals that present no surprises to builders and suppliers. The predictability and uniformity are highly desirable. They have established the level of competency that courts expect construction professionals to display.

When specifications are written using standards and methods that are not uniformly understood and accepted, several undesirable results may occur. First, there may be fewer bidders on the project due to the added time required to research costs from the non-uniform, non-conforming specifications. Second, the probability of an error in the bidding process is increased as a result of the specification of unfamiliar methods. Finally, the specifier who does not use standardized methods of construction assumes a high risk of legal action against himself due to the increased probability of errors after construction is under way.

The Writers

At one time specifications for projects were written by the architects, engineers, or landscape architects who designed the projects. This is still true for some projects. However, it is increasingly probable that the project manual's preparation and the actual writing of the specifications will be done by a full-time, professional specifier. While some specifiers work independently, most work as part of a larger multidisciplinary office. Working closely with the project designers, the specifiers must be experienced in as many aspects of the construction industry as possible. While competency as a technical writer is essential for a specifier, he or she must also know how things are built.

Methods of Specifying

There are different methods of specifying the requirements for a project. Most large construction projects use several methods of specification,

Figure 18-2 Industry trade organizations, such as the American Nursery and Landscape Association, establish their own standards for easy incorporation by specification writers.

although not within a single material or technique description. Specifications are usually technical, performance, or design specific.

Technical specifications are often used for portions of projects that are not easily verified for correctness after their construction has been completed. The specifications seek to provide written descriptions of the technical standards that the work must meet as construction proceeds.

Performance specifications allow the contractor much more freedom than do technical specs. Performance specifications establish the standard of performance or service that must be met by the feature being described. It is then left largely to the contractor to perform or provide the service to the standard established by whatever method he believes best. For example, the specifications might require a cantilevered deck to extend over an embankment and support a given weight and/or wind stress. While it would have to look like the plans described, the actual methods of construction, anchoring, and bracing would be left to the discretion of the contractor.

Performance specifications often result in lower construction costs, since each bidder is inclined to select materials and methods that he has used before and can fine-tune his bid more precisely.

Design specifications are nearly the opposite of performance specifications. All details regarding materials, methods, and performance standards are included in the specifications. Specific products may be cited by brand name and must be used by the contractor. The architect, engineer, or landscape architect is responsible for the finished product, allowing the contractor little or no autonomy on the project. The contractor is specified to carry out the directions of the design professional, down to the last detail.

Sources of Specification Data

Much specification data is available to the specifier from industry associations such as the American Institute for Timber Construction, the Brick Institute of America, and the American Concrete Institute. These organizations monitor the quality of their established products as well as the new products that are periodically introduced. The standards that the industry associations establish are reliable and are used with confidence by specification writers. Contractors must be familiar with these standards if they are to bid and compete successfully in the construction industry.

Additional sources of specification data include product manufacturers; professional associations; and federal, state, and local government agencies. It is not easy for specifiers or contractors to stay current with all of the new products and techniques of construction that enter the industry each year. However, staying up to date is critical.

Closed and Open Specifications

When the specifications require the use of specific products, identified by brand name, model number, manufacturer's name, and/or other unique characteristics, the specifications are termed **proprietary**. When the specifications do not allow for substitution with a similar product, they are termed **closed**. Even when a brand name is not used in the specifications, they are regarded as closed if the product specified can only be obtained from one source. Since there is a suggestion of discrimination when only one supply source can be used, closed specifications are not common in public contracts. They are more commonly used in private contracts when the owner or designer wants to assure a certain quality standard. Closed specifications tend to reduce competitiveness among would-be suppliers.

Open specifications permit substitutions after approval by the design professional or owner. As expected, the proposed substitute must meet the same quality and performance standard of the specified proprietary product. The desire to substitute for the proprietary product may stem from the availability of a comparable product at a lower price, or from the unavailability of the specified material.

When Changes Are Needed

Sometimes it is necessary to make changes in the specifications, drawings, or bidding documents even while the project is still being bid. When that happens, an addendum is prepared in writing and sent immediately to all known bidders. The addendum then becomes part of the project manual and eventually part of the contract documents.

Achievement Review

A. Define the following terms:

1. project manual
2. CSI
3. Masterformat
4. specifications
5. closed specifications
6. open specifications

B. What is the value of uniform material and method specifications?

C. Who prepares specifications, and what qualifications must they possess?

D. List and describe three methods of stating project specifications.

E. What sources do specification writers use for the uniform data they need?

F. In which of the following types of specifications would the described project requirements likely be found?

- technical specifications
- performance specifications
- design specifications

1. Lamp fixtures are to be Starkman Lighting Company Model 1226, rusticated finish, on 12' Starkman Pole Model 1226-D. No substitution permitted.
2. Concrete is to be of a mix that is 1/3 Portland cement, 1/3 coarse sand, 1/3 crushed stone, reinforced with heavy gauge wire.
3. The drive is to be of concrete and of strength adequate to the support of a 1-ton truck.
4. Flagstone is to be set in mortar atop a 6-inch layer of crushed stone that has been spread over a compacted subgrade.
5. The specimen trees in the entry planters are to be transplanted from elsewhere on the property. They are unique, one-of-a kind plants. The contractor is to install them in a way that will assure their survival.

G. Indicate if the following statements are true or false.

1. Specification writers are reluctant to accept the descriptions of product standards provided by industry trade associations.
2. Specifications that are written using nontraditional performance standards are likely to result in a greater than normal number of bidders.
3. If a bidder knows how to save the client money by using a product or technique that deviates from what is described in the specifications, he or she could propose that opportunity to the client if the specifications are of the *open* type.
4. It is not possible to change the specifications once a project has been let out for bidding.
5. Design specifications may also be proprietary.

Suggested Activities

1. Select a single landscape task that you are familiar with, such as installing a shrub or filling a flower planter. Then write a complete description of how to do that task exactly as you would do it. Prepare it for the use of someone else who would be doing the work. Try to be sufficiently thorough so that there will be no need for questions by the person trying to accomplish the task. If possible, have someone try to accomplish the task by reading your specifications. Each time that the person either must ask a question or gets something wrong indicates an omission in the completeness of your specifications.

2. Have two students sit back to back or separated by a divider so that they cannot see each other. Give both a comparable set of construction toys, such as Legos or Tinker Toys. Have one student build a moderately complex structure, unseen by the other student. Then have the student describe verbally to the other student how to construct the same structure having only the spoken explanation of how to do it. When finished, compare the two structures to see how closely they match. This exercise will help to illustrate the importance of both drawings and written specifications in accomplishing a project satisfactorily.

CHAPTER 19

Human Resources: The Needs and the Opportunities

Objectives:

Upon completion of this chapter, you will understand the following:

- the staffing challenges that the landscape industry must address
- the career opportunities available within the industry
- the personal and professional qualities that contribute to career success
- how best to match an individual with a specific company and position

Staffing: The Industry's Number One Problem

Few industries enter the twenty-first century with more growth potential or employment opportunities to offer people of all ages than does the landscape industry. As the diversity of services offered by landscape companies expands, there has been an accompanying need for a broader range of employee skills and training. Today's landscape industry needs horticulturists, artists, business managers, researchers, trainers, mechanics, technicians, builders, salespeople, office administrators, scientists, writers, and more. To attract the quality and quantity of personnel needed, the industry has taken cues from other professional groups to learn what it must do to recruit and retain the employees needed for the growth and development of specific companies and the industry at large. Increasingly, companies across the country are adding human resource specialists to their roster to help them find the best people for their organizations. Many companies support the training of key staff members to make them more effective recruiters during meetings with potential employees.

Yet despite great advances in the way that the industry recruits, its competitive wage and salary scales, and its provision of benefit and retirement packages, there is a severe shortage of people. If any one thing is restraining an even more rapid growth of the national landscape industry, it is the lack of trained personnel. There is no easy explanation of the reason. Nor has there been any definitive research to ferret out the cause of the problem so that a solution could be found. Conjecture suggests the possibility of the following reasons being at least partially responsible.

- *Poor public image* In an age where technology is admired and omnipresent, many perceive landscaping not as a profession, but as a low paying trade. A client willingly accepts a pricey charge for services from his or her physician or attorney, because doctors and lawyers are generally regarded as trained, educated professionals warranting high fees. However, even though the same number of years in school may be required or held by some landscape professionals, many clients do not see them as being comparable to those above. It is unlikely that people possessing a low-level image of the landscape industry will encourage or approve of their son or daughter entering the field.

- *Misleading educational programs* Despite their glossy brochures and trendy titles, many secondary and college "landscape" programs are still centered almost totally around a core curriculum of traditional horticulture. With an almost cycloptic emphasis on plant courses, these programs convey the image of landscaping as a craft, not a business. Other programs take a trade-school approach in which students spend their time building things, operating equipment, and developing an impression of landscaping as a trade, not a profession. Students who aspire to a career that is not just gardenesque or blue collar can be expected to look elsewhere at precisely the time in their lives when they are making career decisions and could be looking at the landscape industry. Too few educational programs present a total image of the landscape industry to their students, enabling them to see it as a business and a viable career option that has now stepped outside the traditional parameters of just working with plants and driving a pickup truck while getting a great suntan.

  Guidance counselors who have not stayed abreast of the industry and the excellent opportunities that it offers to all young people do additional disservice. Their failure to offer landscape careers as options for consideration by the students and recent graduates whom they advise is unfortunate.

- *Limited exposure to the industry* Many young people growing up in the small towns and suburbs of America are unaware of the size of the national landscape industry and the opportunities available. Local landscaping may be done by very small companies, homeowners, or part-time, untrained amateurs. With no occasion to observe an established firm and professional practitioners, a young boy or girl has no incentive to consider becoming a landscaper. One seldom finds landscapers portrayed favorably in the television or motion picture media. Here too it is the attorneys, physicians, and politicians who are the successful role models of so many story lines.

- *Perception as a low-tech industry* In an era where it seems that everyone carries a cell phone, uses e-mail, has a laptop computer, and can play his or her music of choice on a dozen different devices, the landscape industry can seem rather old-fashioned. When even fast-food counter staffs wear headsets and cooks read orders from monitor screens, landscape workers seem to be missing out on all of the fun. While most people would be surprised at how much the industry has embraced technology, the perception of those looking from the outside becomes their misguided reality.

- *Changing work ethic* Closely linked to the perception of the landscape industry as low-tech is the unwillingness of some people to seek careers that require long hours and hard, physical work. Many prefer doing something else that they see as requiring fewer hours and/or more intellectual skill than physical skill.

- *Nontraditional employees have been overlooked* The industry was slow to seek employees who were not young and male. It created and fostered the image of being an industry for men . . . *young* men . . . *strong* young men. While not deliberately overlooking women and older workers, employers made no effort to attract their attention or to provide opportunities for them. That has now changed, but it will require some time for staffing levels to reflect the new workforce.

The Career Opportunities

The methods of recruiting employees for landscape companies range from small ads in the local classifieds to nationwide searches. Budgets for recruiting have a comparable range starting at zero for some companies and soaring to thousands of dollars in other companies. In part, the recruiting technique used by a company is related to the level of employees needed. Positions for local seasonal workers do not warrant the same effort or expense used to recruit a branch manager. While not all-inclusive, the following recruiting techniques are currently used to bring qualified candidates for employment together with some of the leading companies in the landscape industry.

Someone who is interested in employment within the landscape industry would be wise to match the recruitment techniques with the level of position he or she is seeking at that stage of life. For example, a high school student looking for a summer job would most likely find opportunities listed in the classified ads of the local newspaper.

It would be pointless to respond to an ad in a national trade magazine or to consult the Web site of a distant company. The soon-to-graduate college student might attend recruitment fairs and discuss management opportunities with a number of actively recruiting companies, Figure 19-1. An older, experienced worker seeking to relocate or upgrade his or her position might register with a national recruitment firm that is hired by landscape companies to do a widespread search in their behalf.

Qualities Needed For Career Success

As the landscape industry evolves to ever-higher levels of professionalism, companies seek not only to recruit successfully, but also to retain employees once they are hired. There are not enough potential employees to support a revolving-door policy, and studies have shown that it is far more expensive for a company to recruit new employees than to retain existing ones.

One of the keys to employee satisfaction and improved retention has been a more careful eval-

Recruiting Technique	**Target Employee**
Classified ads in local papers	Seasonal, part-time, and/or unskilled laborers
Bonuses to current employees for recruiting their friends	Seasonal and/or unskilled laborers, crew staff
Classified ads in national trade journals	Supervisory personnel with education and experience
Career fairs at high schools	Entry-level, trainee positions
Career fairs at colleges	Entry-level supervisory field and/or management positions
Postings on company Web sites	Varied positions; both entry- and upper-level opportunities
Displays at national conventions	Entry-level supervisory field and/or management positions
Use of green industry recruitment firms	Varied positions at the middle management and higher levels

Figure 19-1 Recruitment fairs, held at schools or during industry conferences, enable company representatives to meet and speak with potential employees. Such events can be valuable first contacts for both the industry representatives and job seekers.

uation of candidates at the time of hiring. The objective is to match candidates with the job that they are prepared for both by training and by personal qualities. Many companies believe that employees' behavioral suitability is more significant to long-term success than their formal training. The belief is that the company can train an employee to do the work, but the basic attributes of a person's personality and behavior cannot be changed. Therefore, it is in the best interests of both the company and an applicant to know what qualities are required in an employee to assure his or her success and satisfaction in the workplace. Some of the qualities that landscape companies seek in employees include the following:

- *Strong work ethic* Landscape companies work long days and long weeks at peak seasons of the year. It is expected that employees understand that and accept it as a condition of employment without complaint.

- *Team player* Whether working on a crew or as a member of management, an employee will be expected to fuse his or her efforts with those of coworkers. It is essential that that fusion be cooperative and supportive, contributing to the accomplishments of the team and leading to the satisfaction of the customer.

- *Compliance with the company culture* Most successful companies have a culture. It may have evolved over time, or it may be spelled out in an employee handbook. Regardless, it becomes the persona and personality of the company. It may be manifested in the attire of the employees, the length or style of their hair, the presence or absence of facial hair, the condition and/or cleanliness of the company vehicles, a shared passion about their bowling league, or a hundred other things that are important to the long-term members of the company staff. The sooner that a new employee can grasp, understand, and accept the company culture, the sooner will he or she be accepted into the fold of the organization.

■ *Evidence of training or prior experience* Documentation of formal schooling or prior work experience is commonly summarized on resumes, in letters of recommendation, or through reference checks. Applicants need to describe their education, training, and experience in a complete and succinct manner that clearly illustrates how their backgrounds match with job performance expectations.

■ *Understanding of service* Landscaping is not a manufacturing business—it is a service industry. It is labor intensive, not product centered. Whether working in a customer's backyard or responding to a customer's questions on the telephone, every member of the landscape company contributes to customer satisfaction by the manner in which he or she performs the job. People who enjoy accommodating the needs of others and who can interact with customers in a positive and friendly manner are well suited for landscape employment, Figure 19-2.

■ *Cultural sensitivity* More than many industries, landscape companies are staffed with a large percentage of employees who are of Hispanic birth or heritage. At the crew level, many of these workers speak little or no English. They may not understand some of the traditions or customs of this country, just as some of theirs are unfamiliar to Americans. Successful employees will find

Figure 19-2 The impressions that are conveyed during the first meeting of a company representative and a qualified potential employee are often critical in helping each one decide if it is worthwhile taking the contact to the next level.

ways to bridge the cultural differences within the workplace. Highly valued are employees who are bilingual, speaking fluently in both English and Spanish.

■ *Computer literacy* Even the smallest landscape companies are using computers. Throughout the industry, computer technology is used to create designs, estimate costs, generate and store employee information, bill customers, and do dozens of other tasks. From the executive office to the field foreman, the computer is omnipresent. No applicant should be without an understanding of computer technology.

■ *Personal and career goals* Most successful companies want to know what their key employees are seeking from their careers. Only then can a company chart a career path for an employee that will assure his or her long-term service to the company. Goals can encompass such diverse objectives as

- type of work that will give the employee the greatest satisfaction
- degree of leadership or responsibility that the employee hopes to have within the company
- location of the branch he or she hopes to manage
- salary expectations
- opportunity to lead the company into a new market
- movement out of the field and into the office or vice versa

It is unlikely that a successful employee can start work for a company with no thought of what he or she is seeking from the job and looking only as far as the next payday. The company is likely to leave that person in the same job for a long time if only because they do not know if or where he or she wants to go.

- *Intellectual curiosity* Graduation from school is only the beginning of what must be a lifelong commitment to learning. Those who desire success in the landscape profession must stay abreast of new techniques, technologies, products, and ways of doing business. Attendance at local and national conferences for the green industry, reading the published results of research and product introduction in trade journals, and seeking periodic advice and assistance from the local Cooperative Extension Service are just some of the ways that landscape professionals continue to learn. Even new applicants should convey their commitment to ongoing learning at the time they are discussing employment with a recruiter.

- *Patience* The landscape industry is conservative. No matter how much education and prior experience a person brings to a new position, most companies want to watch and wait to see how the new employee is performing before making a commitment to that employee's advancement within the company.

Salaries may start lower than the applicant desires, but once the employee is determined to fit well within the company, salary improvement is usually forthcoming.

New employees, especially those who are on their first job out of school, need the same patience. Advancement may not come as fast as desired. The supervisor may not seem as appreciative of the new employee's performance as his or her teachers had. A reprimand from the foreman may seem harsher to the new employee than was intended. The new town may seem unfamiliar and unfriendly to someone who has relocated. If the new employee goes into a new position with a personal commitment to stay the course for at least three years and not pack up and go even if that is the first impulse, then often the *crisis du jour* will not be as stressful.

Successful Matchup: Creating a Win-Win Relationship

The hallmark of successful employment, as measured by both the employee and the employer, is a contented employee performing for the company at a level of competency that permits the company to reward him or her, thereby sustaining the employee's loyalty and continuing performance. To do that requires a careful matching of the employee's training, goals, and personality with the technical and personal requirements for successful accomplishment of the job. Matching the *technical* skills of an employee with the task requirements of a job is the easiest to measure and understand. If a job requires the employee to know and be able to do certain things, an applicant for the position either does or does not have those skills. Granted, there are degrees of competency, enabling one person to perform the job more or less skillfully than another person, and those variances in competency are measurable. Matching the *personality* of a job with the personality of an applicant is not as easily measured. Only recently landscape companies have begun to recognize the importance of matching job needs with employee personalities. For example, landscape sales require someone who enjoys meeting people, who can convey a quality of sincerity and inspire trust when dealing with a client, who can be warm and positive with consistency, and whose appearance, manners, and language skills convey a positive image of the company to clients. It is not the job for someone who prefers to work alone, or who is uncomfortable in a coat and tie or business suit, or who becomes impatient with customer indecisiveness. An employee with those traits might be more successful in a shop or in a field supervisory position where there is less direct contact with the clients and decisions can be made more easily and quickly. As another example, an employee with an assertive personality might be less effective as a team member than as an innovator. He or she might do an excellent job breaking ground in a new market region or developing a new service line for the company where daring and decisiveness are needed. Putting a person with that personality into a position requiring a high level of sensitivity to the concerns and needs of other employees or customers could be a mistake.

Young people as well as experienced older workers will find an increasing number of landscape companies trying to recruit employees who are behaviorally suited to the jobs being filled and to the overall culture of the company. It has made the job of staffing a company more difficult, but it is contributing to improved retention of employees once hired. It is to the advantage of both the industry and those who seek careers within the industry to support this matchup of technical skills and worker personality with the requirements of the job.

Developing Citizenship Skills

Community service activities help members learn to develop and practice good citizenship and leadership abilities needed now and in the future. The focus is on making the community a better place to live and work. Personal development can be analyzed in terms of desirable personal characteristics, accepted social behavior, and good citizenship.

Citizenship development helps members become informed of civic responsibilities, such as voting, paying taxes, and abiding by the laws of society. Community service activities are often designed to develop interested, experienced, and knowledgeable community leaders and citizens. They seek to develop a sense of pride in the community and the initiative to make it a better place to work and live.

Measurable skills are ability to plan and establish community service projects, develop a respect for national symbols and customs, develop an understanding that "if we belong, we pay dues," respect the rights and views of others, and cooperate with others in group activites.

Achievement Review

A. For each of the landscape positions listed below, prepare a comparative listing of (a) those personal characteristics that you believe a successful employee in that position would have and (b) a numerical score or letter grade indicating how you believe you would perform in that job based upon your own personality.

  1. landscape sales
  2. construction field supervisor
  3. maintenance field supervisor
  4. landscape estimator
  5. human resource specialist

6. employee trainer
7. safety officer
8. landscape designer
9. landscape teacher
10. company president

B. Make a list of your personal goals that you will seek to accomplish with your first five years of employment. Explain why each goal is important to you.

C. Write an employment ad for each of the positions below. Try to convey as clearly and concisely as possible what the responsibilities of the position are and what type of person is being sought. Limit each ad to 50 words or less.

1. landscape sales
2. maintenance crew foreman
3. seasonal summer worker for the maintenance crew

D. Write a letter to the Ajax Landscape Company, 123 Main Street, Your town, Your state. In the letter, express your interest in one of the positions and explain why you believe you are suited for the job.

Suggested Activities

1. Prepare a resume that can be used when applying for a job in the landscape industry. Include the following information:

   - name
   - address and telephone number
   - e-mail address
   - educational background
   - work experience, beginning with the most recent and working backward
   - special skills
   - organizations and activities
   - references

2. Invite a human resource specialist, preferably from a landscape company, to speak with the class about the best way to do a job search, including how best to screen the job possibilities, how to get an interview, how to get the most from the interview, and what to do after the interview.

3. Practice interviewing. Have students take turns playing the role of applicant and employer. Focus on questions and responses that will elicit and convey the most information in the least time.

4. Have a discussion concerning students' opinions about the landscape industry. What were their initial impressions before beginning their study? What are their current impressions? What are their parents' impressions? Their friends'? If their impressions are different now, what is responsible for the change?

CHAPTER 20

Installing Landscape Plants

Objectives:

Upon completion of this chapter, you should be able to

- identify the tools used in the installation of landscape plants
- condition soil used in the installation of landscape plants
- describe the advantages and disadvantages of bare-rooted, balled and burlapped, and containerized plant material
- select the best season for transplanting
- outline procedures for the installation of trees, shrubs, groundcovers, bedding plants, and bulbs
- describe the advantages and disadvantages of organic and inorganic mulches
- explain the benefits of antitranspirants
- describe installation problems unique to the American Southwest and Southeast

The Importance of Proper Installation

High-quality landscapes begin with top-quality plant materials. Both depend upon careful installation techniques to assure the survival and growth of the transplanted stock. Landscape contractors joke about $25 plants set into $75 holes. In fact, a great deal of labor and materials are often needed to prepare a hostile planting site for a new plant.

Few sites offer a perfect combination of proper soil texture, fertility, and pH with correct drainage and optimum water and humidity throughout the post-transplant period. All are necessary for successful transplanting.

The Necessary Tools

Since the plant material to be installed includes seeds, bedding plants, groundcovers, and trees and shrubs of all sizes, a wide range of tools must be available to accomplish the installation. The hand tools most commonly used are shown and described in Appendix B.

As plant materials increase in size, hand tools must be supplemented with power tools. The tree spade makes possible the successful transplant of large trees and shrubs, Figure 20-1. It operates on a hydraulic system, with each movement controlled from a set of levers that permits the machine to be operated by anyone, regardless of physical strength. Other power tools helpful in the installation of landscape plants include the power auger, power tiller, tractor, and front-end loader.

The Soil for Installation

It is best to use existing soil of the site to assure the most successful transplanting of trees and shrubs. Research has shown that plant roots adapt faster and grow deeper under those conditions. However, the

Figure 20-1 Large trees can be moved successfully with powered transplanters. There is limited disturbance to the root system. (Courtesy of Vermeer Manufacturing)

soil removed from a planting hole may be unsatisfactory in its unaltered state for use as replacement (or backfill) soil after the new plant is set. It may be too heavy with clay and need the addition of sand and peat moss to provide better aeration and flocculation (aggregation of soil particles). There may be too much sand, requiring addition of humus to retain moisture around the new plant's roots. Landscapers may have to deal with the hard caliche layer if in the Southwest, salt-saturated soil next to roadways and walks, construction debris buried by builders, and the natural stoniness of rocky regions. Rich loam with ideal pH and good drainage is not common on most sites, where the landscaper is usually one of the last developers to be called in.

The soil to be filled into the planting hole must provide a medium in which the root system of the new plant can resume growth and develop fibrous root hairs to absorb water and nutrients for the new plant. If the soil does not drain adequately, the new plant may die from a lack of oxygen. If the backfill is too sandy, the new roots of the plant will stay within its soil ball and not grow out into the new soil.

Correct soil for plant installation has these qualities:

- Loamy texture (near equal mix of sand, peat, and soil)

- Good drainage

- Suitable pH

- Balanced nutrients

These qualities may be attained by blending conditioners such as sand, peat moss, compost, leaf mold, and manure with the soil before backfilling. In some situations, the original soil may be so unsatisfactory that it must be completely replaced, although recent research suggests that this is seldom necessary. In large landscape installations, the conditioning needs should be determined in advance so that necessary quantities of the additives can be ordered and available at the planting site. All members of the planting crew should be instructed in how to prepare the backfill mix. Periodic checks by the crew supervisor will assure that the new soil is mixed correctly and uniformly.

Root Forms of Landscape Plants

Landscape plants are available in a variety of root forms. Bedding plants and groundcovers are usually grown in pressed peat pots or plastic packets that permit the root system to be transplanted intact. Trees and shrubs can be purchased as bare-rooted, balled and burlapped (B & B), or containerized plants. The advantages and disadvantages of each are compared in Figure 20-2.

Which root form is best to use depends upon the season of the year, the availability of stock, the size of the plants at the time of installation, and the budget of the project. Bare-root is a common root form for deciduous shrubs and a few trees that develop new roots quickly after transplanting.

Root Form	Advantages	Disadvantages
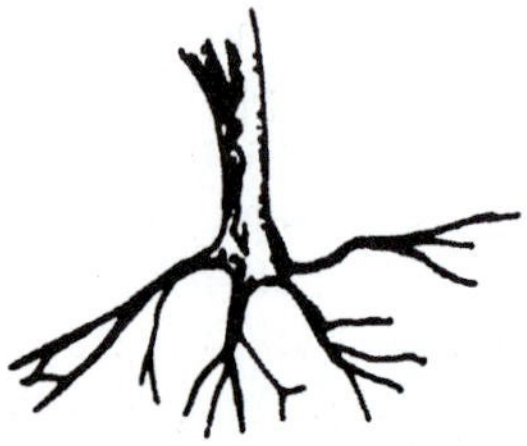 Bare-rooted	• Comparatively inexpensive • Lightweight and easy to transport • Dormant at the time of planting	• Severely reduced root system • Transplant season limited to early spring • Usually small, requiring time to mature
 Balled and burlapped	• Larger material can be transplanted • Less damage to the root system • Can be transplanted throughout spring and fall	• Usually the most expensive • Soil ball adds weight and bulk • For large plants, costly installation equipment is required
 Containerized	• Less expensive than B & B material • Root system intact • Can be transplanted throughout spring, summer, and fall	• Seldom available in large sizes • Can become root-bound if kept in containers too long

Figure 20-2 Root forms of landscape plants

Evergreens are most often balled and burlapped or containerized. Deciduous trees and shrubs may also be obtained in B & B or containerized forms. Vines are usually containerized.

The Time to Transplant

The best season for transplanting depends upon the type of material being planted. Usually, the prime objective is to transplant at a time that will permit good root growth before shoots and leaves develop. For most plants in most parts of the country, that time is early autumn. Then the roots can grow as long as the soil remains unfrozen, while the cool air temperatures encourage the above-ground parts to go dormant. Early spring, when root growth exceeds shoot growth, is the second best season. Summer is not a good season unless containerized material is used, with its intact root system. Winter is not a good season for transplanting in northern regions because the roots cannot grow. In regions where the winter temperatures are milder and the ground doesn't freeze, winter can be a satisfactory alternate season for transplanting.

Flowering bulbs have definite transplant seasons. Hardy bulbs, which bloom in the spring, must be planted in the fall. Tender bulbs, which flower in the summer and will not survive the winter, are planted in the spring, dug up in the fall, and stored indoors over winter.

Annual flowers, purchased as bedding plants, are transplanted in the spring after all danger of frost is past. In regions where the fall season is long and mild, a second planting of cool season annuals may occur after the hot temperatures of summer are past.

Methods of Installation

Trees and Shrubs

Assuming that the plants being installed are suited to the site and in good health, the next most critical factor in their success is proper planting procedure. The ultimate objective of correct installation

is to minimize transplant shock (stress) and return the plant to a normal state of growth as quickly as possible. Differences in technique based on regional geography are minimal. While those differences will be noted here, the basic methods of installation described are nearly the same throughout the country. Variations in technique are more related to the root form of the plants and whether they have a spreading root system or a tap root than anything else.

Bare-Rooted Plants. These plants are likely to be small and deciduous at the time of their transplant. They should be installed in planting pits that are either flat bottomed (for plants with a tap root) or mounded (for plants with a spreading root system), Figure 20-3. Roots should be spread out from the center and over the mound to approximate their natural habit of growth. The planting pit needs to be wide enough to avoid cramping the roots, and the earth mound should be firmly packed to prevent settling. Settling would allow the plant to sink below grade level and lessen its chance of survival. There is usually a soil stain on the bark of the plant to indicate the level at which it was growing before harvest. It should not be planted deeper than that level when transplanted.

Containerized Plants. The first thing to be done when installing containerized plants is to remove the container and determine the condition of the root system. The plant may have been harvested as a bare-root plant and only recently placed in the container, not having sufficient time to produce new roots. It needs to be installed like a bare-root plant. Conversely, the plant may have been in the container too long, causing its roots to grow around themselves. This condition is termed *potbound*, and the plants will have to be either unwound before planting or given several vertical cuts through the root mass to induce new root growth that is oriented properly, Figure 20-4. Another technique used to overcome the **girdling** tendency in some containerized plants is *butterflying*. It requires the soil ball to be split halfway up its center so that the root system can be spread apart

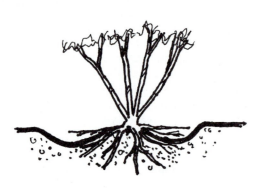

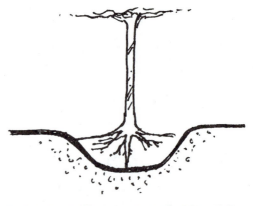

INSTALLING BARE-ROOT PLANTS WITH FIBROUS, SPREADING ROOT SYSTEMS. DISTRIBUTE THE ROOTS OVER A COMPACTED SOIL MOUND.

INSTALLING BARE-ROOT PLANTS WITH A TAP ROOT SYSTEM. THE PIT MUST BE DEEP ENOUGH TO PERMIT THE TAP ROOT'S NATURAL EXTENSION. STAKING MAY BE REQUIRED.

Figure 20-3 Installing bare-root plants

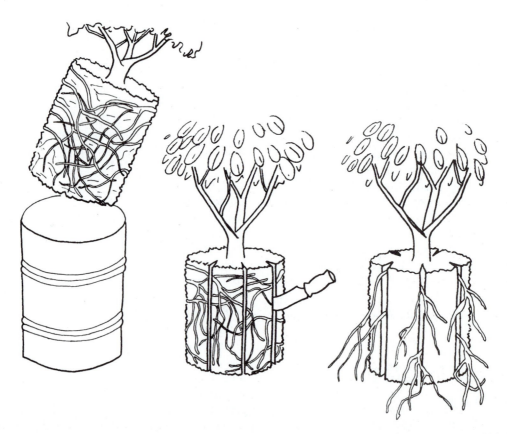

Figure 20-4 Containerized plants can become pot-bound. Before transplanting, they should be removed from the container and the root mass cut vertically at 2-inch intervals to promote new root growth that is oriented properly.

(like butterfly wings) when placed into the planting hole. If the potentially girdling roots are in the top half of the root mass, butterflying will not help. However, where applicable, the technique can promote a more natural root orientation in the new transplant. Ideally, the plants will have been grown in the container for a proper period of time and will be ready to place into their flat-bottomed pit with minimal disturbance to their established root system. Even with containerized plants that are not pot-bound, it is good to loosen the roots a bit before placing them in the pit.

Balled and Burlapped Plants. Nearly all large trees and shrubs are balled and burlapped at the time of their harvest. Some will be wrapped in biodegradable burlap that need not be removed when planted. Some will be wrapped in treated or synthetic burlap that should be removed as much as possible after the plant is set into the pit. Either type of burlap may be secured with non-biodegradable rope or twine that must be removed once the plant is in the pit. Not removing the rope or twine could result in the eventual girdling of the plant, if the ties encircle the trunk of a tree or the crown of a shrub. Girdling prevents nutrients from moving from the roots to the canopy of the plant, causing it to starve and die. Very large balled and burlapped plants may be placed in wire baskets at the nursery to prevent the soil ball from breaking. If it is possible to do without disturbing the soil ball, the basket should be cut away before planting. If that is not possible, then minimally the upper rows of wire should be cut away once the plant is positioned in the planting pit.

The planting pit for balled and burlapped trees and shrubs should be flat bottomed. It should be no deeper than the height of the root ball in order to allow the plant ball to rest on a solid base. There should be no allowance for soft backfill beneath the soil ball—otherwise it will settle, making the transplant too deep, Figure 20-5. In areas of the country where drainage is poor and/or the soil is heavy with clay, it is recommended that the depth of the planting pit be *less* than the height of the soil

ball to permit 3 to 5 inches of the soil ball to extend above grade. Pit width is a much different story. The intent is to assure that the roots grow horizontally out from the plant and eventually into the surrounding undisturbed soil. Accordingly, the planting pit should be at least twice as wide as the soil ball, and on sites where the soil is compacted, the planting hole should be increased to three or more times the diameter of the soil ball. The same applies to the width of trenches used for hedge plantings.

Correct Backfilling. *Backfill* is the soil returned to the planting pit after the plant is set. The soil *interface* is the place where the backfill meets both the soil ball or roots of the transplant and the sides of the planting hole. If the backfill soil is significantly different from either the soil in which the plant has been growing or the surrounding soil in texture, moisture content, aeration, or added amendments, the transplant may be negatively affected. For example, if the backfill is amended to become finer textured than the soil in the root ball, it creates an underground sponge that quickly pulls water away from the root ball and holds it where the plant cannot reach it. If the backfill is more conducive for root growth than the surrounding soil, many root systems will not develop properly. They will stay within the planting hole and not spread wide and deep as they should. Therefore, for most planting situations, the backfilling should either be done using soil removed from the hole or with soil that closely matches the original soil in texture and other attributes.

As backfill is added, care should be taken to avoid leaving air pockets within the planting pit. Bare-root plants should have the soil worked under and into the root mass to eliminate air space. The persons doing the installation should use their feet to tamp down the soil around the soil ball of containerized and B&B materials. Just enough pressure should be applied to eliminate the air pockets while avoiding compaction of the backfill. Balled and burlapped plants should have the fabric being buried tucked down into the pit. If permitted to

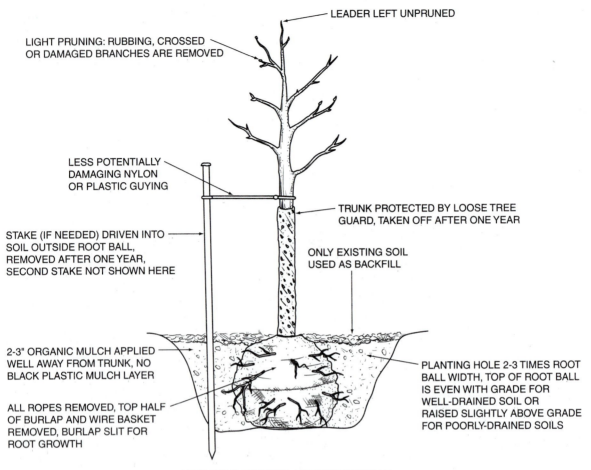

LEADER LEFT UNPRUNED

LIGHT PRUNING: RUBBING, CROSSED OR DAMAGED BRANCHES ARE REMOVED

LESS POTENTIALLY DAMAGING NYLON OR PLASTIC GUYING

STAKE (IF NEEDED) DRIVEN INTO SOIL OUTSIDE ROOT BALL, REMOVED AFTER ONE YEAR, SECOND STAKE NOT SHOWN HERE

TRUNK PROTECTED BY LOOSE TREE GUARD, TAKEN OFF AFTER ONE YEAR

ONLY EXISTING SOIL USED AS BACKFILL

2-3" ORGANIC MULCH APPLIED WELL AWAY FROM TRUNK, NO BLACK PLASTIC MULCH LAYER

ALL ROPES REMOVED, TOP HALF OF BURLAP AND WIRE BASKET REMOVED, BURLAP SLIT FOR ROOT GROWTH

PLANTING HOLE 2-3 TIMES ROOT BALL WIDTH, TOP OF ROOT BALL IS EVEN WITH GRADE FOR WELL-DRAINED SOIL OR RAISED SLIGHTLY ABOVE GRADE FOR POORLY-DRAINED SOILS

NOTE SOLID PEDESTAL HOLDING ROOTBALL

Figure 20-5 Key points in tree and shrub installation

extend above the backfill it will serve as a wick to evaporate water away from the plant roots and promote undesirable drying.

Watering can be done as the backfill is added, but the creation of a muddy pit that will encase the root system in a glazed shell after drying should be avoided. For trees and larger shrubs, the excess soil can be used to create a mounded ring over the root zone to catch and hold water. In large landscapes, such as golf courses, parks, or highway plantings, the soil ring may determine whether the transplants survive during the first year when moisture

is especially critical. In northern regions, these soil rings should be removed at the onset of winter to avoid repeated freezing and thawing that can injure the plants at their base.

Staking and Guying. Once installed, a tree or shrub may need additional safeguards to counter the effects of strong winds, protect it from vandals or maintenance equipment, or stabilize it in sandy soils until it can support itself. Staking or guying may be used to provide those safeguards. **Staking** uses one, two, or three wood or metal stakes driven into the ground parallel to the trunk of a tree and down into solid, undisturbed soil. The stakes should be long enough to extend about 8 inches above their point of attachment to the tree. The point of attachment should be about 6 inches above the highest point where the trunk can be bent by a strong wind yet return to its upright position.

The stakes can be attached to the tree in several ways. The traditional way has used heavy wire looped through links of hose at the point of contact with the tree, then attached to the stakes and tightened by twisting the wire at its center until the desired tautness is attained. Although still used, the wire and hose link technique has been shown to cause injury to some plants when repeated movement by the wind causes the hose coverings to wear away the bark. That may result in a girdling condition. Currently gaining favor and acceptance are commercial ties that replace the wire and hose links. Made of tough, weather resistant webbing, the ties are flat and wider than wire, so they do not gouge the bark. They also have a slight elasticity, providing a more flexible support to the plant than the wire and hose technique. Since the webbing material cannot be twisted like wire, it should be applied with a figure-eight loop between the stake and the tree to allow for flexibility.

Using a single stake is most appropriate when the intent is to prevent a plant from being pushed off center by a strong prevailing wind. In such a case, the stake is positioned on the upwind side of the plant. For the stabilization of small trees, two stakes are often used. They are placed on opposite sides of the trunk and attached to the same point on the tree, Figure 20-6. Securing the trunk at additional levels along the stakes should be avoided because it creates additional stress points when the tree moves in the wind. However, the two stakes can be reinforced at the base with a wooden cross tie to prevent their wobbling and coming loose. On windy sites, three stakes may be used for added stability. Three stakes also offer

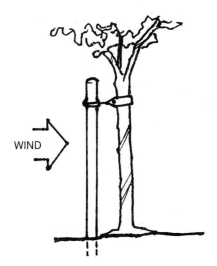

WHEN ONLY A SINGLE STAKE IS USED IT SHOULD BE PLACED ON THE UPWIND SIDE.

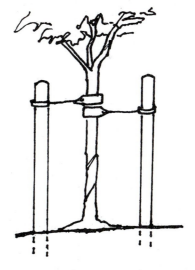

WHEN TWO STAKES ARE USED, THEY ARE PLACED ON OPPOSITE SIDES AND ATTACHED AT THE SAME POINT. A WIDE, FLEXIBLE MATERIAL SHOULD BE USED TO AVOID CUTTING INTO THE TRUNK. STAKES ARE REMOVED AFTER ONE GROWING SEASON.

Figure 20-6 Staking techniques

greater protection from mower damage and may dissuade vandals.

Guying is the stabilization of large trees and multistemmed plants using the hose wrapped wires or strapping material described above, but attached to anchoring devices that are driven into the undisturbed soil. If stakes are used as the anchors, they should be driven at a 45-degree angle to the ground surface and point toward the tree. If driven to point away from the tree, they can become loosened over time, Figure 20-7. Usually three guys are needed to provide the desired stability. For exceptionally large trees, underground anchors, termed *deadmen,* may be used to ensure that the guys do not work loose, Figure 20-8. Support wires can be twisted to the desired tautness or turnbuckles installed to accomplish the same thing. For safety, all exposed wires should be flagged with colorful tape so that no one trips over them. Whether a plant is staked or guyed, all apparatus should be removed after one year to avoid injury and prevent girdling.

Protecting Thin-Barked Trees. Some young trees are especially susceptible to injury during the first year after transplanting if they have thin bark that

can be scalded by the sun or dried out by winter winds. Still other young trees are damaged by the feeding of rabbits and rodents or the gouges of lawn mowers. Vinyl tree guards can be applied after installation to protect against those types of injury. They coil around the trunks from the base upward, Figure 20-9. To avoid girdling, the coils should be removed annually and reapplied if necessary. That way they do not impede the plant's growth.

Groundcovers and Bedding Plants

Both groundcovers and bedding plants are commonly sold in strips of plastic or pressed peat moss pots or in flats. Certain plants such as geraniums or flowering perennials may also be marketed in small clay or plastic pots. Peat pots need not be removed, but other containers must be. Remember that the rim of the peat pot must be buried in the soil to prevent the wick-drying effect.

To install these plants, the entire bed is prepared rather than individual holes. This allows the groundcovers or flowers to be planted with a hand trowel or hoe rather than a spade, Figure 20-10. The soil must provide good drainage and nutrients. It should also be as weed-free as possible at the time of planting. For large areas, a garden tiller may be used to loosen the soil and incorporate the necessary conditioners. The use of a preemergence herbicide prior

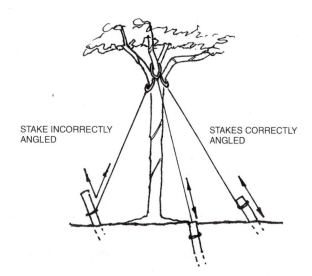

Figure 20-7 Guying a tree: Stakes must be driven at a 45° angle and point toward the tree.

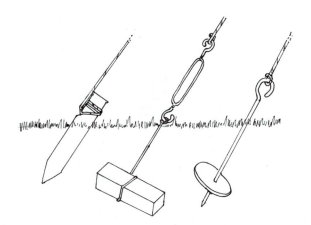

Figure 20-8 Deadmen: Various types of anchoring devices used for guying trees

Figure 20-9 Applying a vinyl tree coil to a new transplant

Figure 20-10 With its plastic pot removed and the bed properly conditioned, this geranium is easily installed using a hand trowel.

to planting can reduce some of the maintenance requirements of the planting.

The spacing of groundcover plants depends upon the species and the speed of coverage desired. Naturally, closer spacing will result in more rapid coverage, but it will also increase the material and installation cost. To assure even and maximum coverage, groundcovers should be installed in a staggered planting pattern, Figure 20-11.

Due to the shallow root system of groundcovers and bedding plants, the plants can dry out easily. Prior to transplanting, each plant must be watered thoroughly. As each is set into the soil, it should be watered again. Thereafter, the new planting must be watered frequently and deeply to establish the plants successfully.

Groundcovers will benefit from mulching to reduce weeds and, most importantly, aid in preventing alternate freezing and thawing of the ground. Such ground activity can result in heaving the groundcovers to the surface where their roots are exposed to the cold and drying air. To be successful, mulch should be applied to a new groundcover planting after the ground has frozen, to prevent premature thawing.

Bulbs

Flowering bulbs require a rich, well-drained soil. They are planted either in flower beds and borders or as masses in the lawn. They may be gently tossed by the handful into open, turfed lawn areas to be planted wherever they land, in an irregularly spaced pattern. Bulbs are planted at differing depths and spacings, depending upon their species. Table 20-1 lists the depths and spacings of some common bulbs.

Bulbs are always set into the ground with the base oriented downward and the shoot pointed

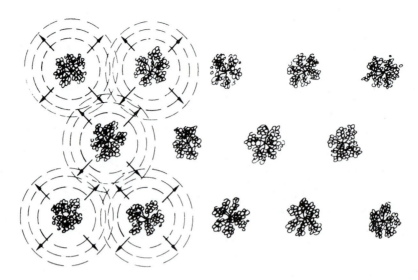

Figure 20-11 Alternating the placement of groundcovers fills space most efficiently.

Table 20-1   A Guide to Bulb Installation

Plant	Depth to Top of Bulb	Spacing
Amaryllis	Leave upper 1/3 of bulb exposed	One bulb per pot
Anemones	2 inches	12 inches
Bulbous iris	2 inches	12 inches
Caladium	2–3 inches in North/1 inch in South	As desired for effect
Calla lily	Leave upper 1/3 of bulb exposed	One bulb per pot
Cannas	3–4 inches	1 1/2–2 feet
Crocus	3 inches	2–4 inches
Daffodil	4–5 inches	6–8 inches
Dahlia	5–6 inches	2–3 feet
Elephant ears	Just below soil surface	As desired for effect
Gladiolus	3–4 inches	6–8 inches
Grape hyacinths	3 inches	2–4 inches
Hyacinths	4–6 inches depending on size	6–8 inches
Lilies	Two to three times the thickness of bulb	1 foot
Paperwhite narcissus	Just below soil surface	One or two bulbs per pot
Ranunculus	2 inches	12 inches
Snowdrops	3 inches	2–4 inches
Summer hyacinths	4 inches	6–8 inches
Tuberous begonia	Just below soil surface	6–8 inches
Tulips	4–5 inches	6–8 inches

upward, Figure 20-12. Many can be installed with a bulb planter.

Other bulb-like structures, called tubers and rhizomes, are installed in mounded holes that permit the structure to be oriented horizontally and the roots directed downward, Figure 20-13. As always, the backfilling step should be done carefully to assure that no air pockets form. Water collecting around the bulbs in air pockets can promote rotting.

Mulching

All plants benefit from mulching after installation. Mulching refers to the application of loose aggregate materials to the surface of a planting bed. The materials may be organic or inorganic. Examples of both types and their advantages and disadvantages are listed in Table 20-2.

The benefits of mulching a newly installed plant are that:

1. Water is retained in the soil around the root system and wilting is avoided.
2. Weed growth is discouraged.
3. The aesthetic appearance of the planting is enhanced.
4. Soil temperature fluctuation is minimized, preventing winter heaving of bulbs and groundcovers. Repeated freezing and thawing of the soil around a plant's base can also damage the bark and permit the entry of pathogens or insects.

For mulches to be effective, they must be applied 3 to 4 inches deep. A shallow layer of mulch does not reduce sunlight enough to discourage weed seed germination, retain moisture, or prevent changes in the surface temperature of the soil. If a more shallow layer of mulch is desired, weeds can still be controlled and water retained by spreading

Figure 20-12 Bulbs are installed either by hand (as shown) or with a bulb planter. They are set into the ground "noses up" and are covered with soil pressed firmly to prevent air pockets. (Courtesy of USDA)

Figure 20-13 This iris rhizome is installed by spreading its roots evenly over a mound of soil, then backfilling to cover the roots, while allowing the leaves and top of the rhizome to remain exposed.

Table 20-2 Characteristics of Common Mulches

Organic Mulches (peat moss, wood chips, shredded bark, chipped corncobs, pine needles)	Inorganic Mulches (marble chips, crushed stone, brick chips, shredded tires)
• reduce soil moisture loss • often contribute slightly to soil nutrition • may alter soil pH • are not a mowing hazard if kicked into the lawn • may be flammable when too dry • may temporarily reduce nitrogen content of soil • require replacement due to biodegradation • may support weed growth as they decompose	• reduce soil moisture loss • do not improve soil nutrition • seldom alter soil pH • are a hazard if thrown by a mower blade • are nonflammable or fire-resistant • have no effect upon nitrogen content of soil • do not biodegrade • may retain excessive amounts of solar heat

weed-barrier fabric around the plant base and adding 1 or 2 inches of mulch to weigh it down. The fabric prevents sunlight from penetrating and promoting weed growth. It also creates a physical barrier that weeds cannot grow through. This mulching technique works well on flat land but is less satisfactory for use on slopes, because rainwater tends to wash the mulch off the fabric surface. It must also be used cautiously on heavy clay soils because the fabric may cause the soil to hold too much water, drowning the new transplant.

Using Antitranspirants

Antitranspirants, also called **antidesiccants**, are chemicals that reduce the amount of water plants lose through transpiration. Antitranspirants are useful because excessive water loss can result in transplant shock. They normally act either to induce closing of the stomata or to cover the stomata with a water-impermeable coating. Several popular brand names are available. The antitranspirants are sprayed onto the plant before and after transplanting. Since most plant stomata are present in

the greatest numbers on the lower surfaces of leaves, the underside of the canopy should receive thorough coverage.

Antitranspirants are of greatest benefit in the transplanting of deciduous trees and shrubs that are in leaf. They are also of benefit to evergreens, especially broad-leaved forms. Any evergreen will benefit from antitranspirants if it is transplanted in the fall, right before the dry winter period.

Problems of Arid Regions

The landscaper installing plants in the American Southwest encounters four distinct problems:

1. The soil quality is generally poor.
2. Irrigation is necessary throughout the year.
3. Higher altitudes can produce extremely hot daytime temperatures and very cool nights.
4. High winds dry out plants quickly and often damage them physically.

Arid soils generally fall into three categories: pure sand or gypsum, adobe, and caliche. Sand lacks both nutrient content and humus. **Adobe** is a heavy, clay-like soil that holds moisture better than sand but needs humus to lighten it and improve its aeration. **Caliche** soils are highly alkaline due to excessive lime content. They have a calcareous hardpan deposit near the surface that blocks drainage, making plant growth impossible. The hardpan layer may lie right at the surface or from several inches to several feet below ground level. The deposits may occur as a granular accumulation or as an impermeable concrete-like layer.

Generally, these are the characteristics of arid soils:

■ Lack humus

■ Require frequent irrigation

■ Are nutritionally poor; nutrients are continually leached out by the irrigation water

■ Are highly alkaline (pHs of 7.5 to 8.5 and higher)

■ Are low in phosphate; phosphate may be rendered unavailable by the high pH

■ Lack iron or contain it in a form unavailable to plants

■ Have a high soluble salt content resulting from alkaline irrigation waters and from manures and fertilizers that do not leach thoroughly

When installing plants in the Southwest, the landscaper must add organic matter to the soils to improve their structure. Organic matter improves the water retention capability of light, sandy soils and breaks up heavy adobe soils. The only way to improve the drainage of caliche soil is to break through it and remove the impermeable layer. The excavated soil can be replaced with a conditioned mix that will support healthy plant growth.

To catch and retain the water so vital yet so limited in arid regions, the planting beds should be recessed several inches below ground level to create a catch basin, Figure 20-14. This method traps and holds applied water, preventing loss through runoff. In addition, organic mulches should be applied to a depth of 4 inches to slow moisture loss and create a cooler growth environment for

the roots. Trunk wraps and whitewash paint are also applied to the trunks of trees to prevent water loss through their thin bark due to sun scald.

Cactus plants are sufficiently different from other plants to warrant special mention. They can be transplanted successfully by following these steps.

1. Before transplanting, mark the north side of the cactus. Orient this side of the plant to the north in its new location. The plant will have developed a thicker layer of protective tissue on its south side to withstand the more intense sunlight.

2. By trenching around the cactus, lift as much as possible of the root system.

3. Brush soil from the roots and dust them with powdered sulfur.

4. Place the cactus in a shaded area where air circulates freely, and allow the roots to heal for a week before replanting.

5. Plant the cactus in dry, well-drained soil. Stake the plant if necessary.

6. Water the plant in three or four weeks, after new growth starts. Thereafter, apply water at monthly intervals.

Whenever possible, native or naturalized plants should be selected for Southwestern landscapes. They have a better chance of surviving the transplant, and they keep maintenance costs down. In situations where the soil is especially unsuitable for planting, there may be little choice but to install the plants above ground in planters.

Planting in the Southeast

While the plants and soils of the American Southeast are definitely different from those of the temperate zoned states, the methods of plant installation do not vary greatly. What differences do exist are related mostly to the uniqueness of certain species, most notably the palms.

Palms are like grasses, only woody-like trees and shrubs. They may have single or multiple trunks.

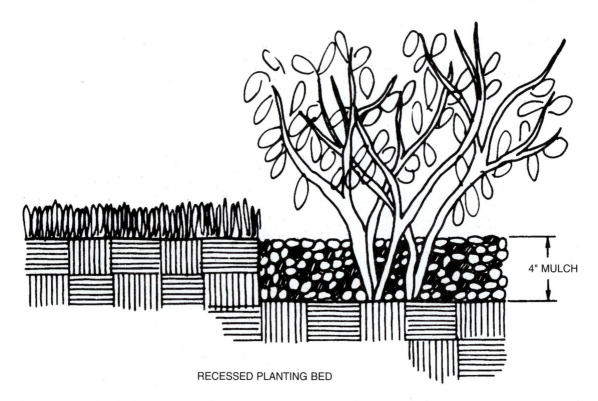

4" MULCH

RECESSED PLANTING BED

Figure 20-14 A recessed planting bed creates a catch basin for moisture.

There are many different types of palms, surviving in a range of hardiness zones and settings as diverse as the seacoast and the desert. Each branch has only a single terminal bud, allowing the plant to grow only at that point. Roots are produced continuously from the base of the trunk, and they are shorter lived than the roots of temperate zone trees and shrubs. The majority of palms that are used in landscapes are grown in nurseries, not the wild. They may be produced in containers or grown in fields. Like other trees and shrubs, palms must have an adequate root and soil ball attached if they are to transplant successfully. However, because of the difference in how and where the roots are produced, the root ball of a palm is seldom as large as that of other transplants. A root ball diameter that is about two feet greater than the diameter of the

trunk is generally accepted as an adequate size. To be certain, landscapers should consult the local industry standards of their geographic region since there is no single industry norm. Large palms can be moved with a tree spade. Smaller plants are handled with ball carts or carried at the root ball. To protect the tender terminal bud, the fronds of a palm are usually tied up and around the end of each branch with biodegradable twine before being harvested and moved. It is important to keep the root ball moist until the installation is accomplished, Figures 20-15 A and B.

Palms should be set at the same depth they were in the field or in their production container. If set too shallow, they may topple over. If set too deep, they may suffer nutrient deficiency. Staking is often needed to stabilize the new transplants.

Figure 20-15 A and B Large palms, with fronds tied, are set into planters using a crane.

The technique used is an exterior support brace, not the devices described earlier in this chapter, Figure 20-16.

Following their installation, the palms should be kept well watered and misted throughout their first season. Where conditions are excessively hot and dry, the fronds remain tied around the terminal bud for most of the first season as well to protect it from drying. The support structure may remain in place for up to a year.

Figure 20-16 Exterior bracing supports the palms. Protective banding prevents injury where braces meet the trunks.

Achievement Review

A. Answer each of the following questions as briefly as possible.

1. Identify the following tools, used in the installation of landscape plants. (See Appendix B for review.)

a.

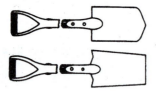

b.

c.

d.

e.

f.

g.

h.

i.

j.

k.

l.

m.

n.

o.

2. List the four qualities of good backfill soil.

3. Indicate whether the following are characteristic of bare-rooted (A), balled and burlapped (B), or containerized (C) plants.
   a. lightweight and easily transported
   b. retain the entire root system
   c. severely reduced root system
   d. usually the most expensive
   e. plants may become pot-bound
   f. permits large plants to be transplanted
   g. transplant season limited to early spring
   h. allows transplanting in any season

4. What is the primary objective when deciding the timing of a plant transplant?

5. Why is early autumn the most desirable transplanting season for most plants?

6. Indicate whether the following are characteristic of inorganic (A) or organic (B) mulches, or both (C).
   a. reduce water loss from the soil
   b. may alter soil pH
   c. may temporarily reduce nitrogen content of soil
   d. hazardous if thrown by a mower blade
   e. do not biodegrade
   f. sometimes are slightly nutritional

7. What benefit is gained by applying anti-transpirants to new transplants?

B. Indicate if the following statements are true or false.

1. The root form of a transplant determines the size and configuration of the planting pit.
2. Balled and burlapped plants are installed in a pit with a mounded bottom.
3. Butterflying is used when installing groundcovers.
4. Burlap around the soil ball should be permitted to extend above the backfill to assure proper aeration to the root system.

5. Wire baskets should be removed entirely before the root ball is set into the planting pit.
6. Where soil drainage is poor, setting the plant above grade level is recommended.
7. Landscape fabric is better suited for use around plants that have been installed as containerized stock than as balled and burlapped stock.
8. For rapid and maximum coverage, groundcovers should be installed in staggered, alternating rows.

9. *Staking* and *guying* are interchangeable terms for the same technique.
10. Backfilling a planting pit with the same soil dug from the pit is recommended over the use of soil conditioned with additives.

C. Installation of plants in the American Southwest is complicated by poor soil and the lack of adequate natural water supplies. What must a landscaper do to counter these threats to plant survival?

Suggested Activities

1. Install a tree in a nearby yard, park, or school campus. Apply the information given in this chapter. If possible, install both a deciduous and an evergreen plant. Stake or guy the trees as appropriate. *NOTE:* Arbor Day is a good occasion to do this activity. There may be a number of planting opportunities nearby.

2. To demonstrate the benefits of landscape fabric in weed control, fill three greenhouse flats with non-sterile soil. Cover one with a thin layer of organic mulch, one with a sheet of landscape fabric and a thin layer of mulch, and leave one without any coverage. Give all the same amount of sunlight and water over a period of several weeks. Observe the number of weeds seen in each flat.

3. To observe the effects of girdling, use a small containerized woody plant for this demonstration, which will require some time for the effects to be noticed. Using a sharp knife, carefully cut through the bark, encircling the main stem or trunk. Remove about a ½-inch wide strip of bark around the entire base. Be careful not to cut any deeper than is necessary to peel off the strip of bark. This duplicates the type of damage done to a plant by rodents or other means. Keep the plant watered. You may even choose to add fertilizer to the soil. Over time you should notice that the plant does not wilt because it can still take up water. However, the leaves will begin to show signs of discoloration and other indications that it is not receiving the nutrients from the soil because part of its vascular system has been destroyed by the girdling. Eventually the plant will starve and die.

Selecting the Proper Grass

Objectives:

Upon completion of this chapter, you should be able to

- list six factors used in the comparison of different turfgrasses
- list the information required by law on grass seed labels
- explain the differences between single-species plantings, single-species blends, and mixtures of species

Turfgrasses are among the oldest plants used for landscaping. They are the most common choice for surfacing the outdoor room. A neatly trimmed lawn of good quality is not only comfortable to walk on, but ideal for many athletic and recreational activities. Because the growing point of the turfgrass is at the crown of the plant, near the soil, it is protected. This permits turfgrasses to be mown and walked upon repeatedly.

Comparison of Turfgrasses

Most turfgrasses used in landscapes are perennial, surviving from one year to the next. Nearly all species reproduce from seed, although several can be reproduced vegetatively, without pollination and seed production. A typical grass plant produces new leaves throughout the growing season. During the growing season, the turfgrasses will increase beyond the number of seeds sown. One of the objectives of good lawn development is to encourage turf growth as quickly and as evenly as possible.

Growth Habits

Grasses have differing growth habits, which result from the three different ways that new shoots are produced, Figure 21-1.

- Rhizome-producing (rhizomatous): A **rhizome** is a horizontal underground stem. New shoots are sent to the surface some distance out from the parent plant. Each new plant develops its own root system and is independent of the parent plant.

- Stolon-producing (stoloniferous): The shoots extending out from the parent plant are above ground. They are called **stolons**. New plants develop independently as described above. Some grasses are both rhizomatous and stoloniferous.

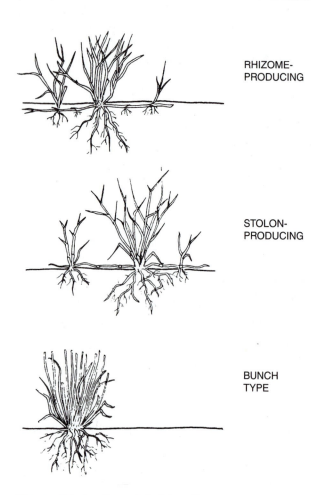

RHIZOME-PRODUCING

STOLON-PRODUCING

BUNCH TYPE

Figure 21-1 Growth habits of grasses

- Bunch-type: New shoots are produced from the sides of the plant, gradually increasing the plant's width.

Rhizome and stolon-producing grasses tend to reproduce more quickly and evenly than bunch-type grasses. Therefore, the bunch-type require more seed and closer spacing in order to cover an area quickly and without clumps.

Texture, Color, and Density

Grass **texture** is mostly a way of describing the width of the grass leaf (blade). The wider the blade is, the coarser will be the texture. Generally, fine-textured grasses are more attractive than coarse-textured grasses. They are also more expensive. The color of a grass and its density will also differ among species. Colors can vary from pastel greens to dark, bluish tones. **Density** refers to the number of leaf shoots that a single plant will produce. It can range from sparse to thick, depending upon the type of grass.

Size of Seed

The size of the seed is another reason for variation in the quality and quantity of grass seed mixes. Fine-textured grasses have very small seeds. Coarse-textured grass seeds usually are much larger. Thus, a pound of fine-textured grass seed contains considerably more seeds than a pound of coarse-textured grass seed.

Because of the greater number of seeds per pound, a pound of fine-textured grass seed plants a larger area of land. For example, a pound of fine-textured Kentucky bluegrass contains approximately 2,000,000 seeds. That number of seeds plants about 500 square feet of lawn. A pound of coarse-textured tall fescue contains 227,000 seeds; therefore, only 166 square feet can be planted with a pound of this particular seed.

Other comparisons can further point out the difference in seed sizes. For example, there are as many seeds in 1 pound of bluegrass as there are in 9 pounds of ryegrass; and as many seeds in a pound of bentgrass as there are in 30 pounds of ryegrass.

Soil and Climatic Tolerance

Most grasses, like almost all other plants, do best in good-quality, well-drained soil. However, every state contains landscape sites that fall short of the ideal conditions preferred for turfgrass success. Some grasses can adapt to a wide range of soil conditions, while others are very limited in their adaptability.

Similarly, some grasses tolerate high humidity and reduced sunlight; others do not. Some thrive in the subtropics and tropics; others are better-suited for temperate and subarctic regions.

Grasses are often grouped into two categories based upon the temperatures at which they grow best:

- Cool-season grasses are favored by daytime temperatures of 60 to 75 degrees F.

- Warm-season grasses are favored by daytime temperatures of 80 to 95 degrees F.

Figure 21-2 shows the peak growth rates of the two types of grasses. Knowledge of the optimum growing temperatures of grasses explains why northern lawns are often brown and dormant in midsummer when the days are very warm. Likewise, warm-season grasses do not really flourish in early spring and late fall when temperatures fall below the optimum temperature.

Common Warm-Season Grasses	Common Cool-Season Grasses
Bermuda grass	Kentucky bluegrass
zoysia grass	red fescue
centipede grass	colonial bentgrass
carpet grass	ryegrass
St. Augustine grass	
Bahia grass	
buffalo grass	

Figure 21-3 illustrates the climatic regions of the continental United States that are favorable for the growth of the grasses listed and others. Any seed purchased for planting in a certain climatic region should be composed of the appropriate grasses.

Use Tolerance

Under the same conditions of use, some grasses will survive and others will quickly wear away. Some will accept heavy use and recover quickly, while others will recover much more slowly. Some can accept the compaction of heavy foot traffic and still look good; yet others may discolor and slow their rates of growth.

Disease and Insect Resistance

Certain grasses possess greater resistance than others to insect and disease pests. The resistance may be natural or may have developed through the efforts of plant breeders. Many grasses are continually being improved by horticultural scientists searching for features such as better color, resistance to drought, shade tolerance, and pest resistance.

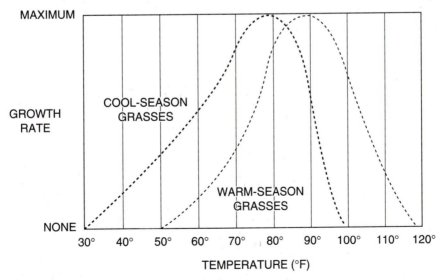

Figure 21-2 Relation of temperature to growth rate in cool-season and warm-season grasses

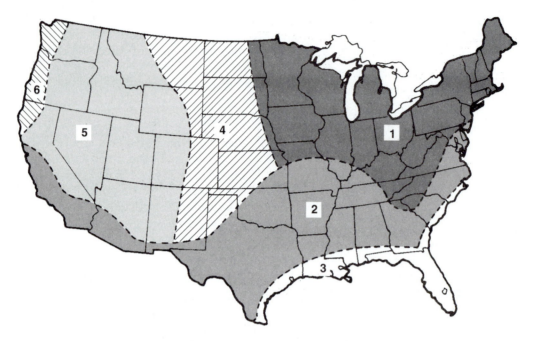

CLIMATIC REGIONS IN WHICH THE
FOLLOWING GRASSES ARE SUITABLE FOR LAWNS:

1. Kentucky bluegrass, red fescue, and colonial bentgrass. Tall fescue, Bermuda, and zoysia grasses in the southern part.
2. Bermuda and zoysia grasses. Centipede, carpet, and St. Augustine grasses in the southern part; tall fescue and Kentucky bluegrass in some northern areas.
3. St. Augustine, Bermuda, zoysia, carpet, and Bahia grasses.
4. Nonirrigated areas: crested wheat, buffalo, and blue grama grasses. Irrigated areas: Kentucky bluegrass and red fescue.
5. Nonirrigated areas: crested wheatgrass. Irrigated areas: Kentucky bluegrass and red fescue.
6. Colonial bentgrass, Kentucky bluegrass, and red fescue.

Figure 21-3 Regions of grass adaptations

In "A Comparison Chart for Turfgrasses" in this chapter, many of the most commonly used turfgrasses are compared.

Purchasing Grass Seed

Grass seed is sold in small quantities through retail outlets such as garden centers, supermarkets, hardware stores, and department stores. It is also sold in bulk amounts through wholesale suppliers.

Professional landscapers usually purchase seed wholesale. However, most clients have purchased packaged seed from retailers in the past. They may not understand why the seed selected by the landscaper is priced higher than expected. Landscapers must be prepared to explain why all grass seeds are not alike and how the quality of seed is measured.

The key to determining the quality of grass seed is the seed analysis label. The **seed analysis label**, which by law must appear on every package of seeds to be sold, gives a breakdown of the contents of the seed package on which it appears. The analysis label may be on the package itself, or, if the seed is being sold in large quantities, on a label tied to the handle of the storage container.

While legal definitions vary somewhat from state to state, most analysis labels contain the following information:

Purity. The percentage, by weight, of pure grass seed. The label must show the percentage by weight of each seed type in the mixture.

Percent Germination. The percentage of the pure seed that was capable of germination (sprouting) on the date tested. The date of testing is very important and must be shown. If much time has passed since the germination test, the seed is older and less likely to germinate satisfactorily.

Crop Seed. The percentage, by weight, of cash crop seeds in the mixture. These are undesirable species for lawns.

Weeds. The percentage, by weight, of weed seeds in the mixture. A seed qualifies as a weed seed if it has not been counted as a pure seed or a crop seed.

Noxious Weeds. Weeds that are extremely undesirable and difficult to eradicate. The number given is usually the number of seeds per pound or per ounce of weed seeds.

Inert Material. The percentage, by weight, of material in the package that will not grow. In low-priced seed mixes, it includes materials such as sand, chaff, or ground corncobs. Inert material is sometimes added to make the seed package look bigger. At other times, the inert material is already present in the seed and is not removed because the cost involved would raise the price of the seed.

Three sample analyses follow. Study the contents of each mixture and determine which would probably cost the most and which the least.

It is likely that Mixture C would be the most expensive. It contains the highest percentage of fine-textured grasses, no coarse grasses, and the lowest percentage of weeds. Mixture A would probably cost the least, since it contains a high

Mixture A

Fine-Textured Grasses	
12.76 percent red fescue	85 percent germ.
6.00 percent Kentucky bluegrass	80 percent germ.
Coarse Grasses	
53.17 percent annual ryegrass	95 percent germ.
25.62 percent perennial ryegrass	90 percent germ.
Other Ingredients	
2.06 percent inert matter	
0.39 percent weeds—no noxious weeds	

Mixture B

Fine-Textured Grasses	
38.03 percent red fescue	80 percent germ.
34.82 percent Kentucky bluegrass	80 percent germ.
Coarse Grasses	
19.09 percent annual ryegrass	85 percent germ.
Other Ingredients	
7.72 percent inert matter	
0.34 percent weeds—no noxious weeds	

Mixture C

Fine-Textured Grasses	
44.30 percent creeping red fescue	85 percent germ.
36.00 percent Merion bluegrass	80 percent germ.
13.54 percent Kentucky bluegrass	85 percent germ.
Coarse Grasses	
None claimed	
Other Ingredients	
5.87 percent inert matter	
0.29 percent weeds—no noxious weeds	

percentage of coarse-textured grasses, the lowest percentage of fine grasses, and the greatest percentage of weeds. None of the mixtures is very poor in quality, since there are no crop or noxious weed seeds claimed by any.

A Comparison Chart for Turfgrasses

Grass Species	Cool Season or Warm Season	Growth Habit	Leaf Texture	Mowing Height/ Inches	Fertilization. Pounds of Nitrogen Per 1,000 Square Feet Per Year
Bahiagrass	Warm	Rhizomatous	Coarse	1 1/2 to 2	1 to 4
Bermudagrass	Warm	Stoloniferous and rhizomatous	Fine	1 to 2	4 to 9
Bentgrass, Colonial	Cool	Bunch-type (with short stolons and rhizomes)	Fine	1/2 to 1	2 to 4
Bentgrass, Creeping	Cool	Stoloniferous	Fine	1/2 or less	4 to 8
Bentgrass, Redtop (see Redtop)					
Bentgrass, Velvet	Cool	Stoloniferous	Fine	1/2 or less	2 to 4
Bluegrass, Annual	Cool	Bunch-type or stoloniferous	Fine	1	2 to 6
Bluegrass, Canada	Cool	Rhizomatous	Medium	Does not mow well	1 or less
Bluegrass, Kentucky	Cool	Rhizomatous	Fine	1 to 2 1/2	2 to 6
Bluegrass, Rough	Cool	Stoloniferous	Fine	1 or less	2 to 4
Bromegrass, Smooth	Cool	Rhizomatous	Coarse	Does not mow well	1 or less
Buffalograss	Warm	Stoloniferous	Fine	1/2 to 1 1/2	1/2 to 2
Carpetgrass, Common	Warm	Stoloniferous	Coarse	1 to 2	1 to 2
Carpetgrass, Tropical	Warm	Stoloniferous	Coarse	1 to 2	1 to 2

Soil Tolerances	Climate Tolerances	Uses	How Established. If Seeded, Pounds Per 1,000 Square Feet
Infertile, acidic, and sandy	Subtropical and tropical	Utility turf; good for use along roadways	Seeded at 6 to 8
Does well on a wide range of soils	Warm temperate and subtropical	Sunny lawn areas; good general purpose turf for athletic fields, parks, home lawns	Plugging or seeded at 1 to 1 1/2
Moderately fertile, acidic, and sandy	Temperate and seacoastal	Areas where intensive cultivation is practical	Seeded at 1/2 to 2
Fertile, acidic, and moist	Subarctic and temperate	Golf greens and other uses where intensive cultivation is practical	Sprigging or seeded at 1/2 to 1 1/2
Moderately fertile, acidic, and sandy	Temperate and seacoastal	Shaded, intensively cultivated areas	Seeded at 1/2 to 1 1/2
Fertile, neutral to slightly acidic	Temperate and cool subtropical	Not planted intentionally; but common in intensively cultivated turfs during spring and fall	Does not apply
Infertile, acidic, and droughty	Subarctic and cool temperate	A soil stabilizer	Seeded at 1 to 2
Fertile, neutral to slightly acidic	Subarctic, temperate, and cool subtropical	Sunny lawn areas; good general purpose turf for athletic fields, parks, and home lawns	Seeded at 1 to 2
Fertile and moist	Subarctic and cool, shaded temperate	Some use on shaded, poorly drained sites	Seeded at 1 to 2
Infertile and droughty	Dry and temperate	A soil stabilizer	Seeded at 1 to 2
Does well on a wide range of soils; tolerant of alkaline soils	Dry temperate and subtropical	Useful in semiarid sites as a general purpose lawn grass	Seeded at 3 to 6
Infertile, acidic, and moist	Subtropical and tropical	Utility turf; good for use along roadways and as a soil stabilizer	Seeded at 1 1/2 to 2 1/2
Infertile, acidic, and moist	Humid subtropical and tropical	Utility turf; good for use along roadways and as a soil stabilizer; can be used as a lawn grass in tropics	Seeded at 1 1/2 to 2 1/2

A Comparison Chart for Turfgrasses *(continued)*

Grass Species	Cool Season or Warm Season	Growth Habit	Leaf Texture	Mowing Height/ Inches	Fertilization. Pounds of Nitrogen Per 1,000 Square Feet Per Year
Centipedegrass	Warm	Stoloniferous	Medium	1 to 2	1 to 2
Fescue, Chewings	Cool	Bunch-type	Fine	1 1/2 to 2	2
Fescue, Creeping Red	Cool	Rhizomatous	Fine	1 1/2 to 2	2
Fescue, Hard	Cool	Bunch-type	Medium	Does not mow well	1 or less
Fescue, Meadow	Cool	Bunch-type	Coarse	1 1/2 to 3	1 or less
Fescue, Sheep	Cool	Bunch-type	Fine	Does not mow well	1 or less
Fescue, Tall	Cool	Bunch-type	Medium to Coarse	1 1/2 to 3	1 to 3
Gramagrass, Blue	Warm	Rhizomatous	Fine	Does not mow well	1 or less
Redtop (a bentgrass)	Cool	Rhizomatous	Coarse	1 1/2 to 3	1 to 2
Ryegrass, Annual	Cool	Bunch-type	Medium	1 1/2 to 2	2 to 4
Ryegrass, Perennial	Cool	Bunch-type	Fine	1 1/2 to 2	2 to 6
St. Augustinegrass	Warm	Stoloniferous	Coarse	1 to 2 1/2	2 to 6
Timothy, Common	Cool	Bunch-type	Coarse	1 to 2	3 to 6
Wheatgrass, Crested	Cool	Bunch-type	Coarse	1 1/2 to 3	1 to 3
Zoysiagrass (Japanese lawngrass)	Warm	Stoloniferous and rhizomatous	Medium	1/2 to 1	2 to 3
Zoysiagrass (Manilagrass)	Warm	Stoloniferous and rhizomatous	Fine	1	2 to 3
Zoysiagrass (Mascarenegrass)	Warm	Stoloniferous and rhizomatous	Fine	Does not mow well	2 to 3

Soil Tolerances	Climate Tolerances	Uses	How Established. If Seeded, Pounds Per 1,000 Square Feet
Infertile, acidic, and sandy	Subtropical and tropical	Utility turf; also usable as a low-use lawn grass	Seeded at 1/4 to 1/2
Infertile, acidic, and droughty	Subarctic and temperate	Shaded sites with poor soil	Seeded at 4 to 8
Infertile, acidic, and droughty	Subarctic and temperate	Shaded sites	Seeded at 3 to 5
Fertile and moist; not tolerant to droughty soil	Moist and temperate	A soil stabilizer	Seeded at 4 to 8
Widely tolerant of all but droughty soils	Moist and temperate	Utility turf; good for use along roadways	Seeded at 4 to 8
Infertile, acidic, well-drained, and droughty	Dry and temperate	A soil stabilizer	Seeded at 3 to 5
Does well on a wide range of soils	Warm temperate and subtropical	Utility turf; good for use along roadways; new cultivars (Brookston, Olympic, and Rebel) good for lawns	Seeded at 4 to 8
Does well on a wide range of soils	Dry and subtropical	Utility turf; good for use along roadways and in arid sites	Seeded at 1 to 2
Does well on a wide range of soils	Subarctic, temperate, and cool subtropical	Utility turf; good for use along roadways and in poorly drained areas	Seeded at 1/2 to 2
Fertile, neutral to slightly acidic and moist	Temperate and subtropical	Useful for quick and temporary lawns in the temperate zone and for winter color in the subtropic zones	Seeded at 4 to 6
Fertile, neutral to slightly acidic and moist	Mild and temperate	Used in mixed species lawns and as an athletic turf	Seeded at 4 to 8
Does well on a wide range of soils	Subtropical and tropical seacoastal	A good lawn grass with excellent shade tolerance	Sprigging
Fertile, slightly acidic, and moist	Subarctic and cool temperate	Utility turf; good for athletic fields in cold regions where preferable species won't survive	Seeded at 1 to 2
Does well on a wide range of soils	Subarctic and cool temperate	Useful as a general purpose turf on droughty sites	Seeded at 3 to 5
Does well on a wide range of soils	Temperate, subtropical, and tropical	Useful as a general purpose turf for home lawns, parks, and golf courses, especially in warmer regions	Plugging
Does well on a wide range of soils	Subtropical and tropical	A good lawn grass	Plugging
Does well on a wide range of soils	Warm subtropical and tropical	A soil stabilizer and groundcover	Plugging

Mixtures, Blends, and Single-Species Lawns

Grass seed is commonly purchased either as a mixture or a blend. It is also available as a single species (such as all Kentucky bluegrass or all Chewings fescue). A mixture combines two or more different species of grass. A blend combines two or more cultivated varieties of a single species. Both mixtures and blends have their places depending upon the site and circumstances. Mixtures are most common in temperate zone landscapes; single-species plantings are more common in subtropical and tropical landscapes.

Mixtures sometimes have the disadvantage of variegated color and texture. This is a result of the different species they contain. They have the advantage of being able to tolerate mixed environmental conditions and can recover from insect and disease pests that would wipe out a single species.

Single-species turf plantings offer a more uniform appearance than mixtures. However, a single-species planting is often unable to adjust to severe changes in environmental conditions. It can also be completely destroyed by a single insect or disease invasion.

Blends attempt to retain the advantages of both mixtures and single-species plantings. If the cultivated varieties of the blend are carefully selected, a blend offers these advantages: uniform color and texture, resistance to damage from environmental changes, resistance to wear, resistance to pest injury, and the varieties in the blend will have similar maintenance needs.

Achievement Review

A. Not all turfgrasses are alike. They can be compared using different factors. Insert the correct factor into each of the following sentences.

  1. Adapting to differences in pH, aeration, fertility levels, humidity, light, and temperatures measures a turf's _____.
  2. Rhizomatous, stoloniferous, and bunch-type are different _____ of grasses.
  3. Blade width, color variation, and the number of shoots per plant are measures of _____.
  4. The ability of turf to withstand the compaction of foot traffic indicates its level of _____.
  5. Certain grasses will suffer pest damage more than other grasses because of their _____.

  6. One pound of fine-textured grass differs from a pound of coarse-textured grass in many ways. One way is in the number and _____ of the seeds.

B. What could cause a very high-quality grass seed purchased in the south to be unsuitable for planting in the north?

C. Of the three seed mixtures A, B, and C shown in this chapter, which mixture is most likely to result in a sparse second-year lawn? Why?

D. Why is a grass seed mixture usually preferable to a pure, single-species seed?

E. List and define the important terms found on a grass seed analysis label.

Suggested Activities

1. Grow some grasses. Start flats or flowerpots of pure grass species in the classroom. Compare fine-leaf and broad-leaf types. If possible, also grow samples of warm-season and cool-season grasses for comparison.

2. Obtain several grass seed mixtures from various sources and in as many price ranges as possible. Rank the mixtures on the basis of package appearance, advertised claims, and brand names. Rank the mixtures again, using the seed analysis labels as the measure. How closely do the package claims match the actual facts about the mixture as shown on the labels? How closely does the price ranking follow the quality ranking?

3. Make a seed count. Weigh 1/4-ounce quantities of a fine-textured grass and a coarse-textured grass. Be as accurate as possible. Count the number of seeds in each measure. Do the fine-textured seeds outnumber the coarse-textured seeds?

   NOTE: Do not use redtop for the coarse-textured grass in this exercise. Its seeds are atypically small for a coarse grass.

CHAPTER 22

Lawn Construction

Objectives:

Upon completion of this chapter, you should be able to

- describe four methods of lawn installation
- outline the steps required for proper lawn construction
- explain how to calibrate a spreader

Selecting the Method of Lawn Installation

There are four methods that may be used to install a turfgrass planting:

- Seeding
- Sodding
- Plugging
- Sprigging and stolonizing

The method selected depends upon the species of grass, the type of landscape site, and how quickly the turf must be established.

Seeding

Seeding is the most common and least expensive method of establishing a lawn. The seed can be applied by hand or with a spreader on small sites. On large sites, a cultipacker seeder (pulled by a tractor) or a hydroseeder (a spraying device that applies seed, water, fertilizer, and mulch at the same time) may be used. The hydroseeder is especially helpful for seeding sloped, uneven areas, Figure 22-1.

Sodding

When a lawn is needed immediately, sodding may be selected as the method of installation. Sod is established turf that is moved from one location to another. A sod cutter is used to cut the sod into strips. These are then lifted, rolled up, and placed onto pallets for transport to the site of the new lawn, Figure 22-2A, B, and C. At the new site, the sod is unrolled onto the conditioned soil bed. The effect is that of instant lawn, Figure 22-3. Sod is produced in special nurseries where it can be grown and harvested efficiently and in large quantities. Sodding is much more

Figure 22-1 Use of the hydroseeder gives rapid stabilization to this steep embankment. (Courtesy of USDA)

Figure 22-2B The sod cutter removes the growing turf along with a thin layer of soil.

Figure 22-2A A sod cutter being used on a sod farm

Figure 22-2C Pallets of fresh sod are carried from the field for transport to landscape sites.

Plugging

Plugging is a common method of installing lawns in the southern sections of the United States. Certain grasses, such as Bermuda, St. Augustine, and zoysia,

costly than seeding. However, the immediacy is important to some clients, and it is necessary on sites where seed might wash away.

Figure 22-3 After careful preparation of the soil, the sod is unrolled at the site of the new lawn.

are not usually reproduced from seed. Instead, they are usually placed into the new lawn as **plugs** of live, growing grass. Since the growing season in the southern regions is longer than elsewhere, the plugs have time to develop into a full lawn. Plugging is a time-consuming means of installing a lawn, which is its major limitation. However, plugging is necessary for many warm-season grasses that are poor seed producers. On large sites, some mechanization of the planting is possible.

Sprigging and Stolonizing

Like plugging, sprigging and stolonizing are more commonly used with warm-season grasses than cool-season grasses. A **sprig** is a piece of grass shoot. It may be a piece of stolon or rhizome or even a lateral shoot. Sprigs do not have soil attached and so are not like plugs or sod. They are planted at intervals into prepared, conditioned soil. Several bushels of sprigs are required to plant 1,000 square feet. If done by hand, the process is slow and tedious. Mechanization can lessen the time required.

Stolonizing is a form of sprigging. The sprigs are broadcast (distributed evenly) over the site and covered lightly with soil. Then they are rolled or disked. Since each sprig is not individually inserted into the soil, this method is faster.

Proper Lawn Construction

If the lawn is to be of the best quality, it must be given every possible chance for success. Proper construction of the lawn is vital. Six steps should be followed by the landscaper to assure a successful beginning for the lawn.

- Plant at the proper time of year
- Provide the proper drainage and grading
- Condition the soil properly
- Apply fresh, good-quality seed, sod, plugs, or sprigs
- Provide adequate moisture to promote rapid establishment of the lawn
- Mow the new lawn to its correct height

Time of Planting

Lawns in southern sections of the country require warm-season grasses. Such grasses grow best in day temperatures of 80 to 95 degrees F. It is most effective to plant them in the spring, just prior to the summer season. In this way, they have the opportunity to become well established before becoming dormant in the winter.

Cooler northern regions require cool-season grasses to yield the most attractive lawns. Bluegrasses and fescues germinate best when temperatures are in the range of 60 to 75 degrees F. These lawns thrive in locations where days are cool and nights are warm. The best planting time for these grasses is early fall or very early spring, prior to the ideal cool season in which they flourish. If cool-season grasses are planted too close to the intensely hot or cold days of summer and winter, they will die or become dormant before becoming well established.

Grading and Draining the New Lawn

Each time the rain falls or a sprinkler is turned on, water moves into the soil and across its surface. **Grading** (leveling land so that it slopes) directs the movement of the surface water. **Drainage** allows the water to move slowly down into the soil to prevent erosion or puddling.

Even lawns that seem flat must slope enough to move water off the surface and away from nearby buildings. If a slight slope does not exist naturally, it may be necessary to construct one. A fall (grade) of between 6 inches and 1 foot over a distance of 100 feet is required for flat land to drain properly. Failure to grade lawns away from buildings can result in flooded cellars and basements.

Drainage of water into and through the soil is important. Without a supply of water to their roots, neither grasses nor any other plants can live. Without water drainage past their roots, turf grasses and other plants can be drowned. Depending upon the soil in the particular lawn area involved, good drainage may require nothing more than mixing sand with the existing soil to allow proper water penetration. In cases where the soil is heavy with clay, a system of drainage tile may be necessary.

If drainage tile is needed, it should be installed after the lawn's grade has been established, but before the surface soil has been conditioned. Regular 4-inch agricultural tile is normally used, placed 18 to 24 inches beneath the surface. Tile lines are spaced approximately 15 feet apart, Figure 22-4. Each of the lateral tiles runs into a larger main drainage tile, usually 6 to 8 inches in diameter. This, in turn, empties into a nearby ditch or storm sewer, Figure 22-5.

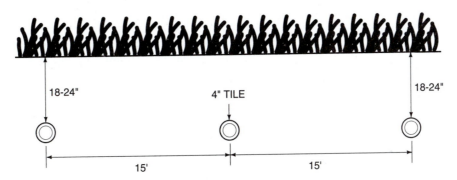

Figure 22-4 Drainage tile installation (Drawing not to scale)

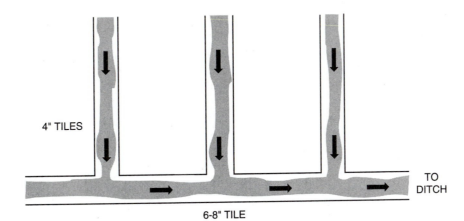

Figure 22-5 The flow of water through tile drainage system (Drawing not to scale)

Where the soil is naturally sandy, no special consideration for drainage may be necessary.

Conditioning the Soil

Proper soil preparation requires an understanding of soil texture and soil pH.

Soil texture is the result of differing amounts of sand, silt, and clay in the composition of the soil. A soil that contains nearly equal amounts of sand, silt, and clay is called a **loam** soil. Loam soils are excellent for planting. Soil textures such as sandy loam, clay loam, and silty clay loam are named for the ingredient or ingredients that make up more than one-third of the composition of the soil. For example, the composition of sandy loam is more than one-third sand. Silt and clay each make up more than one-third of the composition of silty clay loam, and hence, it is less than one-third sand. In conditioning soil for lawn construction, clay, sand, silt, or humus (organic material) may be added to bring the existing soil closer to a medium-loam texture.

Soil pH is a measure of the acidity or alkalinity of soil. A pH measurement of less than 7.0 indicates increasing soil acidity. As the pH increases beyond 7.0, the soil becomes more alkaline or basic. Most turf grasses grow best in soil with a neutral pH (expressed as 7.0) to slightly acidic pH (6.5).

The measurement of soil pH is obtained from a soil test. Soil tests are usually available through county Cooperative Extension Services. Also, pH test kits can be purchased at a reasonable cost, allowing landscapers to make their own determination of pH much more quickly.

If the pH of soil is too acidic, it is usually possible to raise the pH by adding dolomitic limestone. The limestone should be applied in the spring or fall. As the following chart illustrates, the amount applied depends upon the texture of the soil and how far the natural pH is from 6.5 to 7.0.

Where it is necessary to lower the pH of the soil to attain the desired level of 6.5 to 7.0, landscapers commonly use sulfur, aluminum sulfate, or iron sulfate.

The attainment of a suitable texture and the proper pH are very important in the conditioning

Amount of Dolomitic Limestone Applied Per 1,000 Square Feet of Lawn			
Natural Soil	Soil Texture		
	Sandy	Loam	Clay or Silt
pH 4.0	90 lb.	172 lb.	217 lb.
pH 4.5	82 lb.	157 lb.	202 lb.
pH 5.0	67 lb.	127 lb.	150 lb.
pH 5.5	52 lb.	97 lb.	120 lb.
pH 6.0	20 lb.	35 lb.	60 lb.
pH 6.5–7.0	None needed	None needed	None needed

of lawn soil. Equally important is the removal of stones from the surface layer of the soil, the loosening of the soil to a depth of 5 or 6 inches, and the incorporation of organic matter into the soil.

Stones may be removed by hand, by rake, or by machine. If the lawn is to be smoothly surfaced, even the smallest surface stones must be discarded.

Decaying organic matter creates **humus**, a valuable ingredient of soil. Humus aids the soil in moisture retention. It also helps air reach the soil. Organic matter can be added to the soil during its conditioning with materials such as peat moss, well-rotted manure, compost, or digested sewage sludge. The landscaper may choose the material that is most easily available and relatively low in cost.

All necessary soil additives (pH adjusters, organic matter, sand, and fertilizers) can be worked into the soil at the same time. This is done most effectively with a garden tiller, which also loosens the soil surface and breaks the soil into small particles, Figure 22-6. Once the soil has been properly conditioned, it is ready to plant.

Planting the Lawn

Seed. Seed is applied to the prepared soil in a manner that will distribute it evenly. Otherwise, a patchy lawn develops. When applied with a spreader or cultipacker seeder, the seed may be mixed with a carrier material such as sand or topsoil to assure even spreading. The seed or seed/carrier mix is divided into two equal amounts. One part is sown across the lawn in one direction. The

Figure 22-6 A large garden tiller turns the soil while working soil additives into it. For large lawn areas, the tiller is a necessity.

other half is then sown across the lawn at a 90-degree angle to the first half, Figure 22-7.

Placing a light mulch of weed-free straw over the seed helps retain moisture. It also helps to prevent the seed from washing away during watering or rainfall. On a slope that has not been hydroseeded, it is wise to apply erosion netting over the mulched seed to reduce the possibility of the seed washing away, Figure 22-8.

Sod. Sod must be installed as soon after it has been cut as possible. If not, the live grass will be damaged as a result of the excessive temperatures that build up within the rolled or folded strips of sod. Permitting the sod to dry out while awaiting installation can also damage the grass and result in a weak, unsatisfactory lawn.

The soil should be moist before beginning installation of the sod. The individual strips are then laid into place much as a jigsaw puzzle is assembled. The sod should not be stretched to fit, as it will only

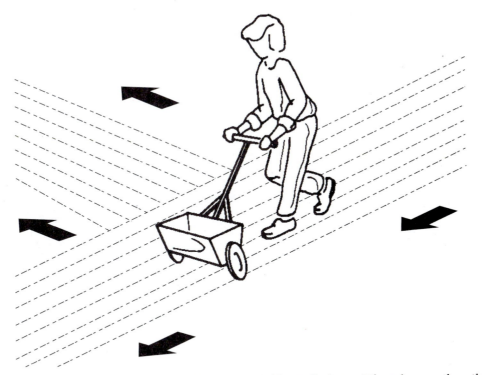

Figure 22-7 Spreader application. Half the material is applied at a 90° angle over the other half.

Figure 22-8 Erosion netting is used here to prevent the grass seed from washing away until it becomes established. Steep slopes such as this one are usually difficult to seed.

shrink back later, leaving gaps in the lawn surface. Instead, each strip should be fitted carefully and tightly against the other strips. Using a flat tamper or roller, the sod should be tamped gently to assure that all parts are touching the soil.

Plugs. Plugs are small squares, rectangles, or circles of sod, cut about 2 inches thick. Their installation is similar to that of groundcovers. They are set into the conditioned soil at regular intervals (12 to 18 inches), in staggered rows to maximize coverage. The top of each plug should be level with the surface of the conditioned soil. The soil should be moist but not wet at the time of installation. This prevents some of the plugs from drying out while others are still being installed.

Sprigs. Sprigs are planted 2 to 3 inches deep in rows 8 to 12 inches apart. In hand installations, rows are not drawn. Instead, the sprigs are distributed as evenly as possible over the prepared soil surface and pushed down into the soil with a stick. As described earlier, stolonizing uses a top-dressing of soil over the sprigs and eliminates the need for individual insertion. The soil should be moist, but not overly wet, when planting begins. If the lawn area is large, planted areas should be mulched and lightly rolled as the installation progresses. Waiting until the entire installation has been completed could result in drying out of the sprigs.

The Importance of Watering. Water is essential to the growth of all plants. As long as grass is dormant in the seed, it needs no water. However, once planted and watered, the seed swells and germinates. At that point, an uninterrupted water supply is very important. The soil surface must not be allowed to dry until the grass is about 2 inches tall. Watering several times a day, every day for a month, may be necessary.

Caution should be taken to keep the new seedlings moist without saturating the soil. Too much moisture can encourage disease development. The use of a lawn sprinkler is a much better method than simply turning a garden hose onto the new grass. With a sprinkler, the water can be applied slowly and evenly.

The First Mowing. The first mowing of a new lawn is an important one. The objective is to encourage horizontal branching of the new grass plants as quickly as possible. This creates a thick (dense) lawn. The first mowing should occur when the new grass has reached a height of 2 1/2 to 3 inches. It should be cut back to a height of 1 1/4 to 1 1/2 inches. Thereafter, different species require differing mowing heights for proper maintenance. For the first mowing, it is a good practice to collect and remove the grass clippings. After that, clipping removal is usually unnecessary unless the grass has grown so tall between mowings that clumps of grass are visible on the lawn. Grasses used for soil stabilization do not require mowing.

Calibrating a Spreader

Two types of spreaders are used to apply seed, fertilizer, and other granular materials to lawns. These are the **rotary spreader** and the **drop**

spreader. The rotary type dispenses the material from a closed hamper onto a rotating plate. It is then propelled outward in a semicircular pattern. The drop spreader dispenses the material through holes in the bottom of the hamper as it is pushed across the lawn, Figure 22-7. In both types, the amount of material applied is controlled by the size of the holes through which the material passes, and by the speed of application. Therefore, the spreader must be calibrated to dispense the material at the rate desired. For materials that are applied often and by the same person, the spreader need only be calibrated once, with the proper setting noted on the control for future reference. Different materials usually require different calibrations, even when the rate of application is the same.

The object of calibration is to measure the amount of material applied to an area of 100 square feet. A paved area such as a driveway or parking lot is an excellent calibration site. Afterwards, the seed or other material can be swept up easily for future use. Covering the area with plastic is also helpful in recollecting the material.

The spreader should be filled with exactly 5 pounds of the material being applied. Selection of a spreader setting near the center of the range is a good point at which to begin.

The material is applied by walking at a normal pace in a straight line. The spreader is shut off while it is being turned around. Each strip should slightly overlap the previous one. When an area of 100 square feet has been covered once, the spreader is shut off. The material remaining in the spreader is then emptied out and weighed. By subtracting the new weight from the original weight, the quantity of material applied per 100 square feet is determined. The spreader can then be adjusted to increase or reduce the rate of application.

Achievement Review

Briefly answer each of the following questions.

1. How does the cost of sodding compare to that of seeding?
2. Which method has a more immediate effect, seeding or sodding?
3. Which lawns are commonly started by plugging, sprigging, and stolonizing?
4. Which method of lawn installation and establishment requires the most time?
5. At what time of year should warm-season grasses be planted?
6. At what time of year should cool-season grasses be planted?
7. At what time of year should bluegrasses and fescues be planted? Why?
8. Why is it important that soil drain properly?
9. What size of agricultural drainage tile is recommended for lawn use, and how is it spaced?
10. Define soil texture.
11. What type of soil is considered ideal for planting?
12. Explain soil pH.
13. What is a neutral pH level?
14. If soil pH is raised, does the soil become more acidic or more alkaline?
15. If a sandy soil has a pH of 5.0, how many pounds of dolomitic limestone per 1,000 square feet are needed to raise the pH level to that required for a lawn?
16. If the soil mentioned in question 15 covers a lawn area of 3,000 square feet, how much limestone should the landscaper purchase?
17. What are the water requirements of a new lawn?
18. At what height should a lawn mower be set for the first mowing of a new lawn?
19. What is meant by the calibration of a spreader?
20. How is a spreader calibrated?

Suggested Activities

1. Invite a Cooperative Extension Service agent to visit the class for a discussion of soil testing. Ask the agent to demonstrate how a soil sample is collected and to explain how landscapers in the state can arrange to have soil tested.

2. Obtain several inexpensive pH testing kits. Bring in soil samples from gardens for testing.

3. Construct a lawn. If materials for proper lawn construction are available at the school, install a new lawn there. If budget restrictions prevent this, volunteer as a class to work for a nearby park, institution, or property owner in return for equipment and materials for the project.

4. Visit a sod farm if one is located in the area.

5. Borrow several spreaders for calibration. (Families of students might be one source.) If there is no budget for seed, substitute sand for demonstration purposes.

CHAPTER 23

Landscape Irrigation

Objectives:

Upon completion of this chapter, you should be able to

- describe historic and current uses of landscape irrigation
- distinguish between sprinkler and trickle irrigation use and understand key irrigation terms
- select and describe irrigation sprinkler heads
- explain precipitation rates and the determination of water needs by geographical area
- describe how irrigation pipe is sized and the available flow and pressure of water are determined
- explain drip tube and emitter placement
- explain simple irrigation designs

Irrigation is the supplying of water to land through artificial means. While the delivery systems have varied and changed over the years, the concept is not new. Once humans ceased to be hunters and gatherers and instead became farmers, the realization that water could be directed and relocated to help plants grow soon followed. Canals and channels that connect to nearby rivers have permitted water to flow across fields and plains yielding bounty from an earth otherwise incapable of bearing food. Runoff from the snowcapped peaks of distant mountains has been collected and carried to farm fields and communities whose lifelines are literally the water pipelines that link their arid locales to the water sources.

As humans became garden builders, not just food producers, irrigation principles were found as applicable to the growth and maintenance of pleasure grounds as they were to agriculture. Princely potentates throughout Europe, Asia, and Africa commonly developed their gardens by redirecting the channels of streams and rivers to bring them into or near their properties. That nearby communities might be either flooded or deprived of their water supplies was often not a concern of the aristocratic garden builder.

Today, the irrigation of landscapes is common. Each year more properties receive designed irrigation systems for their lawns and plantings. These systems are gradually replacing the lawn sprinklers, garden hoses, and watering cans of years past. There are several reasons for the increased interest in landscape irrigation. Turf and ornamental plant research continually provides new information regarding the water requirements of specific plant species. Water needs are no longer guessing games. Automated irrigation systems also save water because they apply only the amount that can be absorbed, so there is no runoff. That savings is important for economic reasons, since water costs are rising. It is also important for environmental reasons. Controlled water application prevents water laden with fertilizers or pesticides from washing into waterways and sewers. Also, in regions of limited rainfall, landscape irrigation assures that none of the precious liquid is wasted.

Types of Landscape Irrigation Systems

Landscape irrigation water can be applied in two different ways. **Sprinkler irrigation** applies water under pressure through a delivery system that delivers the water over the tops of plants, Figure 23-1. **Trickle irrigation** supplies water directly to the root zone of the plants, Figure 23-2. The delivery system is low pressure and may be placed on top of or within the soil. Table 23-1 compares sprinkler and trickle irrigation systems. The table should be studied before proceeding further.

The Terminology of Landscape Irrigation

Since landscape irrigation is a somewhat specialized technical field, it has its own specialized vocabulary. An understanding of these terms can provide a basic comprehension of some of the technology of the field.

Figure 23-1   Sprinkler system

Sprinkler head: The device through which water leaves the pipe and is propelled onto the lawn or plantings.

Spray pattern: The specific distribution pattern of a specific sprinkler head.

Spray head: One of two types of sprinkler heads. Spray heads are stationary and made of two parts, the sprinkler body and the spray nozzle. The nozzles control the amount of water that leaves the spray head as well as the spray pattern.

Rotary heads: The other type of sprinkler head. Rotary heads move in a circle and propel water in tiny streams under pressure. Rotary heads can dis-

Figure 23-2 Water to these containerized plantings is being delivered through a trickle irrigation system. Look carefully to see the thin tubes that carry the water to the emitters located on the surface of the potted soil.

tribute water much further than spray heads are able to.

Impact drive heads: One of two types of rotary sprinkler heads. As the pressurized water flows from the head it bumps a spring-loaded arm that slowly moves the head around in a circle. The repeating bumping of the arm makes this a somewhat noisy form of irrigation.

Gear drive heads: The other type of rotary sprinkler head. The head contains a number of gears that are turned by the water as it leaves the sprinkler. The movement of these internal gears is what causes the head to turn.

Precipitation rate: The amount of water placed over a landscape area. It is measured in inches of water per hour.

Geographical area: An area of the landscape with a specific water need, based upon the needs of the plants growing there and the rate of water absorption by the soil. A landscape may contain numerous and different geographical areas.

Emitter: The device that functions as a sprinkler head for trickle irrigation.

Discharge rate: The amount of water flowing from the irrigation system over a measured period of time. Sprinkler systems are measured and rated in gallons per minute (gpm). Trickle systems are slower and measured in gallons per hour (gph).

Pressure compensating emitters: Not all emitters can maintain the same discharge rate if the pressure of the water changes. Those that can are termed pressure compensating emitters. They are more expensive than noncompensating types.

Micro spray heads: Similar to the spray-head design described earlier, but with a discharge rate more like that of an emitter (in gph). The diameter of water throw is considerably reduced. It is used for small, specialized areas of the landscape such as flower beds and groundcover plantings.

Irrigation pipe: Used for sprinkler irrigation systems, there are two types of pipe currently popular due to their superiority over metal pipe. Polyethylene pipe (PE) and polyvinyl chloride pipe (PVC) are widely used both separately and together.

Schedule rated pipe: A PVC pipe used as a covering or sleeve for thinner, weaker pipe when it passes beneath walks, roads, driveways, or walls. The higher the schedule rating is, the stronger the pipe will be.

Pressure rated pipe: A PVC pipe used to carry the water through an irrigation system. It is produced

Table 23-1 A Comparison of Sprinkler and Trickle Irrigation Systems

Feature	Sprinkler System	Trickle System
Delivery System	Water is sent under pressure through polyethylene or polyvinyl chloride pipe and delivered from either spray heads or rotary heads.	Water is sent under pressure through plastic or poly pipes or tubes and delivered from a small sprinkler head known as an emitter.
Delivery Site	Over the tops of the plant and onto the soil beneath	At the roots of the plants
Water Pressure at the Delivery Site	Varies with each system, but always greater than trickle systems. The pressure must propel the water through the air.	Low pressure resulting from either the small openings in the emitters or the long distance traveled through the pipes.
Speed of Delivery	Faster than trickle systems. Water delivery is measured in gallons per minute.	Slower than sprinkler systems. Water delivery is measured in gallons per hour.
Delivery of Pesticides and Fertilizers	As long as the materials are water-soluble, they can be delivered in the irrigation water. Some waste and/or chemical burning of foliage may result.	As long as the materials are water-soluble, they can be delivered in the irrigation water. Chemical burn is less likely. Plugging of the emitters is possible. Filtration of the water going into the system will help control the problem.
Water Runoff	Possible if the system is not set to deliver only what the soil can absorb	Little or none due to the slow speed of delivery
Weed Growth	Possible, due to the greater area of coverage	Less of a problem due to the more restricted area of coverage
Evaporated Water Loss	Water is lost to evaporation because it is thrown through the air, left atop foliage, and dispersed into soil outside the root area.	Little water loss because the water is never airborne or left on the tops of leaves to evaporate
Insects and Disease	Excessive moisture on foliage and soil can encourage pests to develop.	By eliminating the amount of wet foliage and soil, pest development is discouraged.
Installation Costs	More expensive due to the greater amount of hardware required	Less expensive due to the comparative simplicity of the system
Maintenance	Moderate maintenance. It is quickly apparent when all or parts of the system are malfunctioning.	High maintenance. Because the system cannot be seen working as easily as sprinklers are seen, it requires constant checking.

in four forms. Each form is guaranteed to withstand a particular water pressure of either 125, 160, 200, or 315 pounds per square inch (psi).

Drip tubing: Used for trickle irrigation systems, the thin black tubing carries the water to the emitters. It is usually black in color and made of plastic.

Flow: The movement of water through the irrigation system.

Friction loss: The loss of pressure during water flow that results from the increasing speed of the water, the increasing length of the pipe, and the roughness of the lining of the pipe.

Velocity: The rate of flow, as expressed in feet of movement per second. For landscape irrigation, the ideal velocity of water is 5 feet per second.

Surge pressure: A shock wave type of pressure that results when water velocity is too high and then is suddenly stopped. Excessive surge pressure over time will weaken the pipe and connections within an irrigation system.

Static pressure: The pressure (in psi) that is present in a closed system, when the water is not flowing.

Dynamic pressure: The pressure (in psi) that is present at any one point in a system as a given quantity of water flows past that point. Dynamic water pressure varies within an irrigation system due to the rise and fall of water as the pipes move up and down across a landscape. Friction losses within the pipes also affect dynamic pressure.

Up to this point both sprinkler and trickle irrigation systems have been discussed simultaneously. Now the discussion will focus on each system separately.

Sprinkler Irrigation

Where large quantities of water must be applied in a controlled manner, sprinkler irrigation systems are the best delivery device. Golf courses and other large lawn areas as well as tree and shrub plantings are suitable for the rapid water delivery of sprinkler systems.

Sprinkler Heads

As noted earlier, there are two types of sprinkler heads, the spray head and the rotary. Spray heads are available as pop-up types that are recessed below mower level when not in use. The same water pressure that allows delivery also forces the spray head nozzle above ground level, giving the pop-up its name. Figure 23-3 illustrates several sizes of pop-up spray heads that are available.

Spray heads that do not recess are also available. They are needed for irrigating tall grasses and shrub beds. A plastic riser that may extend several feet above the ground permits a nozzle, specially adapted for shrubs, to deliver water to the plants.

Aside from popping up and down, spray heads have no moving parts. Nozzles are usually purchased separately from the spray head bodies. That permits the system designer to customize the irrigation system for each particular landscape. Nozzles may be selected to distribute water in a full circle, half circle, or quarter circle. They can also deliver

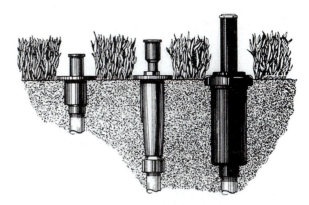

Figure 23-3 Several sizes of the pop-up spray heads are available

water at standard volumes or at low gallonages if site conditions necessitate a slower rate of water application.

Spray heads have their limitations. They can only propel water outward to a distance of 14 to 16 feet from the nozzle before the wind disrupts the spray pattern. A disrupted spray pattern results in non-uniform water application. The overall effect can be inconsistent growth and troublesome wet or dry spots.

Rotary heads have moving parts. They may or may not pop up, depending upon the style, but all rotary heads move in full or partial circles and can propel water further than spray heads. Both the impact drive and gear drive heads are available in sizes that will throw water 110 feet or more from the nozzle. Figures 23-4 and 23-5 illustrate the impact drive and gear drive rotary heads.

In comparison with spray heads, rotary heads do not apply water as quickly. Like spray heads, rotary heads are also affected by the wind.

Precipitation Rates

Because spray heads apply water faster than rotary heads are able to do, it can be said that they have a higher precipitation rate (in inches of water per hour). Regardless of the type of landscape irrigation system selected, it is necessary to match the discharge rate of the system (in gpm) to

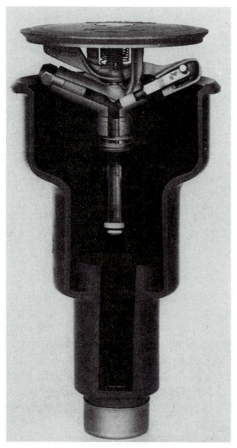

Figure 23-4 Impact drive rotary head

Figure 23-5 The Weather-matic Turbo J2 gear drive head

the precipitation rate that will permit optimum absorption by the plants and meet their water needs, which are usually stated as so many inches of water per week. To convert from gallons per minute through inches per hour to arrive at inches per week and determine irrigation time is not impossible; but there are some necessary steps along the way.

STEP 1: Determine the precipitation rate (PR)

    a. Calculate total gpm by totaling the gpm of all irrigation sprinkler heads in the area of the landscape that is of concern.

    b. Determine the square footage (SF) of the area being covered.

    c. Using a constant figure, 96.3, to aid the conversion, insert the data into the following formula:

$$PR = \frac{\text{Total GPM of an area} \times 96.3}{\text{SF of the area}}$$

STEP 2: Convert the precipitation rate from inches of water applied per hour to inches per minute using the formula:

$$\frac{PR}{60 \text{ minutes}} = \text{Inches of water per minute}$$

STEP 3: Determine how long the sprinklers must run to deliver the plants' weekly water needs.

    a. Learn the water needs of particular plants and plant types. See Table 23-2 for examples.

    b. Divide the plant's weekly water needs by inches of water per minute.

Example

Given: A 1,000 SF shrub installation needing 1.5 inches of water each week is irrigated by a system that delivers 5.2 gpm.

Question: How much irrigation running time is needed to deliver 1.5 inches of water?

Solution:

1. $PR = \dfrac{5.2 \text{ GPM} \times 96.3}{1,000 \text{ SF}}$

    $= .5 \text{ inch per hour}$

2. $= \dfrac{.5 \text{ inches per hour}}{60 \text{ minutes}}$

    $= .008 \text{ inches of water per minute}$

3. $= \dfrac{1.5 \text{ inches}}{.008 \text{ inches/minute}}$

    $= 188 \text{ minutes of operation}$

4. $= \dfrac{188 \text{ minutes}}{60 \text{ minutes}}$

    $= 3 \text{ hours and 8 minutes of operation}$

In the situation described in the example, the sprinkler irrigation would need to operate 3 hours and 8 minutes or 188 minutes each week in the shrub installation to deliver the amount of water needed by the plants.

Geographical Areas

Table 23-2 illustrates that not all plants require the same amount of water for optimum growth. To landscapers, it is no surprise that not all soils are alike either. They differ greatly in their ability to

Table 23-2 A Comparison of Plant Water Needs

Plant Type	Water Needs
Turfgrass	1.5 to 2.0 inches/week
Flowers	1.5 to 2.0 inches/week
Trees and shrubs	1.0 to 1.5 inches/week
Groundcovers	0.5 to 1.0 inches/week

absorb water. Generally, the more clay that a soil contains, the slower is its rate of absorption. The greater the sand content of a soil, the faster will be its rate of absorption. Table 23-3 illustrates some of the characteristics of different types of soil that affect their rate of water absorption.

A landscape that is large enough and of sufficient quality to warrant an irrigation system is likely to contain a mixture of plants. It may also have a mixture of soil conditions. Therefore one operational schedule that will serve all areas of the landscape is unlikely. Instead the landscape must be divided and grouped into geographical areas. Each geographical area will contain plants that have similar water needs and similar soil absorption rates. Additionally, allowance must be made for terrain factors such as slopes or depressions that will speed up or slow down water runoff. Other environmental conditions such as temperatures, drying winds, presence or lack of shade, and the number of competing plants can influence the availability of water to the plant's roots.

Once the landscape site has been analyzed and the number of geographical areas has been determined, such follow-up questions as these must be answered.

1. What sprinkler heads are needed for each geographical area?
2. How many are required and of what size?
3. What is the total gpm required by the sprinkler heads in each geographical area and for the total system?

Table 23-3 Soil Characteristics and Absorption Rates

Soil Type	Components	Water Intake Rate	Water Retention	Drainage	Water Absorption
Clay	Clay, with small amounts of silt and/or sand	Low to very low as the amount of clay increases	High to very high due to reduced air space for drainage	Poor to very poor	0.2 inches per hour (Low)
Loams	Equal or near-equal mixes of sand, silt, and clay	Moderately low to moderately high depending upon the amounts of clay or sand in the loam	Moderately low to moderately high depending upon the amounts of clay or sand in the loam	Good	1.0 to 1.7 inches per hour (Moderate)
Sandy	Sand, with lesser amounts of clay and silt	High to very high	Low to very low	Good to excessive	Greater than 1.7 inches per hour (High)

4. How does the total system need compare to the total water flow available from the source of water?
5. What sizes of pipe are needed, and what layout is best?
6. What static and dynamic water pressures are available?

Only after these questions and others are answered and the irrigation system is made operational will there be enough data to determine things such as the amount of running time for the system. Then, if it is found that a single weekly operation of the system for the calculated amount of time is more than the soil in a geographical area can absorb, it will be necessary to set up a schedule that will apply a lesser amount several times per week. The total amount of water applied will be the same.

Sizing Irrigation Pipe

Cost of the pipe is one of the most expensive factors in the price of an irrigation system. The larger the pipe diameter, the more costly it will be. Therefore, system designers must specify pipe that is large enough to carry the amount of water required by all sprinklers that lead off it, but it need be no larger. Pipe sizes will vary within an irrigation system, being larger in diameter near the water source

and smaller as they are needed further from the source, Figure 23-6. Table 23-4 explains the relationship of irrigation pipe size to the amount of water that can be carried.

Matching Water Flow and Pressure with Pipe Size

To select and specify the most suitable pipe, the irrigation system designer must know the amount of water demanded by the sprinkler heads that are downstream from each section of pipe. Downstream is the term to describe the direction of water flow away from the source. Pipe that is

Table 23-4 Water Flow and Irrigation Pipe Size

If the pipe must carry	Select a pipe size of
1 to 6 gallons per minute	0.50 inch
7 to 10 gpm	0.75 inch
11 to 16 gpm	1.00 inch
17 to 26 gpm	1.25 inches
27 to 35 gpm	1.50 inches
36 to 55 gpm	2.00 inches

PLATE 1 Following construction of this home, a Maryland-area landscape firm began development of the public area setting.

PLATE 2 The entry drive was marked out and forms set where the pavers would be installed.

PLATE 3 Concrete was poured and leveled in sidewalk areas.

PLATE 4 The driveway pavers were set on top of a level and compacted subgrade of stone dust.

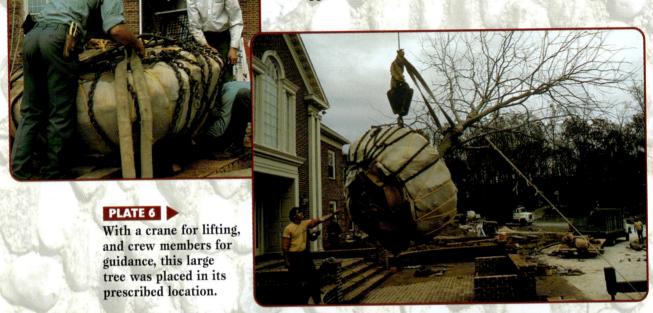

◄ **PLATE 5** Large trees had their soil balls wrapped in chains prior to being lifted into place.

PLATE 6 ►

With a crane for lifting, and crew members for guidance, this large tree was placed in its prescribed location.

▲ **PLATE 7** Once the lawn was established, the public area was complete. The residence was ready to welcome guests.

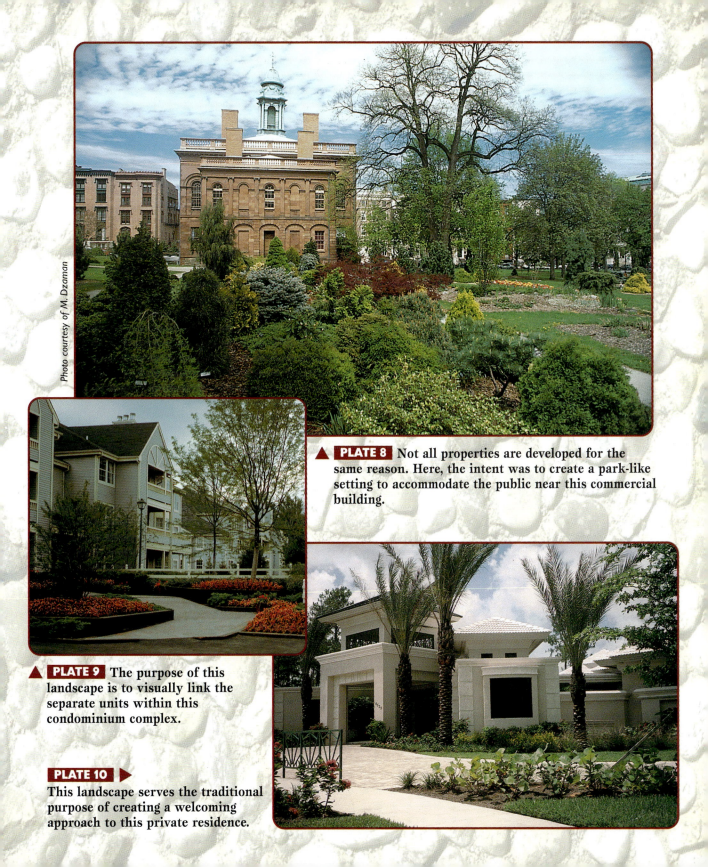

Photo courtesy of M. Dzaman

▲ **PLATE 8** Not all properties are developed for the same reason. Here, the intent was to create a park-like setting to accommodate the public near this commercial building.

▲ **PLATE 9** The purpose of this landscape is to visually link the separate units within this condominium complex.

PLATE 10 ▶

This landscape serves the traditional purpose of creating a welcoming approach to this private residence.

PLATE 11 Urban landscapes are frequently architectural in both style and function. This interior plantscape creates a setting for quiet conversation within a busy hotel lobby.

PLATE 12 ▶
In this multi-level outdoor room, a large deck overlooks the adjacent tennis court.

▼ **PLATE 13** Planted walks, a tree canopy, textured pavers, and comfortable furniture complete this outdoor room.

PLATE 14 Flowers are most dramatic when used in large quantities and in masses of few colors.

PLATE 15 Equally dramatic effects can be achieved by selecting and combining plants with sharp textural contrasts.

PLATE 16 Naturalized effects can be created with bulb flowers that are tossed randomly onto the ground and planted where they fall. Over time, they multiply and create drifts of color such as this planting of daffodils.

PLATE 17 ▶ When used as a small enrichment feature, water is similar to a mirror in the outdoor room. It reflects the light, color, and movement of other nearby components.

▲ **PLATE 18** Used over an extensive area, water becomes a visual adhesive, binding together the many diverse features of a large landscape.

PLATE 19 Floating fountains such as this, not only enrich the landscape visually, but keep the water aerated as well.

PLATE 20 When planning the landscape of a swimming pool area, remember to provide equal amounts of water and patio surface.

▲ **PLATE 21** These twin waterfalls and boat landing are the designed centerpieces of a spectacular hotel lobby near arid Palm Springs, CA.

▼ **PLATE 23** Two repetitions develop the principle of unity here. Identical arching lines are evident in the water, turf, and pavement. Even spacing of the identical planters contributes as well.

▲ **PLATE 22** Unity is created by the repetition of lines, colors, fixtures, and other items. In this resort landscape, the repeating use of the same light fixture applies this principle.

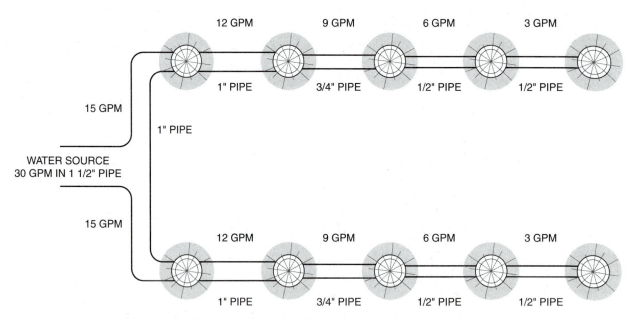

Figure 23-6A Pipe diameter diminishes as it moves downstream.

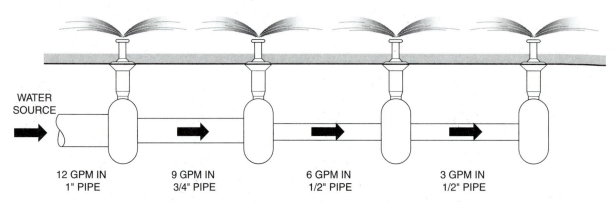

Figure 23-6B The pipe diameter must be able to carry the GPM required.

upstream will be of wider diameter than pipe that is downstream because it must carry a greater volume of water. Figure 23-6A illustrates an irrigation pipe fitted with ten sprinkler heads. Each sprinkler needs 3 gpm to operate. In the illustration the total gpm that must be carried by each section of pipe is noted. In Figure 23-6B are seen the appropriate pipe sizes required to carry the volume of water at specific points.

Calculating Working Water Pressure

Complicating the match-up of water pressure and quantity with the appropriate pipe size is the problem that pressure varies, depending upon whether the water is static (still) or dynamic (moving). Pressure is also affected by the influence of gravity if water drops or rises from one level to another.

Pressure will change as it is released downstream through valves along its route of flow and also in response to the roughness of the inside of the irrigation pipes (friction loss).

Calculating Static Water Pressure

To calculate static water pressure at the system's main valve, the pressure at the main line source must be adjusted by the pressure created by any elevation change that exists between the main line and the valve. This is done by multiplying the elevation change in feet by .433 pounds per square inch. This figure (.433 psi) is the constant weight of a 1-foot tall column of water. It is an accepted constant figure used in irrigation calculations. The following example will help clarify the calculation of static pressure.

Problem:

The water pressure supplied from the municipal source is 60 psi. From the main line, a pipe drops 15 feet to the valve that controls the flow of irrigation water. What will be the static pressure at the valve?

Solution:

Static pressure at valve

= Static pressure at main line
  + elevation change × .433 psi

Static pressure at valve

= 60 psi + 15 (.433 psi)

= 60 psi + 6.50

= 66.5 psi

Calculating Dynamic Water Pressure

To calculate moving water pressure it is necessary either to know or to calculate the friction loss as the water rubs against the inside of the pipe. It is also necessary to calculate the pressure lost as the water passes through valves, as it passes through the water meter, and as it rises and falls due to elevation changes.

As this chapter can only serve to introduce students to the subject of landscape irrigation, certain concepts must be explained with the understanding that further study and supplemental materials are needed if the student is to develop a true competency. Such is the case in describing the determination of friction loss and appropriate pipe sizes. Manufacturers of irrigation pipe provide irrigation system designers with pipe friction loss charts. There are different charts for different classes and types of pipes. Figure 23-7 illustrates a friction loss chart for a common class of plastic pipe. These charts record the friction loss or pressure loss for a particular size and material of pipe or piece of equipment at different flow rates and velocities. The charts also aid the systems designer in calculating the size of pipe needed, based upon the water flow through the pipe section.

Moving water pressure = Static pressure in the main line; minus the pressure loss in the water meter, the friction loss in the pipe, the pressure loss in valves and fitting; plus or minus elevation pressure gains.

In the sample chart various pipe sizes are listed across the top. Beneath each size, two columns list the velocity in feet per second and the friction loss in pounds per square inch for various rates of water flow. The friction losses are for pipe lengths of 100 feet. For pipe lengths of less than 100 feet, the friction losses would be equivalently reduced.

Example

Given: Water flows through 100 feet of 2-inch pipe at the rate of 9 gallons per minute.

Question: What is the velocity of the water and the friction loss?

Answer: Velocity is 0.79 feet per second (fps). Friction loss is 0.06 pounds per square inch (psi).

Given: Water flows through 50 feet of 3/4-inch pipe at the rate of 10 gpm.

Question: What is the velocity of the water and the friction loss?

	3/4"	1"	1 1/4"	1 1/2"	2"	2 1/2"	3"	4"	6"	
SIZE	3/4"	1"	1 1/4"	1 1/2"	2"	2 1/2"	3"	4"	6"	SIZE
OD	1.050	1.315	1.660	1.900	2.375	2.875	3.500	4.500	6.625	OD
ID	.930	1.189	1.502	1.720	2.149	2.601	3.166	4.072	5.993	ID
WALL THK	0.60	0.063	0.079	0.090	0.113	0.137	0.167	0.214	0.316	WALL THK

Flow G.P.M.	Velocity F.P.S.	P.S.I. Loss	Velocity F.P.S.	P.S.I. Loss	Velocity F.P.S.	P.S.I. Loss	Velocity F.P.S.	P.S.I. Loss	Velocity F.P.S.	P.S.I. Loss	Velocity F.P.S.	P.S.I. Loss	Velocity F.P.S.	P.S.I. Loss	Velocity F.P.S.	P.S.I. Loss	Velocity F.P.S.	P.S.I. Loss	Flow G.P.M.
1	0.47	0.06	0.28	0.02	0.18	0.01	0.13	0.00											1
2	0.94	0.22	0.57	0.07	0.36	0.02	0.27	0.01	0.17	0.00									2
3	1.42	0.46	0.86	0.14	0.54	0.04	0.41	0.02	0.26	0.01	0.18	0.00							3
4	1.89	0.79	1.15	0.24	0.72	0.08	0.55	0.04	0.35	0.01	0.24	0.01							4
5	2.36	1.20	1.44	0.36	0.90	0.12	0.68	0.06	0.44	0.02	0.30	0.01							5
6	2.83	1.68	1.73	0.51	1.08	0.16	0.82	0.08	0.53	0.03	0.36	0.01	0.24	0.00					6
7	3.30	2.23	2.02	0.67	1.26	0.22	0.96	0.11	0.61	0.04	0.42	0.01	0.28	0.01					7
8	3.77	2.85	2.30	0.86	1.44	0.28	1.10	0.14	0.70	0.05	0.48	0.02	0.32	0.01					8
9	4.25	3.55	2.59	1.07	1.62	0.34	1.24	0.18	0.79	0.06	0.54	0.02	0.36	0.01					9
10	4.72	4.31	2.88	1.30	1.80	0.42	1.37	0.22	0.88	0.07	0.60	0.03	0.40	0.01					10
11	5.19	5.15	3.17	1.56	1.98	0.50	1.51	0.26	0.97	0.09	0.66	0.03	0.44	0.01					11
12	5.66	6.05	3.46	1.83	2.17	0.59	1.65	0.30	1.06	0.10	0.72	0.04	0.48	0.02	0.29	0.00			12
14	6.60	8.05	4.04	2.43	2.53	0.78	1.93	0.40	1.23	0.14	0.84	0.05	0.56	0.02	0.34	0.01			14
16	7.55	10.30	4.61	3.11	2.89	1.00	2.20	0.52	1.41	0.17	0.96	0.07	0.65	0.03	0.39	0.01			16
18	8.49	12.81	5.19	3.87	3.25	1.24	2.48	0.64	1.59	0.22	1.08	0.09	0.73	0.03	0.44	0.01			18
20	9.43	15.58	5.77	4.71	3.61	1.51	2.75	0.78	1.76	0.26	1.20	0.10	0.81	0.04	0.49	0.01			20
22	10.38	18.58	6.34	5.62	3.97	1.80	3.03	0.93	1.94	0.32	1.32	0.12	0.89	0.05	0.54	0.01			22
24	11.32	21.83	6.92	6.60	4.34	2.12	3.30	1.09	2.12	0.37	1.44	0.15	0.97	0.06	0.59	0.02			24
26	12.27	25.32	7.50	7.65	4.70	2.46	3.58	1.27	2.29	0.43	1.56	0.17	1.05	0.07	0.63	0.02			26
28	13.21	29.04	8.08	8.78	5.06	2.82	3.86	1.46	2.47	0.49	1.68	0.19	1.13	0.07	0.68	0.02			28
30	14.15	33.00	8.65	9.98	5.42	3.20	4.13	1.66	2.65	0.56	1.80	0.22	1.22	0.09	0.73	0.02	0.34	0.00	30
35	16.51	43.91	10.10	13.27	6.32	4.26	4.82	2.20	3.09	0.75	2.11	0.29	1.42	0.11	0.86	0.03	0.39	0.01	35
40	18.87	56.23	11.54	17.00	7.23	5.45	5.51	2.82	3.53	0.95	2.41	0.38	1.62	0.14	0.98	0.04	0.45	0.01	40
45			12.98	21.14	8.13	6.78	6.20	3.51	3.97	1.19	2.71	0.47	1.83	0.18	1.10	0.05	0.51	0.01	45
50			14.42	25.70	9.04	8.24	6.89	4.26	4.41	1.44	3.01	0.57	2.03	0.22	1.23	0.06	0.56	0.01	50
55			15.87	30.66	9.94	9.83	7.58	5.09	4.85	1.72	3.31	0.68	2.23	0.26	1.35	0.08	0.62	0.01	55
60			17.31	36.02	10.85	11.55	8.27	5.97	5.30	2.02	3.61	0.80	2.44	0.31	1.47	0.09	0.68	0.01	60
65			18.75	41.77	11.75	13.40	8.96	6.93	5.74	2.35	3.92	0.93	2.64	0.36	1.59	0.10	0.73	0.02	65
70					12.65	15.37	9.65	7.95	6.18	2.69	4.22	1.06	2.84	0.41	1.72	0.12	0.79	0.02	70
75					13.56	17.47	10.34	9.03	6.62	3.06	4.52	1.21	3.05	0.46	1.84	0.14	0.85	0.02	75
80					14.46	19.68	11.03	10.18	7.06	3.44	4.82	1.36	3.25	0.52	1.96	0.15	0.90	0.02	80
85					15.37	22.02	11.72	11.39	7.50	3.85	5.12	1.52	3.45	0.59	2.09	0.17	0.96	0.03	85
90					16.27	24.48	12.41	12.66	7.95	4.28	5.42	1.69	3.66	0.65	2.21	0.19	1.02	0.03	90
95					17.18	27.06	13.10	13.99	8.39	4.74	5.72	1.87	3.86	0.72	2.33	0.21	1.07	0.03	95
100					18.08	29.76	13.79	15.39	8.83	5.21	6.03	2.06	4.07	0.79	2.46	0.23	1.13	0.04	100
110					19.89	35.50	15.17	18.36	9.71	6.21	6.63	2.45	4.47	0.94	2.70	0.28	1.24	0.04	110
120							16.54	21.57	10.60	7.30	7.23	2.88	4.88	1.11	2.95	0.33	1.36	0.05	120
130							17.92	25.02	11.48	8.47	7.84	3.34	5.29	1.29	3.19	0.38	1.47	0.06	130
140							19.30	28.70	12.36	9.71	8.44	3.84	5.69	1.47	3.44	0.43	1.59	0.07	140
150									13.25	11.04	9.04	4.36	6.10	1.68	3.69	0.49	1.70	0.08	150
160									14.13	12.44	9.64	4.91	6.51	1.89	3.93	0.55	1.81	0.08	160
170									15.01	13.91	10.25	5.50	6.91	2.11	4.18	0.62	1.93	0.09	170
180									15.90	15.47	10.85	6.11	7.32	2.35	4.42	0.69	2.04	0.11	180
190									16.78	17.10	11.45	6.75	7.73	2.60	4.67	0.76	2.15	0.12	190
200									17.66	18.80	12.06	7.43	8.14	2.85	4.92	0.84	2.27	0.13	200
225									19.87	23.38	13.56	9.24	9.15	3.55	5.53	1.04	2.55	0.16	225
250											15.07	11.23	10.17	4.31	6.15	1.27	2.83	0.19	250
275											16.58	13.39	11.19	5.15	6.76	1.51	3.12	0.23	275
300											18.09	15.74	12.21	6.05	7.38	1.78	3.40	0.27	300
325											19.60	18.25	13.22	7.01	7.99	2.06	3.69	0.31	325
350													14.24	8.05	8.61	2.36	3.97	0.36	350
375													15.26	9.14	9.22	2.69	4.25	0.41	375
400													16.28	10.30	9.84	3.03	4.54	0.46	400
425													17.29	11.53	10.45	3.39	4.82	0.52	425
450													18.31	12.81	11.07	3.77	5.11	0.57	450
475													19.33	14.16	11.68	4.16	5.39	0.63	475
500															12.30	4.58	5.67	0.70	500
550															13.53	5.46	6.24	0.83	550
600															14.76	6.42	6.81	0.98	600

Figure 23-7   A friction loss chart for a common class of plastic pipe

Answer: Velocity is 4.72 feet per second. Friction loss = 4.31 psi × .50 = 2.16 pounds per square inch.

It is important to know not only the friction loss but also the water velocity in order to minimize surge pressure in the pipes. Also known as the water hammer effect, surge pressure is likely to occur when the velocity goes above 5 fps. Since velocity and pressure loss increase within a pipe as it is required to carry increasing amounts of water, it is necessary to increase the pipe size when friction loss becomes great or when the velocity approaches the 5 fps that will trigger the water hammer effect.

Example

Question: If a 100-foot-long pipe has 20 gpm flowing through it, what is the smallest size of pipe that could be used?

Answer: From the chart in Figure 23-7, the first velocity below 5 fps is that for 1 1/4-inch pipe, with a velocity of 3.61 fps.

Question: What would be the friction loss in the above pipe?

Answer: 1.51 psi

Given: Water passes through 100 feet of 1 1/4-inch pipe at the rate of 20 gpm. The static pressure in the main line equals 66.5 psi, the friction loss in the 1 1/4-inch pipe is 1.51 psi, and the friction loss as water passes through the valve is 6 psi.

Question: What is the working (moving) water pressure?

Answer: Working water pressure = Static pressure − pressure loss in the valve − friction loss in the pipe = 66.5 psi − 6 psi − 1.51 psi = 58.99 psi.

Once through the valve, the water flows toward the first sprinkler. Again the pipe must be sized to assume enough pressure to pop the sprinkler out of the ground and deliver the desired amount of irrigation water, while avoiding surge pressure.

If the first sprinkler is 25 feet from the valve, reference to the chart finds the friction loss to be 0.38 psi (1.51 for 100 feet of pipe × .25) in 1 1/4-inch pipe with a flow of 20 gpm. Subtracting the friction loss from the working pressure at the valve reveals a working pressure at the first sprinkler of 58.61 psi.

Sprinkler heads operate within ranges of pressure. Different types of sprinkler heads have different ranges of operation. Continuing with the above example, assume that the sprinkler head being used has a pressure range of 35 to 80 psi. The 58.61 psi of pressure at the first sprinkler head is within the operating range, so the head would pop up and deliver its prescribed amount of irrigation water. As an example, assume that the sprinkler heads of the system deliver 10 gpm. After 10 of the 20 gallons of water per minute are used at the first sprinkler, the remaining 10 gallons per minute flow on toward the second sprinkler, spaced 25 feet further away. Again, the pipe must be sized. From the sample chart in Figure 23-7, the smallest pipe size that will carry the 10 gpm with a velocity of less than 5 fps is 3/4-inch pipe. In a real situation, with a higher rate of flow and more sprinkler heads, it is sometimes wise to use a size larger than the smallest pipe possible in order to reduce the friction loss that can reduce working pressure. Conversely, when working pressure needs to be reduced to fit within the pressure range of the sprinklers, the use of a smaller pipe will increase the friction pressure and reduce working pressure in the pipe. A pressure regulator can also be placed along the pipe to reduce water pressure.

Selecting and Locating Sprinklers

Upon leaving the sprinklers under pressure, the irrigation water falls across the landscape in a pattern determined by the type of sprinkler head. Either circular or rectangular patterns are possible. The size of the sprinkler head is partly responsible for the amount of area covered. The other factor controlling area of coverage is the trajectory of the spray. Sprinkler head size is matched to the amount of water being delivered. Trajectory is the path of the water as it is propelled through the air. Flat or low trajectories are used in groundcover and beneath shrubs in planting beds. Higher trajectories are used for lawns and for plantings that are watered from overhead.

Sprinkler Pattern

Rectangular patterns fill prescribed areas that are, obviously, rectangular in shape. Circular patterns may be full or partial circles, in radii ranging from several feet to over a hundred feet. Figure 23-8 illustrates common circular patterns. An irrigation system designer must have a thorough knowledge of the hardware available before he or she can design an efficient system.

Locating the Sprinklers

The objective in the positioning of sprinklers is simple: to obtain an even distribution of water. Uneven irrigation causes inconsistent growth and wet and dry areas within the landscape. Sprinkler heads can be set out in either a square pattern or a triangular pattern, Figure 23-9. Triangular patterns are preferred because they give more even water distribution and waste less water. Both patterns rely upon overlapping of the spray arcs because each sprinkler only delivers the full amount of water to 2/3 of its diameter. The outer third would be drier and the distribution of water uneven were it not for the overlapping.

In a triangular layout, the sprinkler heads are positioned as an equilateral triangle, with equal space between all sprinklers. Where wind is not a predictable factor, the sprinkler arcs should be spaced at a distance that is 60 percent of the diameter of the arc. If wind is predictable and constant, spacing should be reduced. Manufacturers of sprinkler heads will provide recommendations for spacing of the sprinklers depending upon the various wind speeds. These recommendations should not be stretched. To do so will only result in uneven water distribution. In free-form or narrow lawn areas, the spacing may be crowded if necessary, but not stretched.

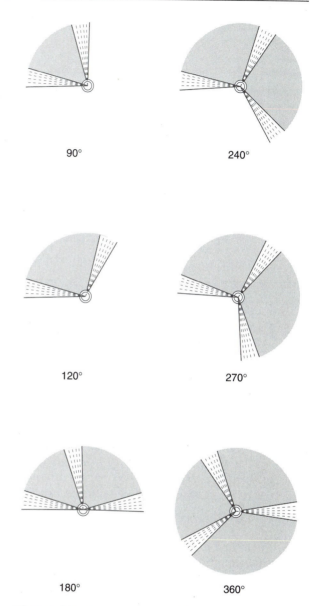

Figure 23-8 Common circular patterns

Problem:

To lay out the sprinkler heads in a square pattern for a rectangular lawn area that is 75 feet × 60 feet. Sprinklers will be used that require 15-foot spacing based upon wind conditions and the manufacturer's recommendations.

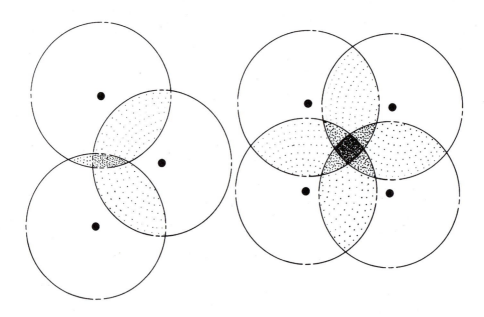

Figure 23-9 Triangular (left) and square (right) spray patterns

Solution:

1. Select the bordering edge that most needs to avoid overspray in order to prevent slippery walks or wet people. Divide the length of the border by the recommended spacing of the sprinkler heads. Round up to the next whole number if necessary.

   75 feet ÷ 15 feet = 5 spaces

2. Divide the other edge length by the recommended spacing of the sprinkler heads. Again round up to the next whole number if necessary.

   60 feet ÷ 15 = 4 spaces

3. Draw a grid across the scaled lawn, providing the calculated number of spaces. Locate sprinklers at all grid line intersections.

4. Select the appropriate sprinkler patterns to provide full coverage without overspray.

NOTE: In the example the border length is divided evenly by the sprinkler spacing. When that is not possible, an additional calculation is needed to determine the actual spacing of the sprinkler heads.

Example

If the border edge had been 79 feet, rather than 75 feet, the edge length would have been first divided by the ideal spacing:

79 feet ÷ 15 feet = 5.27 = 6 (round up to a whole number)

then divided again by that result to determine the actual spacing of the sprinklers:

79 feet ÷ 6 = 13.17 feet

Problem:

To lay out the sprinkler heads for an 80 foot × 65 foot lawn area using a triangular pattern. Again, the sprinklers used will require 15-foot spacing to compensate for wind conditions.

Solution:

1. Select the border edge where overspray must be avoided. Divide the length by the recommended spacing of the sprinkler heads and round up to the next whole number if necessary.

   80 feet ÷ 15 = 5.33 = 6 spaces

2. Divide the border edge length by the number of spaces to determine the actual spacing of the sprinkler heads.

   80 feet ÷ 6 spaces = 13.33 feet

3. Divide the other edge length by the recommended spacing of the sprinkler heads, rounding up to the next whole number if necessary.

   65 feet ÷ 15 = 4.33 = 5 spaces

4. Divide the edge length by the number of spaces to determine the actual spacing of the sprinkler heads.

   65 feet ÷ 5 spaces = 13 feet

5. Draw a grid across the scaled lawn, providing the calculated number of spaces. Then locate the sprinkler heads in a pattern that approximates equilateral triangles as closely as possible.

6. Select the appropriate sprinkler patterns to provide full coverage without overspray.

Trickle Irrigation

For a quick review of the characteristics of trickle irrigation, consult Table 23-1. The system is useful in any situation where slow water application is preferable to sprinkler irrigation due to constraints of space, budget, or limited water supply. Instead of a sprinkler head, water is delivered from an emitter attached to a thin plastic or polyethylene tube. The delivery rate is measured in gallons per hour (gph), and is much slower than the gallons per minute delivery of sprinkler irrigation systems.

Emitters

There are two types of emitters: pressure compensating and non-compensating emitters. The difference is based upon their ability or inability to deliver a constant gph over a wide range of water pressures. In landscapes with rolling topographies, the rise and fall of the land causes drastic inconsistency in the water pressure in the thin tubes. Pressure compensating emitters are designed to allow for that variation and still deliver the intended amount of water and the desired rate. Non-compensating emitters cannot adjust to varied water pressures. If the pressure changes, their delivery rate changes too. In order to compensate for the effects of topographic variation on water pressure, it is common to use pressure regulators to reduce water pressure at points within the trickle system. Doing this permits the non-compensating emitters to deliver water at their intended rate of flow.

Tube and Emitter Placement

Drip tubing is not visually offensive and can be stretched on top of most planting beds or hidden inconspicuously beneath the mulch of the bed. Emitters may be attached directly to the tube or extended away from the tube using micro tubing. For large areas such as shrub plantings, multiple emitters are needed at a plant if sufficient water is to be provided.

In determining the number of emitters needed for a plant, the irrigation designer must consider the soil type and the subsurface wetting pattern that it promotes, the canopy area of the plant, and the percent of the plant's root zone that is to be wetted. It is only necessary to wet 50 percent of the root zone of most plants to provide the benefits of irrigation. Placement of the emitters should avoid direct contact with the trunk or stem of the plant. Failure to do so can result in soil pockets that are too wet for too long, and the health of the plant can be affected.

Subsurface wetting patterns are influenced by the type of soil being irrigated. In sandy soil, with its large particles, the water will move more vertically than horizontally. In fine soils like clay, with

Table 23-5 Soil Types and Emitter Wetting

Soil Type	Area Wetted Per Emitter	Diameter Wetted
Clay soil	65 to 160 square feet	9 to 14 feet
Loam soil	21 to 65 square feet	5 to 9 feet
Sandy soil	5 to 21 square feet	2 1/2 to 5 feet

small soil particles, the movement of water is more horizontal than vertical.

Where the irrigation water contains high concentrations of sodium, chloride, or both, trickle irrigation can encourage a dangerous salt build-up that can kill the plant. Improperly spaced emitters are the usual cause of the problem. Properly spaced, the emitters can both prevent the build-up and encourage the leaching of harmful salts away from the root zone.

Knowing the subsurface wetting patterns is important so that they can be overlapped and enough water applied to leach the salts away. Table 23-5 offers examples of the areas wetted by emitters in different soil types. A designer would need this information to plan for proper emitter placement.

The limitation of an introductory chapter to a complex subject is that it raises more questions in the reader's mind than it can answer. This chapter is not intended as a full coverage of landscape irrigation. It attempts only to introduce the subject to students and offer a few ideas and examples. Texts devoted solely to this subject should be consulted before attempting work with irrigation systems.

The Future

While landscape irrigation is not a technology still in its infancy, neither is it a technology standing still. In most parts of the country, water availability and quality are ongoing concerns. While the demand for landscape irrigation of home and commercial landscapes grows, the water resources needed to supply the systems do not grow. Many

regions, especially in arid areas, are already using water faster than it can be cleansed and purified or replaced. In such regions the mere sight of an irrigation system in operation is, to some people, like waving a red flag in front of a bull. It is seen as a wasteful, aristocratic use of a precious and increasingly expensive resource.

For these and other reasons, the future of landscape irrigation will be shaped or altered by many factors. Among them:

- Turfgrasses, groundcovers, and other landscape plants will be bred, selected, and used on the basis of their low water requirements.

- Passive water conservation methods such as the proper placement of trees for shading buildings, lawns, and plantings will become more topical.

- Computer operation of irrigation systems as well as the use of computers to organize preventive maintenance programs for the system is now a reality. Computer-aided design of complex irrigation systems is also common today. The future will find even greater reliance upon the computer. It is not unlikely that a computer will enable the system to detect its own leaks, pinpoint the location, and send for repair assistance.

- Alternative sources of water for irrigation systems will be sought. Effluent water (treated sewage) and other types of reclaimed water may be possible sources.

- Irrigation systems will become more responsive to the actual need for water rather than merely using a timer that turns on the water whether it is needed or not.

- Systems may have sensors built in that will shut down the system when excessive winds cause inefficiency and waste.

- Systems may be powered by self-contained energy systems such as solar panels.

- The position of water manager will become a valued and professional field of career specialization.

Achievement Review

A. Briefly trace the development of irrigation from its agricultural beginnings to its present use in landscapes. Give examples of landscape irrigation systems in operation within your community.

B. Indicate if the following features are most typical of

(A) sprinkler systems   or   (B) trickle systems

1. Less expensive due to the comparative simplicity of the system
2. Maintenance is high since the system must be checked constantly to be certain it is working
3. Water delivery is over the top of the plants and onto the soil beneath
4. Water is delivered through an emitter
5. Water is delivered through spray heads or rotary heads
6. Water runoff is less due to the slow speed of delivery
7. Water pressure at the delivery site is the greatest

C. Insert the proper term to complete the following definitions.

1. _____ heads move in a circle and propel water in a circle.
2. The specific distribution pattern of a sprinkler head is its _____.
3. _____ heads are driven by a spring-loaded arm that responds to water pressure and bumps the sprinkler around in a circle.
4. The amount of water placed over a landscape area by a sprinkler is the _____.
5. The _____ is the delivery device for trickle irrigation.
6. The amount of water flowing from an irrigation system over a measured period of time is the _____ rate of the system.
7. The _____ of PVC pipe is a measure of its strength.

8. A PVC pipe used to carry water through an irrigation system is termed _____ pipe.
9. _____ is the loss of pressure during water flow resulting from the increasing speed of water, the increasing pipe length, and the roughness of the lining of the pipe.
10. Velocity is the rate of _____, as expressed in feet per second.
11. When the water velocity is too high and is suddenly stopped, water hammer or _____ pressure can result.
12. The pressure present in a closed system, when the water is not flowing, is termed _____ pressure.
13. As water flows past a point and responds to rises or falls within the system, its pressure becomes _____.

D. Indicate if the following characteristics of sprinkler heads apply to

a. spray heads   c. both
b. rotary heads   d. neither

1. The heads are available in pop-up styles.
2. They have no moving parts.
3. They are able to propel water across the greatest distance.
4. Water is distributed in full or partial circles.
5. Water is applied most rapidly.
6. Their spray pattern is affected by the wind.

E. State whether the following statements are (T) true or (F) false.

1. Spray heads have higher precipitation rate than rotary heads.
2. The discharge rate of the irrigation system must be matched with the rate at which plants can absorb the water.
3. Discharge rate is measured in gallons per minute.
4. Plant water needs are measured in gallons per week.

5. It is not possible to convert from gallons per minute to inches per week.

6. Not all plants require the same amount of water for optimum growth.

7. Sandy soil has a slower rate of absorption than clay soil.

8. Loams drain the best of all soil types.

9. Geographical areas contain a variety of plants and soils with a diversity of water needs and absorption rates.

F. As concisely as possible, answer the following questions in essay form.

1. Why does the size of irrigation pipe matter?

2. How do the terms *upstream* and *downstream* relate to water volume in an irrigation pipe?

3. What factors affect working water pressure?

4. What is a friction loss chart?

5. What relationship exists between water hammer, a water velocity of 5 fps, and the size of irrigation pipe?

G. Fill in the blanks to make the following paragraph read correctly.

In a trickle irrigation system, the water is delivered through _____. There are two types of these. They are pressure compensating and _____. Pressure compensating emitters deliver a constant rate of water over a wide range of _____. The use of _____ enables non-compensating emitters to do the same thing. In determining the number of emitters to specify, an irrigation designer must consider the _____ type and the _____ pattern, as well as the canopy of the plant and the percent of the _____ system that is to be wetted. It is only necessary to wet _____ percent of the root system of most plants to provide proper irrigation. Should the irrigation water contain high concentrations of sodium, chloride, or both, dangerous levels of _____ may build up in the soil.

CHAPTER 24

Interior Plantscaping

Objectives:

Upon completion of this chapter, you should be able to

- discuss the current status of the interior plantscape industry
- list problems unique to the interior use of plants
- describe the role of light quality, intensity, and duration
- list the characteristics of a good growing medium
- explain the steps in proper installation, watering, and drainage
- describe the interdisciplinary relationship required between architects, landscape architects, interior plantscapers, and maintenance professionals

Containerizing Plants, Past and Present

The human need and pleasure in having plants within our homes and places of business and play are rooted in antiquity. The ancient Greeks, Romans, and Egyptians commonly grew plants in containers to provide herbs for cooking, fragrance, and/or medicinal uses. To some extent most cultures, primitive or civilized, have used plants to add color, texture, softness, and a sense of life to their static, fabricated interior environments.

In America, the residential use of interior plantings has expanded from houseplants and holiday flowers to large trees and exotic desert and jungle species. The soaring popularity of interior plants began in the mid-1970s. America currently leads the world in the development and appreciation of interior plantscapes. Commercially, it is now expected that every new shopping mall, office building, hospital, or hotel will have plants featured prominently in the interior public areas. Most knowledgeable employers are aware of the increased productivity and reduced absenteeism of employees who have living, green plants as part of their working environment. Therefore, interior plantings are not limited to public areas but are now evident within the nation's workplaces as well, Figures 24-1 and 24-2. Architects and interior decorators recognize that plants are as much a part of the indoor room as furniture, carpeting, and wall

Figure 24-1 Interior plantscaping has expanded throughout the country. It enhances shopping malls and office complexes by providing a natural garden atmosphere.

coverings. Properly selected and displayed, interior plantings duplicate the role and contribution of fine art to the mood and ambiance of a room.

So explosive was the demand for interior plantings in public buildings that the technology necessary to assure the plants' survival was not always able to keep pace. Some of the early efforts, heralded for their aesthetics, became corporate embarrassments a few months later. The lush green lobbies were replaced with dead stalks and leaf-littered carpets. Because the architects and decorators were unaccustomed to working with living materials, their failure to consider such foreign requirements as light, water, drainage, and fertilizer is understandable. Now they appreciate the differences between fabrics and philodendrons, and the mistakes of the past are less likely to be repeated.

The past decade saw the technology needed to sustain plants indoors catch up to the public demand for lavish plantings. The profession of interior plantscaping is still evolving and is still very labor-intensive. Certainly the next few years will find additional advances being made in the way that interior plantings are prepared for installation, irrigated, and drained. New species will be tried. Some will do well and gain public acceptance, while others will not adjust to the change from outdoors to indoors. Thus, while the profession is no longer in its infancy, it has not yet reached maturity. It is entering into a new stage of rapid and exciting growth, and learning from past mistakes.

The Materials

As the profession has developed and changed, so has the array of plants used to create the interiorscapes. Yesterday's windowsill full of ivy and geraniums has been supplemented with containerized trees, in-ground installations, and lavish, changing displays of seasonal color.

Figure 24-2 Interior plantscaping can enhance the architecture of a building by softening lines, creating sheltered spaces, enframing views within the area, and focusing attention on key features of the architecture.

Tropical foliage plants are the most popular and successful indoor plants for long-term installations. This is because they do not require the period of cool temperature dormancy that often makes temperate zone plants unsatisfactory. Originally collected from the jungles of South America, Africa, and other exotic locales, these plants are now grown in the greenhouses and nurseries of Florida, California, Texas, New York, Ohio, Pennsylvania, and Latin America, Figure 24-3.

Figure 24-3 Tropical foliage plants are now grown in greenhouses and nursery fields.

Other plants finding their way into the residential and commercial interior plantscapes of America include subtropical plants such as cacti, traditional and short-lived flowering plants such as poinsettias and mums, and even turfgrasses, Figure 24-4.

Uniqueness of Interior Plantscapes

The transplanting of any plant from its production site to the plantscape site involves some risk even if it is an exterior plant. When the plant is intended for interior use, the relocation initiates problems that no exterior landscaper faces. Among them:

- A drastic reduction in the quality and intensity of light.

- Reduction and constriction of the plant's root system.

- Replacement of natural rainfall by dependency upon humans for watering.

- A reduction in nutrient requirements and a potential for build-up of soluble salts (fertilizers).

Figure 24-4 Poinsettia varieties grown for the winter holiday season

- A lack of air movement and rainfall, allowing dust to accumulate on leaves, often plugging stomata and reducing photosynthesis.

- Potential damage by air conditioners, central heating systems, cleaning chemicals, water additives, and other irritants.

Under these conditions, plants may sustain themselves, but they seldom grow. Because of this, it is unnecessary to space plants for expansion, but it is necessary to install them in a manner that permits the replacement of dead or unsightly individuals as needed.

Light and Interior Plantings

Everyone knows that plants require light to survive and grow. Not everyone knows that there are differences in the kinds and sources of light. Also, few but horticulturists consider the amount of light that plants require. Yet despite how little some people know about the survival needs of interior plantings, expectations for survival of the plantings are always high and are frequently disappointed.

How long will plants live indoors? The answer depends upon the plant and the quality of installation and maintenance. However, in interior design as elsewhere, nothing lasts forever. Carpets wear thin, furniture nicks and sags; walls require fresh paint. Plants must be regarded as perishable furnishings as well. If installed correctly into a properly designed setting and maintained properly, the plantscape will serve satisfactorily for a time period that will unquestionably justify its cost. Then it will require at least partial replacement.

Light Intensity

Human activities do not require as much light as that required for the growth of plants. Even modern homes and buildings with extensive windows and skylights are unable to provide a light intensity equaling the outdoors. Glass filters the sun's light to the extent that an unshaded greenhouse still reduces the intensity of sunlight by at least 15 percent.

To understand light intensity requires knowing how light is measured. Light intensity is expressed in the units of **lux** or **footcandles**. A lux is the amount of illumination received on a surface that is 1 meter from a standard light source known as unity. A lux is an international measurement comparable in use to the metric system. In the United States, the footcandle unit is more commonly used and understood. One footcandle (fc) is equal to the amount of light produced by a standard candle at a distance of one foot. Direct-reading meters are manufactured that measure light intensity in footcandles up to 10,000. That is the intensity of sunlight on a typical clear, summer day in the temperate zone. In the south, intensity can approach 20,000 fc. A light meter is the only way to measure light intensity accurately and should be the first piece of equipment purchased by a beginning interior plantscaper, Figure 24-5.

The challenge of bringing plants accustomed to outdoor light intensities approaching 20,000 fc into a home or shopping mall is best appreciated through several examples. The average residential

Figure 24-5 Light meter calibrated in foot candles

living room has a light intensity of 10 to 1,000 fc by day and as few as 5 fc at night. A good reading light provides 20 to 30 fc. A word processing operator may have 40 to 50 fc of illumination on the keyboard's surface. The average shopping mall provides 20 to 30 fc of light in pedestrian circulation areas and up to 100 fc in sales areas.

The distance between the light source and the plant will affect the amount of light that actually reaches the plant, so that must be taken into account when measuring light intensity for an interior planting. Obviously, tall trees reach closer to an overhead light source than do shorter vining plants. Therefore, when a light meter is used to take readings of an area's light intensity, the readings should be taken at the top of the plants that are or will be growing there.

Acclimatization

Long before interior plantings are installed, the plants must be prepared to survive in their new, reduced light setting. The adjustment of an outdoor plant to interior conditions is known as **acclimatization**. Proper acclimatization changes the plant's physiology (biological functioning) and morphology (physical structure). While there is still much to be learned about the adjustment of plants from outdoor to indoor growing sites, it is to the credit of forward-looking industry groups such as the Florida Foliage

Association and the Interior Plantscape Division of the Associated Landscape Contractors of America that our present body of knowledge exists.

Acclimatization is directed at the adjustment of four vital factors that determine the survival of all plants: light, nutrients, moisture, and temperature. Under field or greenhouse conditions, the plants' metabolisms have been at or near optimum levels. As the time of harvest, transplanting, and interior installation nears, the metabolic activity must be slowed to minimal levels that permit survival and maintain an attractive appearance. Extensive additional growth is seldom an objective.

Light Intensity Acclimatization. The objective of light intensity acclimatization is to reduce the plant's light needs to a level where **photosynthesis** (the production of food) just slightly exceeds **respiration** (the use of food reserves for growth and maintenance). The point of exact balance is known as the **light compensation point** (LCP). Since a plant's ability to capture the energy of light declines as the plant ages, light levels in the interiorscape must be slightly above the LCP of the plant for long-term survival of the plant.

Light intensity acclimatization should begin at the greenhouse or nursery, prior to shipment to the interior site. Plant leaves that are produced in high light intensities are smaller and thicker than those produced under reduced light conditions. The smaller, thicker leaf reduces the potential of radiation damage from the sun. That danger does not exist in the interior setting. During the acclimatization period that may take as long as six months, the new leaves become thinner and larger to permit most efficient capture of the light energy that will sustain them in their interior location. The change in the leaf structure is both morphological and physiological.

Light intensity is reduced gradually over a period of several weeks or months. Each change reduces the light by 50 percent until the desired intensity (usually 100 to 200 fc) is reached. The acclimatization process cannot be rushed without a severe reaction by the plant. That reaction may range from partial leaf drop to death.

Nutrient Acclimatization. Critical to the survival of the plant in its new interior setting is a healthy root system that is in balance with the leafy greenery of the plant. As the plant's ability to photosynthesize is scaled down by the reduced light intensity, so too is the amount of nutrients that need to be taken up by the roots in support of the plant's growth. Thorough soil **leaching** (flushing with water until it flows freely from the bottom of the container) at the beginning of the acclimatization and occasionally afterwards will prevent a build-up of soluble salts (fertilizers). Soluble salt accumulations can damage the root system.

Following the leaching process, later fertilizations are less frequent than when the plant was growing under high light intensity. The reduction in fertilization must be coordinated with the reduction in light as closely as possible for successful acclimatization.

Moisture Acclimatization. Both the frequency of watering and the high humidity levels of the production area are gradually reduced as the plant is prepared for relocation. The interior site will be very stressful to the plant since office buildings and malls are kept at low humidity levels for greater human comfort.

Temperature Acclimatization. Production area temperatures are usually higher than human comfort levels in order to promote more rapid plant growth. During acclimatization, temperatures are gradually reduced to the 65 to 75 degree F range that is common to most interior areas.

Light Quality

Once acclimatized to the reduced light intensity of the interior site, the interior plantscape may still prove unsatisfactory if the quality of light is incorrect. **Light quality** is the color of light emitted by a particular source. The sun emits all colors of light, some of which the human eye can perceive and others that are unseen by humans but beneficial to plants. The green-yellow light most comfortable for humans is of little use in photosynthesis by plants. They depend on light from the blue and red bands of the visible light spectrum. Visible light is only a narrow region of the radiant light spectrum, Figure 24-6.

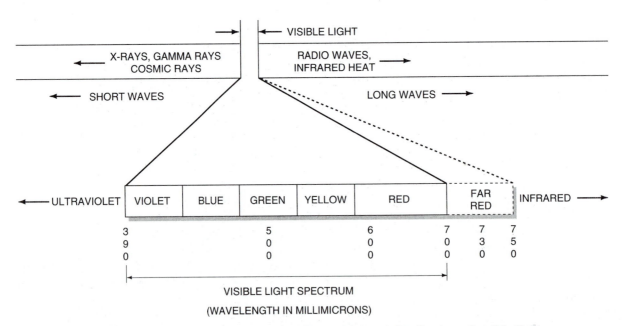

Figure 24-6 Electromagnetic spectrum and special distribution of visible light

As long as both humans and plants can derive their light energy from the sun, the needs of each are satisfied. Indoors, however, where light energy is usually created or supplemented by artificial means, the quality of light can vary considerably. Light selected solely for its benefit to plants may cause the skin tones of human beings to appear ashen and deathly. In similar fashion, an interior decorator may specify a quality of lighting that gives human complexion a healthy glow while making nearby plantings appear brown and dead. Not all lamps provide the same quality of light, nor does any lamp provide a constant quality or intensity as the distance between the lamp and the plants changes. Categories and examples of lamps that have some use in interior plantscaping are shown in Table 24-1. Table 24–2 compares the lamps in all areas important to interior plant survival. You should study these tables thoroughly before proceeding further.

Table 24-1 Lamps for Interior Plant Illumination

Tungsten filament incandescent lamps
- Standard (the familiar household lightbulb)
- Reflector (spot or floodlights)
- Parabolic aluminized reflector (a weather-resistant type of floodlight with a more precise beam)
- Incandescent plant lamps (not proven to be any better than the standard incandescent)

Fluorescent lamps
- Cool white
- Warm white
- Plant lamps
- Wide spectrum plant lamps

High-intensity discharge lamps
- Mercury
- Metal halide
- High-pressure sodium

Interior plantings are seldom illuminated from a single source or type of light. For example, consider the plantscape of a typical office. The plants may be permanently set or in movable planters. Their functions may be to serve as room dividers, establish a mood, or relieve a cluttered desktop. The ceiling may be 8 or 10 feet high. Side windows or a skylight may admit some natural light. In such a setting, cool white fluorescent lighting would be ideal for both general lighting and the maintenance of the plants. People, plants, and furnishings look natural beneath cool white light due to its excellent color rendition. If additional task lighting is needed, small desk lights should be used. Special effects such as shadows or textured highlights can be created with incandescent lights installed beneath the plants and directed upward. (These are called uplights.) Some benefit will accrue to the plants from the addition of lighting at the base. However, if supplemental lighting is needed for photosynthesis, it is most efficient when applied from overhead.

A shopping mall presents a different situation. The corridors may have ceilings too high to permit the use of fluorescent lamps, which are not good for illumination when the ceiling is much beyond 10 feet. There may be skylights as well as decorative architectural lighting. Overall illumination by mercury or metal halide lamps would be best. As in an office, supplemental or decorative lighting might also be desirable for special effects or the health of the plants. When uplights are used, they should be installed directly into the planters and waterproofed. When supplemental lights are added for overhead illumination, they should be positioned to light the plants fully without shining in the eyes of viewers.

In both examples the total light radiation may be from several sources, some natural and some artificial. Add additional, perhaps uncontrollable, sources of light, such as store windows, electrical signs, and reflective surfaces, and the task of determining the total quality and quantity of illumination becomes complex.

Determining Total Radiation. Dr. H. M. Cathey, director of the U.S. National Arboretum, and Lowell

Table 24-2 A Comparison of Artificial Lighting Sources for Interior Plantscapes

Lamp Type	How Light Is Produced	Quality of Light Produced	Percent of Visible Light Radiation	Color Rendition	Initial Cost	Operating Cost
Incandescent (all types)	Current flows through a tungsten filament heating it and making it glow.	High in red light; low in blue light	7–11	Good	Low	High
Cool White Fluorescent	Phosphor coating inside the glass tube is acted upon by radiation from a mercury arc.	High in blue and yellow-green light; low in red light	22	Good (blends with natural daylight)	Moderate	Moderate
Warm White Fluorescent	Phosphor coating inside the glass tube is acted upon by radiation from a mercury arc.	Low in blue and green light; more yellow and red light	22	Poor (blends with incandescent light)	Moderate	Moderate
Fluorescent Plant Growth Lamps	Same as other fluorescents. Special phosphors transmit most light energy in blue and red light regions of the spectrum.	High in red and blue light; low in yellow-green light	22	Average (enhances red and blue colors; darkens green colors)	Moderate	Moderate
Wide Spectrum Plant Growth	Same as other fluorescents. Special phosphors transmit most light energy in blue and red light regions of the spectrum.	Less blue and red than standard plant growth lamps; more far-red and yellow-green light	22	Average (favors red and blue colors; darkens green colors)	Moderate	Moderate
Mercury Deluxe white model, for interior plants)	An electric arc is passed through mercury vapor.	High in yellow-green light; less red and blue light, but still usable for plant growth	13	Poor (favors blue and green colors)	High	Moderate
Metal Halide	Similar to mercury lamps but with metal gas additives to produce a different spectrum	High in yellow-green light; less red and blue light, but still usable for plant growth	20–23	Good (similar to CW fluorescent)	High	Low

Life of the Lamp	Placement Height Above Plants	Plant Responses	Major Advantages	Major Disadvantages
750 to 2,000 hours	At least 3 feet to avoid foliage burn	Plants become long and spindly with pale foliage. Flowering is promoted, and senescence is accelerated.	• Good for special lighting effects • Compact source of light • Simple installation	• Energy inefficient; too much lost as heat • Light does not distribute evenly over a surface. • Glass blackens with time and light output is reduced. • Frequent replacement is needed.
Up to 20,000 hours	10 feet or less	Plants stay short and compact. Side shoots develop. Flowering extends over a longer period.	• Energy efficient • Heat is radiated over the length of the lamp, allowing closer proximity to plant foliage. • Light distributed more evenly over a flat surface	• Light does not focus well. • They are difficult to start when line voltage drops or humidity is high. • Installation is expensive. • Special fixtures are needed.
Up to 20,000 hours	10 feet or less	Same as CW fluorescent	• Same as CW fluorescent	• Same as CW fluorescent
Up to 20,000 hours	10 feet or less	Rich green foliage color. Large leaf size. Side shoots develop. Plants stay short. Flowering is delayed.	• Same as CW fluorescent • Light emission is from the region of the spectrum most important to photosynthesis.	• Same as CW fluorescent • Greater expense with little increase in benefit to the plants
Up to 20,000 hours	10 feet or less	Stems elongate. Side shoots are suppressed. Flowering is promoted. Plants age rapidly.	• Same as CW fluorescent	• Same as CW fluorescent • Growth may not be desired. • Poor color rendition on nonplant materials
Up to 24,000 hours	10–15 feet or more	Plants respond in a manner similar to CW fluorescent.	• Long life; useful for inaccessible fixtures • Medium energy efficiency	• Not interchangeable with other lamps • Warm-up time required
Up to 20,000 hours	10–15 feet or more	Plants respond in a manner similar to CW fluorescent.	• High energy efficiency, surpassing the mercury lamp • Good for both plant and general lighting	• Warm-up time required • Color and light quality change with operating hours.

Table 24-2 A Comparison of Artificial Lighting Sources for Interior Plantscapes

Lamp Type	How Light Is Produced	Quality of Light Produced	Percent of Visible Light Radiation	Color Rendition	Initial Cost	Operating Cost
High-Pressure Sodium	Sodium is vaporized into an arc.	High in yellow-orange-red light	25–27	Poor (similar to WW fluorescent)	High	Low
Low-Pressure Sodium	Sodium is vaporized into an arc.	High in yellow-orange-red light	31–35	Poor	High	Low

Campbell, an agricultural engineer at the U.S. Department of Agriculture, have developed conversion factors that enable light meter measurements of light from different sources to be converted to a common basis of measurement. The unit of measurement is watts per square meter, Table 24-3.

By taking independent light meter readings of the footcandles of illumination that fall upon the plants from each light source during the brightest periods of the day and during various times of the year, then multiplying by the conversion factor for that particular light source, the amount of radiation is determined. All readings should be taken at plant level. Each reading must be taken when the other light sources are off. Then the interior plantscaper can add the separate measurements together to determine if the proper total illumination is being provided. Cathey and Campbell report that most interior plantings can be maintained in a healthy state if given a minimum of 9 watts per square meter of illumination for 12 hours each day.

Natural Light

Most important of all light sources for interior plantscapes is natural sunlight when it can be planned for and depended upon. Each footcandle

of illumination that nature provides is one less that has to be provided and paid for with artificial lighting. In our energy-conscious society, such savings are worth planning for. However, knowledge of how to maximize the benefits of natural light is vital; otherwise more heat energy is lost through inefficient windows than is gained in light energy.

Table 24-3 Conversion Factors for Determining Watts Per Square Meter for Various Light Sources

Light Source	Conversion Factor
Daylight	0.055
High-pressure sodium lamps	0.034
Low-pressure sodium lamps	0.022
Metal halide lamps	0.034
Cool white fluorescent lamps	0.030
Warm white fluorescent lamps	0.030
Fluorescent grow lamps	0.044
Incandescent lamps	0.090
Mercury incandescent lamps	0.070

(continued)

Life of the Lamp	Placement Height Above Plants	Plant Responses	Major Advantages	Major Advantages
Up to 24,000 hours	10–15 feet or more	Typical red-light plant responses; similar to fluorescent plant growth lamps when compared on equal energy	• High energy efficiency. When combined with blue light sources (such as metal halide), they provide good lighting for plants and people. • Long life	• Yellow color makes them unsatisfactory for general indoor lighting by themselves.
Up to 18,000 hours	10 feet or less	Plants respond in a manner similar to HP sodium	• The most energy efficient lamps available	• Yellow color makes them unsuitable where color rendition is important. They must be used at night after closing.

Natural light is most helpful when it offers high levels of illumination throughout most of the year. Traditionally sunny areas like the Southwest can make better use of natural light than areas like the Northeast, where clouds block the sun and snow blankets the skylights through much of the winter season.

Sunlight entering from overhead is of greater use in the illumination of interior plantings than light entering from the side, although both are helpful. In neither situation will the natural light be as intense as outside light. It will be significantly reduced by the glass glazing through which it passes and the distance it travels between the point of entry and the leaf surface. Little usable light passes more than 15 feet beyond glass, so skylights in high lobby areas or shopping malls are of no benefit to plantings beneath them. Nevertheless, they can be of great benefit in single-story buildings with lower ceilings. In a similar fashion, a large interior plantscape may derive little benefit from side lighting, since usable light enters at a 45-degree angle, and plants must be placed within that narrow beam if they are to benefit, Figure 24-7. They cannot be too close to the glass or the foliage may burn, however. In a smaller room, as in a residence, natural side light can be of great value.

When skylights are used, they must be designed to permit the most light to enter while insulating against as much winter heat loss as possible. Within the limits allowed by heating and structural engineering, the ceiling well through which the light passes should be as wide and shallow as possible,

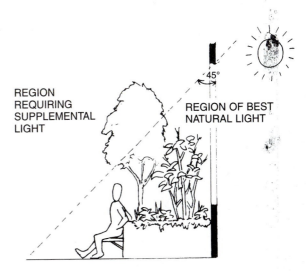

Figure 24-7 Interior plantings receive the best quality of natural light within a 45° arc of the side windows.

with the sides painted a reflective white and beveled outward at 45-degree angles, Figure 24-8. As skylights become more narrow and the ceiling well deeper, there is less area through which the sunlight can enter and a narrower focus of illumination on the surface below. The difference between wide, shallow skylights and narrow, deep ones is similar to the difference in illumination between a floodlight and a spotlight. With natural light and interior plantings, the wide floodlight effect is most desirable.

Selecting the Correct Lighting

In summary, no single recipe for correct lighting can be given. There will be varied settings, needs, and objectives to accommodate. Plants will seldom be the only consideration in the selection of lamps and the quality of illumination. When both plants and people are to be considered, a lamp should be selected that provides the yellow-green visible light needed to render human complexion, clothing, and furnishings attractive, while still providing sufficient blue and red light to allow photosynthesis to

exceed respiration in the plants. The cool white fluorescent lamp is ideal for such a situation provided that the ceiling is not too high and that growth of the plants is not an objective. If growth is desired, additional incandescent lighting can be focused directly on the plants. The use of more expensive growth lamps is unnecessary since they have not been proven superior to the ordinary cool white fluorescent in maintaining plant health, and they do not render color well. Any natural light that can be used advantageously will reduce the cost of lighting the interior planting.

As for light intensity, not all plants available from growers have been carefully studied to determine the minimum at which they will survive attractively. The ALCA, the FFA, and the colleges of agriculture in the major foliage production states should be queried by anyone responsible for creating an interior plantscape. The many books that purport to give lighting specifications and that stock the shelves of libraries nationwide are satisfactory for homeowners but not for professional use. Their information is often dated, usually based upon green-

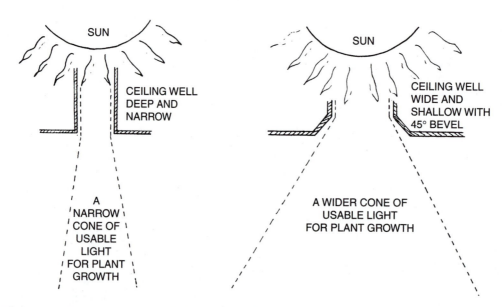

Figure 24-8 Cross-sectional view of skylights and the influence of their design on the amount of light available to plants.

house lighting, and may be questionable in its accuracy. Recommendations should be sought that give lighting minimums in footcandles, not in general terms such as high, medium, and low. When plants are installed without documented knowledge of their lighting requirements, the situation is risky at best and the plants must be watched carefully to determine if additional lighting is needed.

To assure that lighting is of the right intensity and duration, a simple timing device may be necessary. The lights must shine on the leaves for enough time each day (12 hours minimum) to allow adequate photosynthesis to occur. Should the hours of lighting have to be reduced for some reason, the intensity of the lighting must be increased to compensate, Figure 24-9. It does not seem to matter whether foliage plants receive their needed lighting over a short or long period as long as the cumulative photosynthetic activity balances and slightly exceeds respiration. With flowering plants, day length often plays a critical role in determining if and when the blossoms will appear. Since interior plantings are currently valued most for their foliage, day length is of limited importance.

The Growing Medium

No less important than light to the successful acclimatization of interior plants is the provision of a proper growing medium. The roots must be

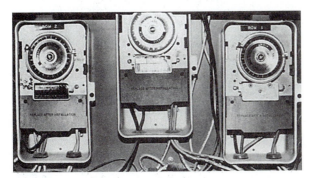

Figure 24-9 A timing system can be used to provide supplemental light for an interior plantscape.

placed in an environment that provides structural support, allows the roots to absorb water, and provides essential minerals. Further, the growing medium must allow rapid drainage of water past the root zone and provide the correct pH for growth.

The medium that serves when plants are growing in a nursery field or production container is likely to be inappropriate for an interior installation. Natural field soil may be:

- too heavy to permit rapid drainage

- too heavy for the floor to support if the container is large

- inconsistent in composition, making standardized maintenance of separate planters difficult

- infested with insects, pathogens, or weeds

Although pasteurized natural soil may be a component of the growing medium for interior plantings, additives will probably be needed. It is even possible that the growing medium selected for the interior plantscape will have no natural soil in it for reasons of improved drainage, hygiene, pH balance, or nutritional consistency. The use of synthetic soils, whose composition is as controlled as a cake recipe, is becoming the rule rather than the exception.

Cornell University and the University of California have been leaders in the formulation of synthetic soils for interior plant production. The Cornell mixes are made from vermiculite or perlite and sphagnum moss. The University of California (U.C.) mixes are made from fine dune sand and sphagnum moss. The ratio of components varies with the species of plants and the maintenance program to be followed.

In addition to the U.C. and Cornell mixes (called the *peat-lite* mixes) there are bark mixes, composed of pine bark, sand, and sphagnum moss. These are especially acidic growing media and may require buffering (with dolomitic and hydrated lime) to sustain healthy plants.

All of the synthetic soils are mixed and sold commercially. In large interior installations, it may be more economical to mix the medium at the site rather than purchase the premixed commercial products. All decisions about the growing medium should be made before the planters are filled and the plants installed. Once planted, errors in the medium's composition are difficult to correct without removing the plant.

Because the artificial media have low nutritional value for the plant, complete fertilizers are required as well as periodic application of the minor elements. Regular soil testing is necessary to assure that the fertilization program is correct.

Installing the Plants

The acclimatization of a plant's root system involves establishment of the correct relationship between roots and foliage. Outdoors, a sizable root system is needed to supply adequate amounts of water and minerals to the leaves for near-maximal photosynthesis. Indoors, photosynthesis is reduced to survival-maintenance levels, so the root system need not be as large. Thus, one of the first steps at the time of installation is to remove the production medium from around the roots and prune away excess roots, Figure 24-10.

Several methods can be used for setting plants into an interior plantscape, depending upon whether the plants are to be placed in ground beds or raised planters. The anticipated frequency of plant replacement will partially determine the method of installation. For example, large trees are often permanently planted in the floor of an enclosed shopping mall because they are too large and expensive to replace, Figure 24-11. The growing medium around them must permit good drainage and retain nutrients, while drainage tiles beneath the plant carry away excess water. Plants in such a setting must not be subjected to detergents and waxes used for floor maintenance, so special design provisions, such as raised edging, may be necessary.

Plants that are in raised planters or that require frequent replacement are usually not installed permanently. Instead, they are often planted in a growing medium within a nursery container with good drainage, Figure 24-12. The containerized plant is then placed in a support medium of peat,

Figure 24-10 Excess roots are pruned off the foliage plant before transplanting to establish a more balanced relationship between the foliage and the root mass.

Figure 24-11 The grate around this permanently installed fig tree allows watering and fertilizing while protecting the root system from being trampled or compacted.

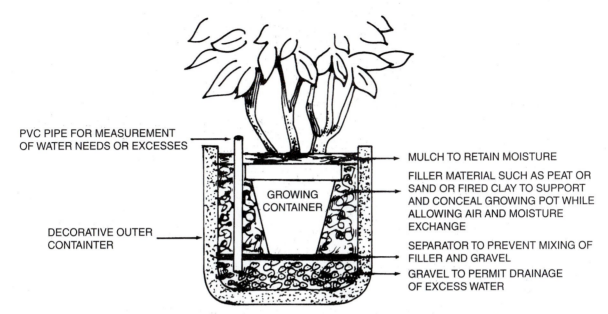

PVC PIPE FOR MEASUREMENT OF WATER NEEDS OR EXCESSES

GROWING CONTAINER

DECORATIVE OUTER CONTAINTER

MULCH TO RETAIN MOISTURE

FILLER MATERIAL SUCH AS PEAT OR SAND OR FIRED CLAY TO SUPPORT AND CONCEAL GROWING POT WHILE ALLOWING AIR AND MOISTURE EXCHANGE

SEPARATOR TO PREVENT MIXING OF FILLER AND GRAVEL

GRAVEL TO PERMIT DRAINAGE OF EXCESS WATER

Figure 24-12 Installation of a containerized plant within an indoor planter

perlite, sand, or similar well-drained material. The support medium serves to retain the plants as well as permitting excess water to drain away from the root zone and insulating the root system against abrupt temperature or moisture changes. A separating sheet (usually of fiberglass or a rot-resistant fine-meshed material) is placed between the support medium and the coarse gravel lining the bottom of the planter. The gravel is needed to facilitate drainage, and without the separator the growing medium and support medium would gradually wash into the gravel and plug it. A plastic tube inserted into the planter permits dipstick testing to determine if the planter is being overwatered. It also provides a means of pumping out excess water if the planting is endangered by overwatering. Mulch on the surface of the planter serves to discourage moisture loss, provides a decorative appearance, and conceals the rims of the plant containers.

The same techniques used for planting containerized plants can be applied to in-ground plant-

ings. If the planter drains directly into the building's drainage system, then the vertical plastic tube is unnecessary. All else remains the same except that there is no outer container.

Watering and Drainage

As noted previously, the ability of the growing medium to drain off excess water is critical. More interior plant deaths result from overwatering than from any other cause. All planters must permit the removal of standing water. The layer of coarse gravel already referred to is one method. Other methods include setting containerized plants on top of inverted pots within the larger planter and incorporating drains and spigots into planter bottoms. Where drainage from the planter base is planned, additional planning must assure that carpets, tiles, and other floor surfacings are not damaged by the runoff water. Sitting the plants on gravel beds into which the water can drain and then evaporate is one method.

Hand in hand with planning for drainage goes provision for proper watering. Some interior plantings require continuous moisture; others do better if permitted to dry out between regular waterings. Obviously the two types of plants would not coexist compatibly in the same planter.

An interior plantscape must be watered according to a schedule and not according to the judgment of a custodian or other unqualified individual. The right watering frequency is most easily determined in controlled environments such as enclosed shopping centers. It is most difficult in locations where environmental variables are not stable, such as by open windows or in drafts near doors.

The need to water a planting can be determined by feeling the soil and observing its color. A grey surface color and failure of soil particles to adhere to the fingers indicate dryness. Moisture meters are also available for a more carefully controlled reading of the growing medium's water content.

Although some automated watering systems exist, there is surprisingly limited use of them in large installations. Nationwide, interior plantscapes utilizing hundreds of plants are all watered by hand. Most definitely, the technology of watering is still in the developmental stages.

When water is applied, it must be in a quantity adequate to wet the soil deeply, not shallowly. Shallow watering encourages shallow rooting and increases the vulnerability of the plants to damage from drying. Deep watering promotes deep, healthy rooting while providing the soil leaching necessary to prevent soluble salt build-up.

The quality of water used on the interior planting may vary with the location. The most likely source will be municipal water lines. Most public drinking water contains chlorine and often fluoride, as germicides and tooth-decay deterrents, respectively. Neither additive will harm plants under normal conditions. Although chlorine is potentially harmful, the amounts used in drinking water are dissipated by aeration as the water bubbles from the faucet or hose nozzle. More heavily chlorinated swimming

pool or fountain water can damage plants and should never be used as a watering source.

NOTE: Plants grown around enclosed pools require good air exchange in the room or the chlorine gas from the pool may damage them.

Water is a source of soluble salts. Water that has been softened by means of cation-exchange softeners may be dangerously high in sodium, which can be toxic to plants. In buildings with such water softeners, alternate sources of water should be sought for the plantings. In regions of the country where the need for water conservation causes recycled water to be used on interior plantings, a chemical analysis of the water should be made to determine if any toxic chemicals are present that could damage the planting.

Other Concerns

As with any planting, certain routine procedures are needed to keep the plants healthy and attractive. The problems that are common to plants grown outdoors are similar to those that trouble plants grown indoors. In addition, indoor plantings often encounter stresses not common in outdoor landscapes.

Fertilization

Fertilization is needed to provide the mineral elements required for photosynthesis. Interior plantings need a complete fertilizer, but not as often as outdoor plantings, because the rate of plant growth is greatly reduced indoors. The ratio of the complete fertilizer should be fairly uniform; for example, 1-1-1 or 2-1-2. Too much nitrogen may lead to spindly, succulent, and unattractive vegetative growth. Trace elements will also need to be applied, especially if the growing medium contains no real soil. Frequent soil testing is necessary, regardless of whether the growing medium is true soil or synthetic. Excessive fertilizing follows only slightly behind overwatering as a major reason why interior plants fail to survive satisfactorily.

Either organic or inorganic fertilizers will work well, but the organics are generally slower in releasing their minerals for use by the plants. This means that soluble salt buildup is less frequently a problem. However, the odor of decomposing organic fertilizers may not be welcomed in such settings as shopping malls or library lobbies. Therefore, the use of organic fertilizers may not be practical in all situations.

Inorganic fertilizers are applied to large interior plantscapes in liquid form. This is faster and easier than applying dry granulars to each pot. When there are hundreds of pots and planters to maintain, speed is vital.

Humidity

Humidity is seldom a problem for plants growing outdoors, but it can cause problems for interior plantscapes. Because of the drying effects of central heating and air conditioning, interior plantings must adapt to an air environment that may contain half or less of the relative humidity outside. Interior humidities of 40 percent or less are common.

Preparation for the dry air must begin during the acclimatization process. Gradual drying of the plant's atmosphere will usually allow the plant to survive after transplanting indoors. In some cases, attempts to increase the humidity around plantings by misting the foliage during the day have proven to be of little or no value. Misting may also cause damage to carpeting or furnishings. Proper acclimatization is the best solution at present.

Air Pollution

Air pollution cannot be escaped by bringing the landscape indoors. Pollutants in the exhaust of cars and trucks, the smoke from cigarettes, the chlorine gas escaping from swimming pools, and the chemical soup that passes for air in our major metropolitan areas are all harmful to plants when sufficiently concentrated. Some plant species are more susceptible than others.

Good ventilation is an important element in the health of interior plantings. Proper ventilation will carry away chlorine vapors, fumes from smokers and chemical cleaning agents, or the ethylene that may be present if a building is heated by some form of hydrocarbon combustion (fossil fuel).

Ethylene can also damage plants as they are shipped to the site. This toxin is present in vehicle exhaust and, if permitted to seep into the cabin where plants are stored, may result in injury. If packed tightly or in restrictive packaging, the plants also may injure themselves, since their own tissue produces ethylene. Because of the potential harm from ethylene, plants should be unpacked and ventilated immediately upon arrival. At the time the plants are purchased, the grower or shipper should guarantee that the transport vehicles will be ventilated and sealed against exhaust fumes.

Dust

Dust is an air pollutant different from the others in that it is a particulate, not a vapor. When the leaves of a plant are coated with dust, they are not only unattractive, but gas exchange may be reduced due to plugged stomata. Air filtration reduces the amount of dust. Regular cleaning of the plants can also prevent dust buildup. Most plants in a residential interior can be rinsed off under the shower or set outside during a rainfall to wash away the dust. Commercial plantings can be kept clean with regular feather dusting and periodic washing. Cleaning should be a regular task within a total maintenance program for the interior plantscape.

Pruning

Pruning will not be extensive in an interior planting because of the plants' reduced rate of growth. Most pruning will be done to keep the plants shaped for an attractive appearance. Broken or damaged branches will also require removal. If the plants are not intended to grow, in order to avoid crowding, the roots as well as the foliage must be pruned back. Excessive root growth in containers can result in strangulation of the root system.

For pruning the plantings, hand pruners will be suitable for most of the herbaceous material and much of the woody material. Lopping shears will be helpful with larger materials, and pruning saws may be needed for indoor trees.

Interior plants should be pruned to a shape that suggests the appearance of a full canopy. Due to the restricted lighting in an interior planting, a full canopy seldom forms, however, and the plant's full branching structure may be visible. The pruning must enhance and take advantage of this sparse foliage covering.

Repotting

Repotting of plants is necessary in plantings where growth is allowed. Containers need to be removed from the planter and from the plant, the next larger size selected, and the excess space filled with growing medium. If roots are matted, they should be loosened before repotting. If they have started to grow around themselves, the large and excess roots should be pruned away. The pot should not be filled to the rim with the growing medium. An inch of unfilled space below the pot rim will give water room to flow into the container.

Insects and Diseases

Insects and diseases are not as common to interior plantings as to exterior ones, but they do occur. Insect problems are more common than diseases.

The initial pest presence may be introduced by the plants themselves as they arrive from the grower. Insects or pathogenic inoculum may be present in the foliage, roots, soil, or containers. All should be checked carefully upon arrival, at an area away from the installation site. The same careful check should be made each time replacement plants arrive. Obviously a reputable grower is a first defense against pests.

The most common pest problems of interior foliage plants are:

- Aphids
- Mealybugs
- Spider mites
- White flies
- Scale
- Thrips
- Nematodes

- Root mealybugs
- Root rots
- Leaf spots
- Anthracnose
- Mildews
- Blights

These pests and others are not uncontrollable when modern pesticides and methods of application can be used against them. However, the interior location and the presence of people make the use of sprays, dusts, and fumigants difficult. Whenever there is a possibility of people making contact with a pesticide, it is dangerous to use it. Even furnishings and carpeting can be damaged by many of the corrosive or oil-based chemicals. Where practical, plants can be wrapped in loose plastic bags and sprayed within the bag. This helps reduce the drift and the danger to people and furnishings.

Control measures are restricted to:

- Chemical pesticides approved for application indoors

- Removal of infected plant parts

- Washing away of insects and inoculum from plant foliage

- Replacement of plants with healthy new ones

Vandalism and Abuse

Vandalism and abuse to interior plantings may necessitate replacement earlier than expected.

Certain locations will bring certain predictable types of abuse. Plantings in cocktail lounges or nightclubs may have alcoholic drinks poured into their soil. Those in college snack bars and dorms may have cigarette holes burned through their leaves. Plantings in shopping malls may be mutilated by home propagators who take cuttings faster than the plants can replace them. Sometimes entire plants are stolen from the planters. The planters themselves are used as litter bins for paper cups, cigarettes, chewing gum, and assorted other debris.

The only real defense against such damage and abuse is public education and cooperation. By keeping the plantings attractive and well maintained, they are less likely to be deliberately damaged. Replacing or repairing damaged plants as soon as they are noticed displays the concern of the owners for an attractive planting that all can enjoy.

Grouping Compatible Species

Grouping compatible species simplifies the maintenance of the interior plantscape. Within any one planter the species selected should have the same requirements for light, moisture, fertilization, and soil mix. While a large plantscape may effectively combine tropical species, desert species, and sometimes even temperate species, they should be in separate planters. When plant replacement is necessary, the new plants should be compatible with existing ones.

The Interdisciplinary Team

The popularity of interior plantings has grown faster than the ability of any one profession to stay abreast of it all. A successful interior planting in a commercial building requires the expertise of interior plantscapers, plant growers, interior decorators, landscape architects, architects, maintenance professionals, and building management. Each professional brings his or her point of view to the project. The architect sees the plants as archi-

tectural features, not as living organisms. Building management personnel want the plants to attract customers and please employees, but they too do not have a horticultural sense of plant needs. Maintenance professionals regard the plants as something to be dusted, watered, and fertilized. They are concerned about the proximity of water, the difficulty of changing lamps, and the ease of reaching the plants. Each has concerns and contributions to the planning process that the others need to know.

The success of the plantscape is measured by its appearance and health. The interior plantscaper should be consulted while the building is still in the planning stage, as errors in lighting quality and intensity can be difficult to correct later. Drainage of planters directly into the building's drainage pipes requires that plantings be sited permanently near the pipes. Watering of the plantings necessitates a nearby supply of water to which a hose can be attached. A shopping mall or office lobby should have appropriate water outlets every 50 feet along the wall and preferably within each planter. Otherwise hoses will be stretched across walkways, endangering pedestrians and creating puddles. Maintenance of the plantings should be the responsibility of a contracted professional plant maintenance firm, which should work closely with the architect to assure sufficient water outlets, storage space, accessibility for equipment, and so on. Failure to involve the maintenance firm in the early stages of planning can result in an unattractive plantscape shortly after installation. The managers of the building must be made to understand why specific lamps are needed for plant survival even if less expensive ones are available.

In short, fewer problems will develop for the interior planting if all professionals whose work affects it work together from the beginning. At present, such interdisciplinary cooperation is the exception rather than the rule. The failure to use the team approach is often a case of architects not realizing their own limitations with horticultural materials.

The Future

No career field in ornamental horticulture holds greater promise than interior plantscaping. The current technology and knowledge of growing plants indoors is comparable to automobile technology when Henry Ford and the Wright brothers began working in their professions. Fresh approaches are needed to move us beyond ficus trees and hanging baskets. More research is needed to introduce new species and varieties suitable for interior use. Also needed is better data on how to acclimatize plants with less foliage drop and transplant shock. Further study of soil mixes, fertilizer needs, and lighting requirements is needed to replace the guesswork of today. Finally, more professional plant maintenance firms are needed. They need dependable studies of such matters as the time required to dust and clean various leaf sizes and textures. Buildings need to be designed as carefully for their planted occupants as for their human ones. Automated watering and fertilization systems are needed to ease current labor-intensive methods. There is ample work for young people who wish to train themselves for it.

Achievement Review

A. Indicate if the following statements are true or false.

   1. Interior use of plants is a new concept.
   2. Architects and interior decorators are trained to write correct specifications for interior plantscapes.
   3. Tropical plants have proven to be better suited for interior use than temperate plants.
   4. Interior plantings require less fertilizer than outdoor plantings.
   5. Soluble salts originate from fertilizers and water.
   6. Clear glass transmits 100 percent of the sunlight that shines upon it.
   7. Light intensity is measured in nanometers.
   8. Light wavelength is measured in foot-candles.
   9. Most plants will survive after being acclimatized to a light intensity of 100 to 200 fc.

B. Select the answer from the choices offered for each question.

   1. The most efficient natural light for interior plantscapes enters from _____.
      a. the sides of buildings
      b. skylights 20 feet overhead
      c. skylights 15 feet or less overhead
      d. incandescent bulbs

   2. The most efficient skylight has _____.
      a. a shallow, wide well with vertical sides
      b. a shallow, wide well with 45-degree beveled sides
      c. a deep, narrow well with vertical sides
      d. a deep, narrow well with 45-degree beveled sides

   3. In most single-story settings the best lamp for lighting an indoor planting is _____.
      a. an incandescent
      b. a cool white fluorescent
      c. a warm fluorescent
      d. a fluorescent plant growth lamp

   4. Interior plantings illuminated with 100 to 200 fc of light require _____ hours of lighting each day.
      a. eight
      b. ten
      c. twelve
      d. fourteen

5. The most important reason for using a synthetic soil in an interior planting rather than natural soil is _____.
   a. drainage
   b. nutrient content
   c. structural support
   d. cost

6. An interior planting that reflects seasonal changes through use of such plants as daffodils, mums, and poinsettias, would be best if designed _____.
   a. as an in-ground planting
   b. as a containerized planting

7. Deep watering of interior plantings will _____.
   a. encourage deep rooting
   b. leach away soluble salts
   c. do both of these
   d. do none of these

8. Watering of interior plantscapes is best done _____.
   a. irregularly
   b. weekly
   c. when convenient
   d. according to an established schedule

9. The two major reasons why interior plants die are overwatering and _____.
   a. too much light
   b. too much fertilizer
   c. too little fertilizer
   d. incorrect pH

10. Chlorine gas, dust, and ethylene are all possible _____ affecting interior plantings.
    a. pesticides
    b. air pollutants
    c. pathogens
    d. toxins

C. Answer each of the following questions as briefly as possible.

1. Place *X*s where appropriate to compare the different lamp types.

Characteristic	Incandescent	Cool White Fluorescent	Fluorescent Plant Lamps	Mercury	Metal Halide	High-Pressure Sodium
High in red light						
Low in red light						
High in blue light						
Low in blue light						
Good color rendition						
High initial cost						
Low initial cost						
Moderate initial cost						
Low operating cost						
Moderate operating cost						

2. Label the parts of this containerized plant.

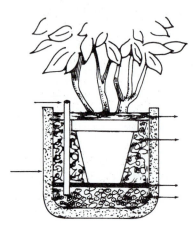

3. Indicate whether the following are characteristic of interior or exterior plantings.
   a. Rapid growth is encouraged.
   b. Pruning is minimal.
   c. Dusting of foliage is needed.
   d. High-analysis fertilizers are used regularly.
   e. Ventilation of the planting site is needed.
   f. Plants must be acclimatized to low humidity.
   g. Plants are most affected by insects and diseases.
   h. Plants suffer most from vandalism and abuse.

Suggested Activities

1. To demonstrate the cultural differences between interior and exterior landscape plants and why plants intended for outdoor use in the temperate zone cannot be used in indoors, do the following. Select a well-lit windowsill or shelf that can hold several plants. Select two small containerized temperate zone exterior woody plants such as a juniper and a rose. Also select a common houseplant, such as a philodendron. Place all three in a setting that is most favorable for the houseplant. Give them identical care over an extended period of time and observe their appearance and apparent health over time.

2. To gain an appreciation of the effect interior plants have on people, position two identical benches in a single public area such as a lobby, but keep them separated far enough that people cannot share in a common conversation when seated on them. Try to standardize the views from both benches and any other factors that might make one location preferable to the other. Then create a simple, pleasant planting near and around one bench. Leave the other bench without plants. Over a period of one or two weeks chart the number of people who elect to sit on one bench or the other. The preference for the plantscaped bench is predicted.

3. The above project could also be done with one bench having a plantscape of real plants and one done using artificial plants. The data should be interesting.

SECTION 3

Landscape Maintenance

CHAPTER 25
Maintaining Landscape Plants

Objectives:

Upon completion of this chapter, you should be able to

- water, fertilize, edge, and mulch tree and shrub plantings
- prune trees and shrubs correctly
- maintain annual and perennial flower plantings

Sustained Care of Plantings

As important as the design and proper installation of landscape plantings is their ongoing maintenance. Many gardens do not attain the appearance envisioned by the landscape designer until the plants have time to mature. Aiding those plants in their healthy maturation requires attentive and knowledgeable maintenance. Also, some designs require specific plant effects such as clipped formal hedges or espalier training to create the garden as envisioned. In both instances, the value of skillful landscape maintenance is apparent. A maintenance program generally will include the following tasks: watering, fertilizing, mulching, edging, pruning, pest control, and winterization.

Watering

Depending on the region of the country, supplemental watering may be an infrequent task or so regular that it requires an automatic irrigation system. The need for a good water supply to the plantings has been discussed in earlier chapters. Here the subject is the objectives of proper watering, its frequency, and its quantity.

Initially, watering must promote deep root development by the plant to establish it securely in its location. Later, watering must keep the plant healthy and growing actively even during dry summer weather. Much winter damage to evergreens can be avoided if the plants are kept well-watered throughout the summer and autumn.

Not all plants require the same amounts of water. Neither will all plant root systems grow to the same depth in the soil. While nearly all landscape trees and shrubs will die if kept too long in either arid or waterlogged soil, certain species are especially sensitive to sites that are too dry or too wet.

Infrequent and deep watering is preferable to frequent, shallow water. Enough water should be applied to wet the soil to a depth of 12 to 16 inches. In the Southeast and Southwest, supplemental water may be required nearly every day. In other regions, supplemental watering may be weekly or even less frequent.

Fertilization

Trees will grow without fertilization in most soils once they are established. However, they will grow with greater health and vigor if they are fertilized annually. Shrubs will respond to proper fertilization with lush growth, greater resistance to pests, and less winter damage.

For shrubs growing in cultivated beds, fertilizer may be applied in early spring. Depending on the plants involved, each 100 square feet of bed area should receive between one and three pounds of a low-analysis, complete fertilizer. The fertilizer should be distributed uniformly over the soil beneath the shrubs, with most of the fertilizer under the outer edge of the shrub where the fibrous roots that absorb the nutrients are located. The fertilizer should not be allowed to touch the foliage, or **foliar burn** (a reaction to the chemicals) may result. If the soil is dry, the fertilizer should be worked into the soil with a hoe. If the weather has not been abnormally dry, the fertilizer can be left untilled, and the next rainfall or irrigation will wash it into the soil.

Trees are fertilized in different ways depending on the species, the age of the plant, the adjacent plants and terrain, and the equipment available to do the job. Small trees may be fertilized to promote their growth and ensure their health. Mature trees may be fertilized to sustain their health but with no concern for size expansion. Where trees stand in open lawn areas, fertilizer can be applied to the surface using the same spreader that would be used for lawn fertilization. In settings where trees are crowded by structures or other plants, or

are on sloped terrain where runoff would prevent movement into the soil, fertilization may be by direct incorporation into the soil or directly into the tree. With only a few exceptions, the method used makes no difference in the trees' reaction. As long as the fertilization provides a correctly balanced nutritional supplement, the delivery system is not significant.

Most important is delivering the fertilizer where it can be taken up by the tree. The take-up of nutrients by trees occurs at the outer extremes of the root zone and within the top 6 to 8 inches of the soil. The outer edge of the root zone in established trees was long believed to correspond to the edge of the foliage canopy, termed the *drip line*. Recent knowledge indicates that tree root systems commonly extend far beyond the drip line. Therefore, placing fertilizer close to the trunk or limiting its application just to the drip line may miss the most absorbent roots entirely.

If fertilizer is being applied to the soil in holes drilled with augers or by injections made with high pressure hydraulic sprayers (Figure 25-1), the holes or injections should extend from halfway between the trunk and drip line to a point approximately a

Figure 25-1 High-pressure injection places fertilizer directly at the zone of absorption.

quarter of the radius outside the drip line. Drilled holes are best for dry fertilizer application. They should be 6 to 8 inches deep and spaced 18 to 24 inches apart throughout the zone of application. A dry mixture of 50 percent high analysis, complete fertilizer and 50 percent sand as a carrier, is poured into the holes.

High-pressure injection uses concentrated liquid fertilizers. The spacing is greater than for dry applications, with the objective being to apply about a gallon of fertilizer with each injection and enough injections to deliver approximately 200 gallons of fertilizer within each 1,000 square feet of the zone of application.

Direct application of nutrients into a tree's vascular system is possible using the same technology as that used to apply systemic pesticides (see Chapter 26). However, from a purely practical standpoint, it is only of relevance when a tree is in need of micronutrients that are unavailable to the tree for specific reasons such as local soil conditions.

The one most common mistake in the fertilization of landscape plants is the application of fertilizer too late in the growing season. The result is often a flush of vegetative growth in response to the nitrogen that leaves the plant ill-prepared for winter. Great damage can result to plants from well-meaning but ill-timed fertilization.

Mulching

The objectives, advantages, and disadvantages of mulching, and examples of commonly available products were outlined in Chapter 20. For extended maintenance of landscape plantings, mulches require replacement. Organic mulches decompose, forming humus *and* an ideal medium for the germination of newly deposited weed seeds unless mulch is replaced annually. Inorganic mulches do not decompose but decline in appearance if not freshened periodically. *Caution*: Old mulch should be removed before new is added. Otherwise, the soil level over the roots is gradually deepened and the plants may die. Also, to reduce the possibility of stem and trunk tissue being killed by the heat of

decomposing mulch or by bacterial and fungal infection promoted by the moist mulch, the mulch should not be piled against the trunk or crown. It should be kept 3 to 5 inches back from shrub crowns and the trunks of young trees. Older trees need at least 8 to 12 inches of mulch-free area.

One important use of mulch is as a protective divider to prevent lawn mower damage to the base of plants, Figure 25-2. A gouge from a lawn mower creates a site for pathogen or insect invasion of the plant. If hit repeatedly, a tree may become partially or completely girdled. In large landscapes with numerous trees and an understaffed maintenance crew who rely on riding mowers for grass cutting, frequent injury to trees results from attempts to mow as close as possible, thereby eliminating hand trimming. A ring of mulch around the base of each tree can protect the tree, speed the overall mowing time, and create a neat appearance, Figure 25–3.

Figure 25-2 Typical damage resulting from a lawn mower when grass is allowed to grow next to a tree

Figure 25-3 A ring of mulch separates the tree from potential mower injury.

Edging

The term **edging** has a double use. As a verb, edging refers to cutting a sharp line of separation, usually between a planting and the adjacent lawn. Reference may be made to edging a bed or edging the lawn along a walk. The term can also be used to describe a product, usually a steel or plastic strip that can be installed as a physical separator between planting beds and lawns or between lawns and paved areas.

To edge a bed requires an edging tool or a flat-back spade. The edger or spade is dug into the ground to a depth of 6 to 8 inches and a wedge of sod removed, Figure 25–4. The process continues along the edge of the bed or the paved area. Cutting through the sod sharply and vertically discourages the roots of the turf from growing into the bed. If the landscape is large, edging can be accelerated by using a power edger.

The installation of edging material against the cut edge further discourages the horizontal spread of grass roots into the bed. It also retains a sharp turf line along walks and drives. The best edging materials are firm, not easily bent and crushed. The corrugated foil edging promoted to the home garden market is not satisfactory for professional use and should be avoided. Heavy-gauged steel and polyvinyl-chloride edging is available from numerous manufacturers. Satisfactory edges can also be created with wood, bricks, and other modular manufactured materials.

In regions where winter heaving of the soil is common, it is best to select an edging material that can be anchored, Figure 25-5. Most anchoring techniques are only moderately effective, however, and resetting heaved edging is a common spring activity in northern landscapes. It is worth the effort, considering the advantages that the material offers in retention of the bedline and separation of the mulch from the lawn mower.

Pest Control

Chapter 26 deals with plant pests, the symptoms of disease and injury that they create, and the principles used to control them. Emphasis is given to the decisions that a landscaper or arborist must make to apply the knowledge of pest control to the maintenance and management of a landscape site. It becomes a three-part judgment call: What is the potential or actual extent of pest damage? What is the quality expectation and/or tolerance level of the client? What are the environmental ramifications of the proposed method of control?

Whether a professional landscaper or arborist, a property manager, or a home gardener answers the questions, they must be answered. In many cases, the control methods used will be restricted or guided by local laws and regulations. Many states and communities ban or restrict the use of certain pesticides or limit the right to apply them to registered pesticide applicators. Those that are

Figure 25-4 A sharply beveled edge around a planting bed will slow rhizomatous and stoloniferous grasses and stop bunch grasses.

Figure 25-5 Plastic edging should be anchored at the time of installation to prevent winter heaving. Here a metal rod is driven through the edging into the soil and clipped over the lip of the edging.

available for general purchase are increasingly limited and have diluted concentrations.

Every state has a college of agriculture that is responsible for approving the use of every pesticide used within the state. Scientists at the colleges are also responsible for determining the appropriate product and formulation for use against a particular disease, weed, or insect on a particular host at a particular time of year or stage of development. It is increasingly probable that their recommendations will incorporate the latest biological control measures. Their recommendations are updated and published each year for use by the states' professionals. Copies of the recommendations are available directly from the state universities and often from local Cooperative Extension offices. Everyone who uses pesticides must realize that although pest problems may be similar in different states and in different regions within a single state, the environmental conditions can vary in subtle ways not always apparent.

Therefore, if a pesticide is not approved for use within a state or region, it is dangerous, unethical, and illegal to purchase the product elsewhere and apply it in the restricted zone. It is also equally dangerous and illegal to increase the dosage or frequency of application beyond that which is recommended.

Managing the landscape in a manner that optimizes plant health is the best way to control pests. Avoiding the use of plants in locations that do not fit their cultural needs will also avoid the costs of labor and materials needed to control the predictable invasion of weeds that will flourish in those locations or the insects and diseases that will find their way to the weakened plants placed in unsuitable locations.

Regular scouting of the landscape to monitor the presence, numbers, and stage of development of pests is another mandatory management task in order to determine what controls are needed, where they are needed, and when to apply them for maximum effectiveness.

Pruning Trees and Shrubs

Pruning is the removal of a portion of a plant to improve its appearance and health and to control its growth and shape. It is easily done, but not so easily done correctly. Each time a bud or branch is removed from a plant, it has a short-term and long-term effect. The short-term effect is the way the plant looks immediately after pruning, and perhaps through the remainder of the growing season. The long-term effect is the way the plant appears after several seasons of growth without the part that has been pruned.

The Tools

Landscape pruning tools are available in a range of sizes and qualities. Whether a hand tool or power tool, all pruning tools share a common attribute: *they are sharp* and must be used carefully with a vigilant regard for the safety of the user and others nearby. The tools used by home gardeners, landscape professionals, and aborists include:

- Hand pruners—Hand pruners are used to cut branches up to about 1/4 inch in diameter. There are two styles in use. One is like a pair of scissors in that it has two sharpened edges (a blade and a curved anvil). The other style has one sharpened blade and a flat, straight anvil that it cuts against. Professional horticulturists generally prefer the scissor-type hand pruner because it gives a more precise cut and is less likely to damage the plant tissue by squeezing against an unsharpened anvil.

- Lopper pruners—Also available as scissor or straight anvil types, loppers are used to cut branches having diameters up to about 1/2 inch. Loppers have longer handles than hand pruners. That allows the user greater leverage and extended reach. It also can result in damage to the tool and the plant if the long handles are used to twist off a partially cut branch. If the loppers cannot make a quick, clean cut through a branch, a pruning saw should be used instead.

- Hedge shears—These are the most task-specific of all pruning tools. They are used to shear and shape hedges, with some secondary applicability to the shearing of plants used in formal settings. Hedge shears are available as a scissor-like hand tool or as a wand-like power tool.

- Hand pruning saw—When a branch exceeds 1/2 inch in diameter, it is important to use a saw, not a hand pruner or lopper. Pruning saws are designed to cut live wood, unlike carpenter saws that are made for dried wood. Hand pruning saws with straight blades cut best when pushed away from the user. Curved blade saws cut best when pulled toward the user. The circumstances of use will usually dictate the proper choice of saw.

- Pole saw—This tool is simply a hand pruning saw or a lopper with a long handle. It permits the user to reach up into a tree for the removal of the same sized branches that are removed at lower levels with regular handsaws or loppers.

Caution: It is difficult to see electric wires that may be in the canopy of trees. Do not cut until you are certain of what is being severed.

- Chainsaw—This is a very dangerous tool! It may be powered by gasoline or electricity and is used to remove branches that cannot be cut with the hand tools described. Razor sharp teeth attached to a chain cut through the wood. Protective clothing and a thorough knowledge of all safety measures regarding the use and maintenance of the chainsaw are essential.

- Other power pruners—Professional landscapers and arborists have available an expanded assortment of tools that do the same tasks described for the hand tools but have the benefits of engine power to increase their strength and cut size capability. Powered forms of pole pruners and loppers are available. These, too, are highly dangerous to use.

Parts of a Tree

The following parts of a tree are important for an understanding of proper tree pruning, Figure 25-6. The *lead branch* of a tree is its most important branch. It is dominant over the other branches called the *scaffold branches*. The lead branch usually cannot be removed without destroying the distinctive shape of the tree. This is especially true in young trees.

The scaffold branches create the *canopy*, or foliage, of the tree. The amount of shade cast by the canopy is directly related to the number of scaffold branches and the size of the leaves. When it becomes necessary to remove a branch from a tree, removal usually occurs at a *crotch*, the point at which a branch meets the trunk of the tree or another, larger branch.

It is always desirable to leave the strongest branches and remove the weakest. Where the crotch union is wide (approaching a right angle), the branch is strong because there has been no

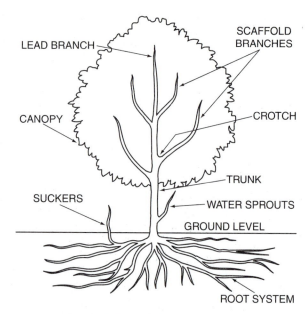

Figure 25-6 The parts of a tree

crowding and pinching of the new wood produced each year by the trunk and branches as they expand. Where the crotch union is narrow, the branch is weak due to a pinch point forming where the expanding trunk meets the expanding branch, Figure 25-7. Growth in that area becomes compressed and dwarfed, and the branch may snap off at that point during a heavy wind or in response to the weight of a person climbing on it.

Two other types of branches often found on trees are *suckers* and *water sprouts*. Suckers originate from the underground root system. Water sprouts develop along the trunk and branches of a tree. Neither is desirable for an attractive tree, and both should be removed.

Parts of a Shrub

A shrub is a multistemmed plant, Figure 25-8. Its branches and twigs differ in age, with the best flower and fruit production usually on the younger branches. The younger branches are usually distinguished by a lighter color, thinner bark, and smaller

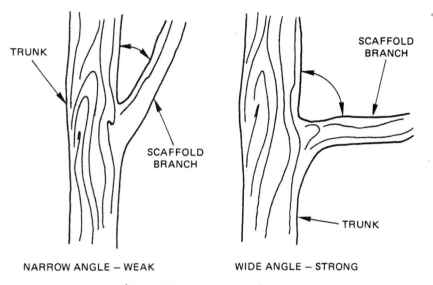

Figure 25-7 Tree crotch structure

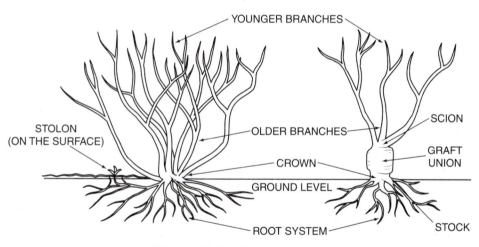

Figure 25-8 The parts of a shrub

diameter. The point at which the branches and the root system of a shrub meet is the *crown*. New branches originate at the crown, causing the shrub to grow wider. New shoots may arise from existing roots or from prostrate stems, termed *stolons,* to create new shrubs from the parent plant. In some grafted plants, the *graft union* may be seen at or near the crown. Shoots originating from the *stock* (or root portion) of a grafted plant are cut away since the quality of their flowers, fruit, and foliage is inferior. Only shoots originating from the scion (or shoot portion) are allowed to develop.

The Proper Time to Prune

Landscapers who design and install as well as maintain landscapes usually prefer to prune when they have little other work. This distributes their work and income more evenly throughout the year. Some plants can accept this off-season attention and remain unaffected by it. Other species accept pruning only during certain periods of the year.

There are advantages and disadvantages to pruning in every season. Since seasons vary greatly from region to region, the following can be used only as a general guide to the timing of pruning.

Winter Pruning. Winter pruning gives the landscaper off-season work. It also allows a view of the plant unblocked by foliage. Broken branches are easily seen, as are older and crossed branches. The major disadvantage of winter pruning is that without foliage it is difficult to detect dead branches. Because of this, plants can become seriously misshapen if the wrong branches are removed. An additional disadvantage is the damage that can be done through cracking frozen plant parts.

Summer Pruning. Summer pruning also provides work during a slower season for the landscaper. An advantage of summer pruning is that it allows time for all but very large wounds to heal before the arrival of winter. The major limitation of summer pruning is that problems of plants may be concealed by their full foliage. Branches that should be removed are often difficult to see. Especially with trees, it is difficult to shape the branching pattern unless all the limbs are visible.

Autumn Pruning. Pruning during the autumn season may conflict with more profitable tasks for the landscaper. In terms of the health of the plant, autumn pruning is acceptable as long as it is done early enough to allow cuts to heal before winter. Autumn pruning should not be attempted on plants that bloom very early in the spring, however. These early bloomers produce their flower buds the preceding fall. Thus, fall pruning cuts away the flower buds and destroys the spring show. Autumn pruning should be reserved for plants that bloom in late spring or summer, producing their buds in the spring of the year.

Spring Pruning. Since spring is the major planting season, most landscapers do not welcome pruning requests unless maintenance is their principal business. However, most plants are most successfully pruned during the spring. As buds begin to swell, giving evidence of life, it is clear which are the live and dead branches. Furthermore, there is little foliage to block the view of the complete plant. Spring pruning provides the plant with maximum time for wounds to heal. In addition, the unfolding leaves conceal the fresh cuts from the viewer's eye.

If the plant is an early spring bloomer, it is best to prune it immediately after flowering. Plants that have a high sap pressure in the early spring, such as maples, birches, walnut, and poinsettias, should not be pruned until summer or fall, when the sap pressure is lower. Otherwise, the excessive exudation becomes unsightly.

Parts of the Plant to Prune

The reason for pruning will determine the limbs and branches to be removed from a tree or shrub. If the objective is to remove diseased portions, the cut should be made through healthy wood between the trunk or crowns and the infected part (Figure 25-9). The cut should never be made through the diseased wood or very close to it. This contaminates the pruning tool, which may transmit the disease to healthy parts pruned later.

If the objective is to improve the overall health and appearance of the plant, branches growing into the center of the plant should be removed. Limbs and twigs that grow across other branches can crowd the plant and cause sites of infection to form by rubbing abrasions through the bark. If more than one limb originates at the tree crotch, the strongest should be left and the others removed, Figure 25-10. Major structural limbs and twigs must be left so that no holes appear in the plant. Often overlooked is the fact that many secondary branches can stem from one older branch. The removal of one branch

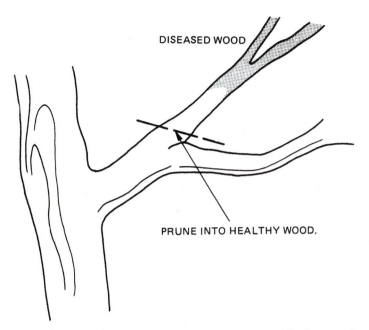

DISEASED WOOD

PRUNE INTO HEALTHY WOOD.

Figure 25-9 The correct way to remove diseased limbs or twigs

from a young tree can result in an older tree missing an entire side.

If the purpose of the pruning is to create denser foliage, as with evergreens, the center shoot is shortened or removed. This encourages the lateral buds to grow and create two shoots where there had been only one, Figure 25-11.

Pruning Methods

The method of pruning a tree or shrub depends on the size and number of branches to be removed. Limbs are pruned from trees with a technique called **jump-cutting**. This method allows a scaffold limb to be removed without taking a long slice of bark with it when it falls. A jump-cut requires three cuts for the safe removal of a limb, Figure 25-12. The final cut should remove the stub of the limb close to the trunk, but not directly against it. Until recently, the wound would then be covered with a wound paint to make it less conspicuous until the plant has time to heal. However, recent research has suggested that wound paints may actually

delay healing of plant tissue. Where this is suspected, the landscaper should not use them.

When shrubs are pruned, one of two techniques is used. **Thinning out** is the removal of a shrub branch at or near the crown or its point of origin. It is the major means of removing old wood from a shrub while retaining the desired shape and size. **Heading back** is the shortening, rather than total removal, of a twig. It is a means of reducing the size of a shrub. In cases where shrubs have become tall and sparse, a combination of thinning out and heading back can rejuvenate an old planting, Figures 25-13 and 25-14.

In heading back, the location of the cut is important, Figure 25-15. If too much wood is left above the bud, the twig will die from the point of the cut back to the bud, but the cut may not heal quickly enough to prevent insect and pathogen entry. Also, the woody stub itself may decay later. A cut below the bud will cause the bud to dry out and possibly die. The cut should be made just above the bud and parallel to the direction in which the bud is pointing. The cut should be close

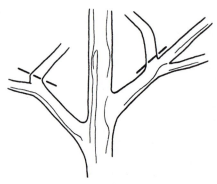

REMOVE BRANCHES GROWING TOWARD
THE CENTER OF THE PLANT.

DORMANT WINTER BUDS

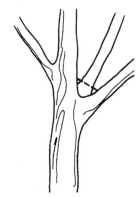

REMOVE EXTRA BRANCHES AT
TREE CROTCH

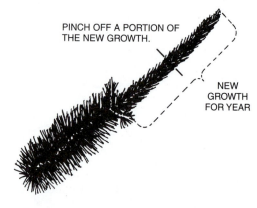

PINCH OFF A PORTION OF
THE NEW GROWTH.

NEW
GROWTH
FOR YEAR

IN THE SPRING, THE DORMANT CENTRAL BUD HAS
THE GREATEST GROWTH UNLESS PINCHED BACK.

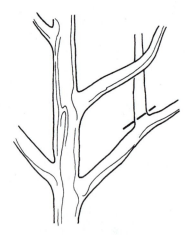

REMOVE BRANCHES GROWING
ACROSS OTHER BRANCHES

Figure 25-10 Selecting branches to be pruned

AFTER PINCHING, ALL BUDS ARE ABLE TO
GROW. THE RESULT IS A FULLER PLANT.

Figure 25-11 Evergreens are pruned in the
spring if denser foliage is desired.

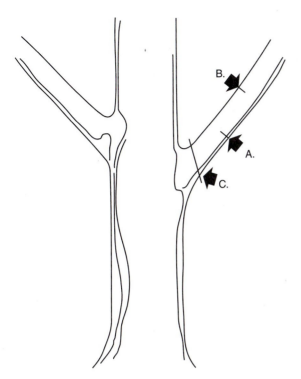

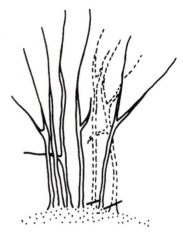

THINNING OUT. As its name implies, this method involves selection of an appropriate number of strong, well-located stems and removal at the ground level of all others. This is the preferred method for keeping shrubs open and in their desired shrub size and form. With most shrubs, it is an annual task; with others, it is required twice a year.

Figure 25-12 The removal of large limbs using the technique of jump-cutting. The cut at A allows the limb to snap off after a cut at B without stripping bark from the trunk as it falls. The final cut at C removes the stub.

enough to the living tissue to heal over quickly but not so close to the bud that it promotes drying. The direction in which the branch of a plant grows can be guided by good pruning techniques. Branches growing into the plant can be discouraged by the selection of an outward-pointing bud when heading back, Figure 25-16. If the twig has an opposite bud arrangement, the unnecessary one is removed.

How to Prune Hedges

The creation of a hedge requires close spacing of the shrubs at the time of planting and a special type of pruning. The landscaper must shear the plant so

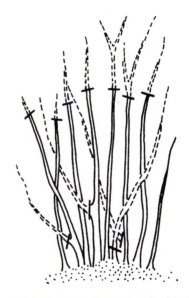

HEADING BACK. This method involves trimming back terminal growth to maintain desired shrub size and form. It encourages more compact foliage development by allowing development of lateral growth. This is the preferred method for controlling the size and shape of shrubs and for maintaining hedges.

Figure 25-13 The techniques of thinning out and heading back

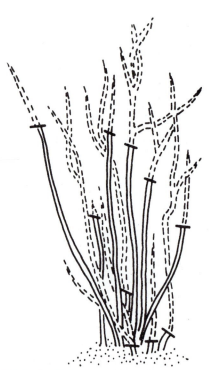

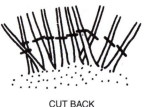

CUT BACK

SELECT SIX OR MORE WELL-PLACED VIG-
OROUS SHOOTS.

HEAD BACK

GRADUAL RENEWAL This pruning method involves removal of all mature wood over a three- to five-year period. Approximately one-third of the mature wood is removed each season. This is the preferred method for shrubs that have not been recently pruned and are somewhat overgrown.

COMPLETE RENEWAL This method involves complete removal of all stems at the crown or ground level. Two to three months later the suckers or new growth that emerges is thinned to the desired number of stems. These, in turn, are headed back to encourage lateral branching. Unpruned, seriously overgrown, or severely damaged shrubs are prime prospects for this treatment.

Figure 25-14 Two techniques used to rejuvenate old shrubs

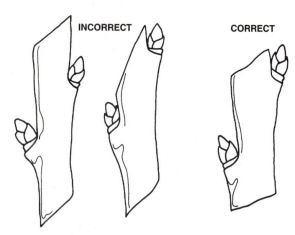

Figure 25-15 Where to prune the twig

that it becomes as dense as possible. This is usually done with hedge shears. The hedge shears easily cut through the soft new growth of spring, the season when most hedges are pruned. For especially large hedges, electric or gasoline powered shears are available. However, practice and skill are required for the satisfactory use of power shears. Damage can occur quickly if the landscaper does not keep the shears under control.

A properly pruned hedge is level on top and tapered on the sides, Figure 25-17. It is important that sunlight be able to reach the lower portion of the hedge if it is to stay full. Otherwise it becomes leggy and top-heavy in appearance.

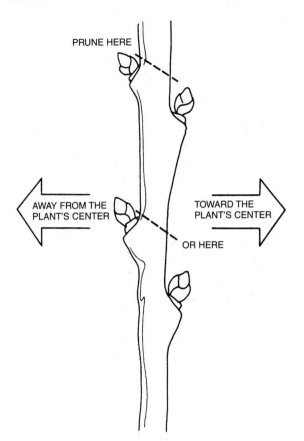

Figure 25-16 Twigs should be pruned to leave an outward-pointing bud.

National Pruning Standard

Recent cooperation between members of the arborist, landscape, and nursery industries and governmental organizations has resulted in the development of national pruning standards. These standards are known as the American National Standards Institute (ANSI) Standards and seek to establish a level of acceptable performance for all professionals engaged in the care of woody plants.

Flower Plantings

No components of the landscape require more maintenance time than flower plantings. This is why landscapes with a low budget for maintenance

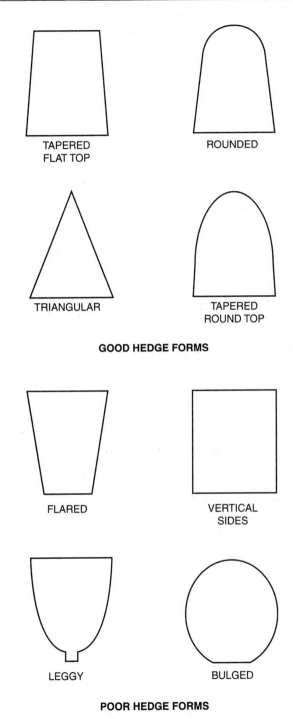

Figure 25-17 Good and poor hedge forms as seen from the side

must minimize the use of flowers. Reasons for the high maintenance costs are that:

- weeds must be pulled by hand or controlled with costly selective herbicides

- flowers are more susceptible to insects and diseases that mar their aesthetic appearance, therefore requiring additional expenditures for pesticides

- flowers tend to go to seed or get leggy with age if not pinched back frequently during the growing season (this subject is discussed later in the chapter)

- many perennials bloom only once a year but must be cared for throughout the growing season to ensure a good flower display the following year

- some perennials, notably the tender bulbs, must be dug up, put into storage for the winter, and set out again each spring

Weed Control

Several good herbicides are approved for use in flower plantings. If applied for preemergence *and* postemergence control, a weed-free flower planting can be attained. One danger of herbicides in flower beds is that the chemicals are always selective—that is, they kill grasses or broad-leaved weeds, but seldom both. Since flower plantings can have characteristics of both, damage can be done to some flowers even when others are uninjured.

Most herbicides used in flower plantings lose their effectiveness if the soil is disturbed after application. In such cases, landscapers must avoid cultivating the soil surface if they wish the full benefit of the herbicide.

Watering

All flowers should be watered frequently and deeply during dry periods. Their shallow roots quickly react to drought conditions, and they reach a critical wilting point much sooner than the woody plants of the landscape. Flowers planted beneath trees and shrubs must compete with the woody plants for surface water, and they will dry out faster than flowers not in such competitive locations.

Fertilization

Annuals can be fertilized in midsummer with a low-analysis fertilizer to keep them lush and healthy. Bulbs should be fertilized immediately after flowering with a high-phosphorus fertilizer such as bone meal. Nonbulbous perennials grow best if fertilized in the early spring. Summer or fall fertilization of perennials can harm the plants by keeping them too succulent as winter approaches.

Dead-heading

Many summer annuals must be continually tended to remove dead flower blossoms, a practice termed **dead-heading**. Plants such as petunias and geraniums are especially demanding. Combining deadheading with routine maintenance can reduce labor costs.

Pinching

Annuals are most likely to benefit from pinching back, but certain perennials such as hardy mums will also do better if pinched. *Pinching* removes the terminal shoot on each branch of the flower and allows the lateral shoots to develop, thereby creating a fuller plant. *Soft pinching* is done with the thumb and forefinger and removes the terminal bud or, at the most, the terminal and the first set of laterals. *Hard pinching* may shorten each stem by one third or more. Most flowers benefit by a hard pinch soon after being set out, followed by one or two soft pinches during the summer. With perennials such as mums, whose flower bud initiation is tied to a photoperiod response, the last pinch should not be after mid-July if a good flower display is to be seen in the autumn.

Flowers such as petunias or rose moss that are valued for their profuse blossoms, can be kept from getting leggy and going to seed by severely cutting them back about midsummer. A tool as indelicate as a pair of grass clippers can be used to provide a hard pinch if the planting is extensive. A period of several weeks with few flowers will follow until new reproductive growth begins. However, the fresh look and new flowers that result will carry the annuals right into the fall season.

Achievement Review

A. From the choices given, select the answer that best completes each of the following statements.

1. The main objective when watering trees and shrubs is to _____.
   a. keep the humidity high
   b. promote deep root development
   c. keep the foliage clean and moist
   d. prevent wilting

2. The proper time to fertilize shrubs is in _____.
   a. early spring
   b. late summer
   c. early fall
   d. winter

3. Trees should be fertilized _____.
   a. near their anchor roots
   b. at the nursery before digging
   c. near their base
   d. beneath the drip line of the canopy

4. The purpose of edging a bed is to _____.
   a. divert surface water
   b. discourage turf roots from entering the bed
   c. hold the mulch within the bed
   d. improve drainage

B. Answer each of the following questions as briefly as possible.

1. You discover a leaf spot disease on the English ivy plantings that are the major groundcover of a large landscape. How can you determine what control to use, what products to select, and when and how much to apply?

2. Complete the following statements.
   a. Removing a portion of a plant for better appearance, improved health, controlled growth, or attainment of a desired shape is termed _____.

   b. The most important branch on a tree is the _____.

   c. Scaffold branches create _____ of the tree.

   d. The point at which a branch meets the trunk or a larger branch is termed the _____.

   e. Undesirable branches originating from the root system of a tree are termed _____.

   f. Undesirable branches originating from the trunk of a tree are termed _____.

   g. The point of a shrub at which the branches and root system meet is termed the _____.

   h. Generally, the younger branches in a shrub will have a _____ color.

   i. Shoots originating from the stock of a grafted plant are _____ compared to those originating from the scion.

   j. Flower and fruit production occur most profusely on _____ branches.

3. Place *X*s in the following chart to compare the advantages and disadvantages of pruning during different seasons.

Characteristic	Spring	Summer	Autumn	Winter
A full view of the plant's branching structure is possible.				
View of the plant's branching structure is blocked.				
Dead branches cannot be detected.				
The timing may conflict with more profitable activities.				
The most time is allowed for cuts to heal.				
This is the best season to prune for the good of most plants.				
This is the best season to prune shrubs that flower in early spring.				
Plants may be damaged by unintentional breakage.				

4. Which three branches shown in this drawing should be removed, and why?

5. What is the difference between thinning out and heading back in the pruning of shrubs?

C. Write a short essay on the maintenance of flower plantings. Outline the care needed by both annuals and perennials to keep them healthy and attractive throughout the growing season. Assume the environmental conditions to be those in your own area.

CHAPTER 26

Plant Injuries: Identification and Care

Objectives:

Upon completion of this chapter, you should be able to

- describe the common causes of injuries to landscape plants
- distinguish between those that are caused by insects and diseases and those caused by other things
- explain the ways that insects and diseases are spread
- describe the most common symptoms of plant injury
- classify weeds in several different ways and explain how they injure landscape plantings
- apply the principles of control to plant pests
- explain the different formulations of chemical pesticides
- describe integrated pest management

Plant Injuries and Their Causes

Anything that impairs the healthy growth and maturation of a plant may be regarded as an injurious agent. Some injurious agents cannot be transmitted from plant to plant. Others can be transmitted and are regarded as either *infectious* or *infestious*. An infected plant has the injurious agent active within it. An infested plant has the injurious agent active on its surface. Some causes of plant injury are members of the plant or animal kingdom and are biological in character. Others are environmental or circumstantial and nonbiological in nature.

When an injurious agent is biological and either infectious or infestious, it also is parasitic. A **parasite** is an organism that cannot manufacture its own food. So *yes*, humans are parasites, as are rodents, deer, rabbits, and other animals that can cause injury to plants, but they are not involved with plants at the cellular level. Therefore, in a discussion of plant parasites the term is usually applied to **insects, bacteria, fungi, viruses,** and **nematodes**. Every one of these parasites has its own branch of science devoted to its study. This chapter can only touch on the high points of what is known about them and how they affect the growth and development of landscape plants. In the same way, our consideration of weeds will focus narrowly on how they affect landscapes, not on their biology. Table 26-1 illustrates the various categorizations of the injurious agents of landscapes.

Different Forms at Different Times

Just as plants change their appearance at different seasons and at different stages of their lives, insects and the agents of plant disease do, too. Most people understand that butterflies were once caterpillars, and some may know that the grub worms found in their lawns will later become the June bugs that buzz around their porch lights. Insects are among nature's most extraordinary animals, changing their appearances and forms dramatically as they grow. That change is known as **metamorphosis**. Depending upon the species of insect, it may have as many as four different stages of development. Not all stages during the life cycle of an insect are injurious to plants. For effective control of insect damage, a landscaper must know

- the insects likely to affect particular plant species

- the stage(s) of the insects' lives that do damage to the plants

- how to recognize the presence of insects

- how to match a type of injury with a specific insect

The bacteria, fungi, viruses, and nematodes that cause plant diseases are collectively termed **pathogens.** Many pathogens, especially fungi, also have different stages in their development. While neither as visually dramatic nor as readily visible to the eye, the changes during the life cycles of pathogens determine when and how they do the most damage to plants. Because they are so small,

their presence can go unnoticed until the plants begin to show signs of injury. For effective control of disease damage, a landscaper must know:

- the pathogens to which the landscape's plants are most susceptible

- the environmental conditions most suitable for pathogen development

- the potential sources of pathogens in the landscape

- how to recognize the presence of pathogens and the onset of disease

- how to diagnose a specific disease

Weeds injure landscape plantings both visually and culturally. They mar the attractive and tidy appearance of the design. Most significantly, they compete with the desirable plants for water, nutrients, and even sunlight. While weeds do not have the complex multiple life stages of insects and pathogens, they do have times during their growth from seed to maturity when they are more easily controlled, and a landscaper must know those times of vulnerability.

Most of the damage caused by larger animals is a result of their feeding and not related to their stage of growth. It may be more predictable at certain seasons of the year when other food sources are limited, such as the winter months. It may also occur in late evening or in remote areas where there is less likelihood of the animals being frightened away by people.

How Insects and Diseases Are Spread

It is not necessary to understand every biological nuance of insects, pathogens, and weeds to understand how they enter the landscape and how they move from plant to plant and area to area. In most cases, they get assistance. Often, it is the landscape personnel and/or the clients and their guests who provide that assistance.

Table 26-1 Injurious Agents

	Infectious	Infestious
Biological Agents of Injury		
Insects	Occasionally	Yes
Mites	No	Yes
Fungi	Yes	Occasionally
Bacteria	Yes	No
Viruses	Yes	No
Nematodes	Yes	Yes
Rodents, rabbits, deer	No	No
Slugs and snails	No	Yes
Weeds	No	No
Parasitic plants	Occasionally	Yes
Other Causes of Injury		
Snow, ice, wind, sun scald, hail	No	No
Mowers and other mechanical tools	No	No
Vandalism	No	No
Nutrient deficiency	No	No
Fertilizer burn	No	No
Chemical injury	No	No
Drought	No	No
Poor drainage	No	No
Incorrect light exposure	No	No

Plant pests can be brought into a landscape setting on the plants that are purchased from a nursery or other supplier whose production hygiene was faulty. This is especially probable when plants for the landscape are collected from the wild, accepted from friends, or obtained from any source that does not guarantee its stock to be pest-free. They can also be carried in by the wind or bodies of water that pass through the landscape site. Animals such as dogs and cats make nice vehicles for pests, as

do the cars, trucks, bicycles, and equipment of people who live, visit, or work at the property. Weeds can be introduced by the means described and are also frequently contained in soil additives, mulch, or seed.

Once on the site, the pests can be spread from plant to plant and from one area to another by natural means such as the wind and splattering rain. They can also be carried by people's hands or garden tools that become contaminated during their use on an infected or infested plant. Some insects can also crawl and/or fly, so their movement through a planting, once infested, is easy. Because insects and pathogens reproduce often and in large numbers, it takes very little of either when transferred to become a sizable presence in short order.

The Symptoms of Injury

When we look at a plant and determine that something is wrong, we are witnessing the plant's response to some type of irritant. The responses of plants to insects and pathogenic irritants are termed **symptoms**. Some symptoms are common to numerous insects and diseases; other symptoms are almost unique to certain pests. The symptoms of an insect or pathogen's activity in or on a plant may change as its life cycle advances or as the severity of the plant's response increases. The sum of all the symptoms expressed by a host plant from the time it is initially infected until it either recovers or dies is known as the **symptom complex**.

Infection of a plant by pests is not a static condition. The symptoms expressed early in the infection may be quite different from those expressed later. For any one pest, however, the symptom complex is usually specific. To diagnose a disease or insect problem correctly requires recognizing the specific changes each major pathogen or insect can create in a host.

Symptoms may be influenced by an assortment of factors including the species of the host plant, the environment, the quantity of infectious material **(inoculum)** present, the insect population, and the stage of development of the insect or pathogen. Furthermore, symptoms can result from other causes such as a damaging environment, improperly applied chemicals, animal injury, and mechanical damage. For this reason, it is often necessary to isolate and identify the specific agent of plant injury or to consider other possible sources of irritation before the cause of plant symptoms can be established.

While specific symptoms cover a wide range often separated only by subtleties, collectively they can be grouped into major categories that permit description and comparison, Figure 26-1.

Wilting

Plants may wilt from lack of water. If such a symptom is environmental in its cause, the plant will recover when water is applied. If pathogens attack the plant's *vascular system*, which permits it to move water up its stem, or if insects destroy the plant's root system, the wilting will be permanent. When fungi invade the tender stem tissue of a

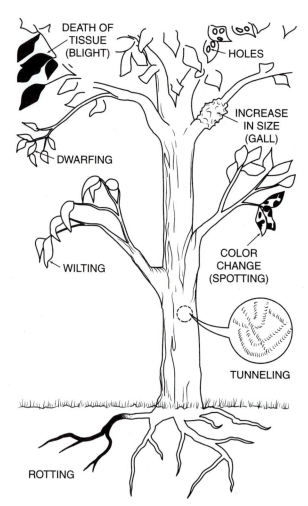

Figure 26-1 Symptoms of plant injury

young seedling, the disease *damping-off* develops, the seedling wilts and drops over to die.

Color Changes

Color changes may be widespread through most of the plant or localized as spots, rings, and lesions. When a plant does not receive enough light to enable it to make food efficiently through the process of **photosynthesis,** it is in a state known as **etiolation**. If adequate light is provided in time, the plant will return to full health. When a plant turns yellow, but does not lack light, the symptom is termed **chlorosis**. Chlorosis is probably the most common of all color changes. It can be caused by insects, pathogens, or environmental problems.

Rotting

Rotting results from a destruction of cells within the host plant, causing a release of the cellular fluids. It may be accompanied by a strong, often foul odor. Rots may be dry or soft and occur in the roots, leaves, stems, buds, or fruits of the host plants. They are usually the result of plant diseases or freezing.

Death of Tissue

When plant tissue dries out and dies, it usually turns brown or black and is said to be necrotic. The dead tissue, or **necrosis,** may be located on leaves as *spots,* or centered in young buds. It may also be extensive, encompassing entire branches as in *blights.* Insects, pathogens, or environmental factors can cause necrosis. It ranks with chlorosis as a highly common symptom. It is often the final symptom in the symptom complex.

Dwarfing

All or part of a plant can be reduced in size as a result of insects, nematodes, and other pathogens, especially viruses. Dwarfing may result from the reduction of water uptake at a time when new tissue is expanding. Insects or nematodes (microscopic worms) in the root system of a plant may create such dwarfing. The symptom is caused by either a reduction in the size of cells produced by the host plant or by a reduction in the number of cells that it produces.

Increase in Size

Plant parts may become malformed in response to insect or pathogen irritants. As cells increase in size or in number, symptoms are expressed as *galls, witches brooms, swollen roots, abnormal shoot growth, scabs,* and *fasciations.*

Tunneling

Insects may bore into plant trunks, chew their way within leaves, or burrow up and down the stems. Borers and leaf miners commonly create such injuries in plants.

Holes

Leaves riddled with holes are symptomatic of both insect and pathogenic causes. Insects usually cause holes by feeding on the leaf tissue. When caused by pathogenic activity, the holes are usually preceded by necrotic lesions and spots that represent dead cells. As the tissue dies in these localized areas, the necrotic spots drop out, creating the holes.

Determining the Cause

Hundreds of combinations and variations of symptoms can result from the many possible causal agents, reactions of individual host plants, and unpredictable modifications caused by the environment. If the quick and accurate diagnosis of what is troubling a particular host plant in the landscape sounds difficult, be assured it is!

Some diseases, such as powdery mildew, are common to many plants, and once you become familiar with it, you will recognize it every time you see it. Some insects, like the aphids, attack many different plants, so they soon become familiar to landscape professionals responsible for keeping a property healthy. It isn't necessary to recognize and have a complete knowledge of every insect and plant disease in nature, because most landscapes within a geographic region will be affected by a limited number of pests that recur year after year. So it becomes important to learn to recognize them and learn how to protect the landscape against them. Still, early in the symptom complex it can be difficult to determine whether the problem you are seeing is caused by an insect, some type of pathogen, or some cultural or environmental problem.

Look for Signs

Insects leave tell-tale signs of their presence, and so do many pathogens and other parasites. There may be the actual insect there. That can make diagnosis fairly easy. However, some insects look like part of the plant they are infesting. Scale insects are an example: some are easily mistaken for bumps on the bark, and others resemble white lint and appear harmless. Aphids often have soft green bodies and cluster around the young tissue at the end of a twig where they can go unnoticed until the damage caused by their feeding begins to show.

At times, other evidence of insects will be the landscaper's only clue to their presence. There may be egg masses on the underside of leaves or in the junction where leaves meet the branches. Some insects will leave slimy trails or traces of webs as a sign of their activity. It can also be baffling to observe plants that have been chewed on, yet find no insect presence. That usually is evidence of *nocturnal* insects or slugs that feed at night and hide by day. Usually turning over nearby rocks or logs will reveal the pests.

Pathogens are usually more difficult to recognize because they are so much smaller. Often only a single cell in size, the signs of their presence are different from those of the insects or larger parasites. Bacteria and viruses are completely invisible to the naked eye. Only after the host plant begins to display symptoms of infection is their presence apparent and their identification possible. Fungi are larger and more complex, although still often microscopic. Some can be seen growing on the surface of leaves and other plant parts. An example is, again, the mildews. What the eye sees as the

white powdery material on the leaves is actually the vegetative body of the fungus. Other fungi can be seen when they begin to reproduce and send their fruiting structures to the surface of the plant where they become visible to the eye. The bright orange spots of rust diseases are actually fruiting structures of the fungus that is growing inside the host plant. A simple magnifying glass will reveal the fungus, its fruiting structure, and its seed-like contents **(spores)** that enable it to spread.

Cultural and environmental injuries do not display the same types of clues to their cause, but there is evidence nonetheless. When the soil in which the plants are growing becomes deficient in one of the nutrients needed for healthy growth, symptoms of that deficiency will begin to show. Necrosis and chlorosis or other color change are among the most common symptoms of nutrient deficiency, but others may be displayed as well. Landscapers should be sufficiently knowledgeable of the soils in their region to anticipate the likely nutrient needs of the plants they install and maintain on a recurring basis. An overall knowledge of the symptoms of nutrient deficiency is also important and is certain to be a part of the educational background of well-trained landscape professionals.

Nutrient deficiencies may appear as patterned symptoms, such as yellowing between the veins of the leaves. They may also appear as localized symptoms, perhaps affecting only the tips of the plant or only the lower, mature leaves. At other times, the symptoms will be widespread with all parts of the plant showing the color change or growth reduction.

Injuries that result from improper culture, such as mowing too short, over-fertilizing, or not providing the proper amount of moisture, are recognizable by the time of their occurrence—they will become apparent shortly after the cultural malpractice has happened.

Environmental injury is often the easiest to diagnose, as when the damage is found only on the side of the plant that is most exposed to strong winter wind or hot summer sun. If the only plants injured are those closest to a walk or roadway that is heavily salted during the winter, then determining the cause of the problem is not difficult. Other types of environmental injury are not as immediately apparent. The symptoms of polluted air will appear slowly and may be evident only on certain plants. Temperature extremes, such as frost or sunburn, may be long gone before the symptoms of their damage are manifested, leaving the late-arriving landscaper to wonder about the cause.

It is not the purpose of this text to identify specific insects, diseases, nutrient deficiencies, or other causes of injuries to landscape plants (there are many excellent resources available to accomplish that). Rather, this chapter is intended to illustrate the need for proper and complete training for those who choose to practice as landscape professionals—the ability to recognize the cause of a plant ailment or to anticipate the potential for injury in advance can make the difference between a successful landscape and a disappointing one.

Weeds

A **weed** may be defined as *a plant having no positive economic value and/or growing where it is not desired.* Unlike insects and pathogens, weeds do not rely on a host plant for food. Instead, weeds compete with other plants for the materials that both need to grow and thrive. In addition, weeds often serve as alternate hosts, providing sites for the overwintering of insects or pathogenic inoculum. Some fungi need weed species in which to produce one or more spore forms as part of a complex life cycle.

Classification of Weeds

Although much larger than insects or pathogens, weeds share one attribute with them: a variety of life cycles. A landscaper who must control weeds needs to be able to identify the specific weed and understand its life cycle.

Annuals. These weeds complete their lives in one year. *Summer annuals* germinate in the spring, grow most actively during the summer months, produce their seeds in the autumn, and then die.

Examples include crabgrass, foxtail, spurge, and prostrate knotweed. *Winter annuals* germinate in the fall and grow into small plants before going dormant and overwintering. They then resume growth in the spring and complete their lives during the early summer. Examples of such weeds are wild garlic, chickweed, and henbit.

Perennials. These weeds live for several years and may produce seeds numerous times during their extended lifespan. Examples include dandelion, quackgrass, clover, and nutsedge.

Weeds have evolved to survive. They are prolific seed producers, with some species capable of producing more than a million seeds per parent plant. As with nonweed plants, some weed seeds germinate quickly and easily, while others remain dormant in the soil for some time. Weed seeds are usually distributed throughout a garden's soil, so turning over the earth may bring to the surface seeds that have been buried and dormant for several years.

The Principles of Control

The control of insects, pathogens, and weeds in a landscape depends upon how successfully the landscaper applies the four basic principles of control.

Exclusion

Exclusion is the first principle of control. It includes all the measures designed to keep a pest from becoming established in an area. The measures vary depending on the type of pest. The most logical application of this principle in landscaping is the use of plant material produced by a nursery that certifies its stock to be disease- and insect-free. To be doubly safe, a member of the landscape staff who is knowledgeable of plant pests should carefully inspect all nursery stock when it is delivered and before it is installed at a project site. Any material displaying symptoms of infection or signs of infestation should be rejected and returned to the grower. Another way to apply the principle of exclusion is to avoid materials that are certain to be contaminated, such as cheap grass seed mixtures or nonsterile mulches, such as hay or nonfumigated straw. Telephone companies or arborists often turn their cut wood into chips and then offer the material to anyone who will take it as a means of disposal. Landscapers who then spread it as mulch run the risk of introducing pest contaminants into a landscape that was formerly pest-free.

Eradication

Eradication is the principle that seeks to remove or eliminate pests that are already in, on, or near plants in infested areas. The measures of eradication attempt to reduce the quantity of pathogenic inoculum, insects and their eggs, or weeds and their seeds.

- *Isolating and destroying individual infected plants* is time-consuming and expensive, especially on large landscape sites, such as parks, cemeteries, and commercial properties. If done as part of another routine maintenance task, such as pruning, then it can be cost-effective. In smaller, residential landscapes with comparatively fewer plants, the eradication of infected plants is easier because they will be more quickly noticed.

- *Hand-pulling and cultivation* are effective in the control of some weeds, especially if they have not yet produced their seeds. The technique usually does not remove all of the root system however, so the weeds may return.

- *Destruction of alternative hosts* such as weeds can also aid pest control. The alternative hosts may allow completion of the life cycle of a fungus or harbor insects that will later invade the garden.

- *Plant rotation and soil treatments* are both methods of eradicating pests that persist in the soil. Many pathogens overwinter as dormant spores on plant litter in the soil. If the same plants are used in that location the next year, as is often the case with flower plantings, the pathogen will be waiting to strike again. By

using flowers that are not susceptible to the previous year's disease, the landscape can avoid another year of injury.

Treating the soil with heat or chemicals before planting can give protection to flower plantings and can be especially effective in more controllable areas such as raised beds and potted flowers.

- *Destroying host parts* that display evidence of insect or disease damage can reduce the amount of inoculum available for dissemination to nearby hosts. This is the benefit received when diseased limbs are cut from a tree, permitting the healthy parts to remain.

- *Removing infested refuse* eliminates a site for the overwintering of pathogens or insects. Collecting grass clippings and raking fallen leaves applies this principle and can be accomplished as part of regular maintenance routines.

- *Chemical sprays, dusts, and drenches* are generally the most expensive methods of eradication. The products, collectively termed **eradicants,** strive to kill the pathogen before it can infect the host, kill the insect before it can do much damage or reproduce, or kill the weed before or shortly after it emerges.

Protection

Protection is the principle of control that sets up a barrier between the host plants and the pests to which they are susceptible. It is a shielding endeavor that can be accomplished either through manipulating the plants' growing environment or by applying chemicals.

- *Manipulating the environment* is an attempt to create conditions for growth that are more favorable to the host plant than to the pest. Most pathogens thrive in conditions of high humidity and moisture. Many diseases can be avoided by watering the landscape early in the day so that water can evaporate quickly from the foliage, rather than in the evening when leaves would stay wet all night. Also, by providing and maintaining nutrient, pH, and other cultural conditions that are more favorable to the host plant and less favorable to known pests, the landscaper can protect the plantings of the garden. A healthy, vigorously growing plant is one of the best defenses against pests.

- *Chemical sprays and dusts* can be applied to seeds, bulbs, foliage, and wounds of plants to place a barrier between the host and the insect or pathogens. In this case, it is essential that the chemical be applied before the pests arrive. It then kills them after they arrive.

Resistance

Resistance is the fourth principle of control. It is an attempt to change a plant either physically or genetically so that it will suffer less from diseases and/or insects. Actual creation of the resistance is not something done by landscapers. It either occurs naturally where it is observed and then reproduced by plant breeders, or it is developed over time by plant geneticists and the resistant plants are eventually released for commercial and public use. However, it is the landscaper's selection of plant varieties known to be resistant to the pests of an area that makes application of this principle valid. Resistance is seldom total and may not last forever. All plants have some degree of susceptibility, and that susceptibility often increases with time. The loss or lessening of resistance to a pest does not necessarily mean that the host plant has changed. It may mean that the pest has altered in a way that allows it to infect the plant. Considering the reproductive potential of most pests and the rapidity of their life cycles, it is not surprising that resistance is often overcome by the natural mutation of insects and pathogens.

Pesticides

Pesticides are chemicals used to kill organisms that injure desirable host plants. If directed against insects, they are called *insecticides;* against fungi,

they are called *fungicides;* against nematodes, *nematicides,* against bacteria, *antibiotics* or *bactericides,* and when directed against weeds, they are called **herbicides**. Pesticides are one of the weapons in a landscaper's arsenal of defenses against the pests of the landscape. Specific recommendations of what product to use against what pest can be obtained from a variety of sources. The chemical companies that develop and manufacture the products are certain to advertise them in trade magazines and through other media outlets. The final determination of what is recommended and safe for use in a particular state or region within a state is usually the responsibility of the scientists at the college of agriculture in each state. Landscapers should consult the list of approved chemical pesticide products for their state before purchasing and using a new product. That information is often available through the Cooperative Extension Service, which makes the latest research and publications of the colleges of agriculture available to the public, including industry practitioners such as landscape professionals.

Understanding chemical pesticides requires knowing what they are and what they are not. What they are is *poisonous*, not only to pests, but to animals and people as well. What they are not is *medicinal*. The belief that pesticides are medicines for ailing plants is misguided and implies a curative quality that is lacking. An infected plant can seldom be cured. Necrotic tissue cannot regain its life, holes chewed in leaves will not restore themselves, and galls will not diminish.

To be most effective, pesticides need to be on the plant before the pathogen or insect invader arrives so it dies promptly on its arrival. These pesticides can be regarded as **protectants**. If the pathogen is already on the plant host and about to infect or the insects are already feeding on the plant, then the pesticides must kill them immediately. Such pesticides are termed **eradicants**.

Herbicides also have varying forms of action. If the chemical kills all green plants, it is termed a *nonselective herbicide*. Those that kill some kinds of plants and not others are *selective herbicides*. These assorted products are also characterized by whether they kill on direct contact with the weeds or after the chemicals have been absorbed by the weeds, making the products *systemic*. Herbicides may kill the weed before they emerge from the soil (preemergence) or after they emerge from the soil (postemergence).

Pesticide Formulations

Whether the pesticide is a fungicide, nematicide, insecticide, or herbicide, it is usually available in several different formulations. The choice of formulation is based on:

- the size of the landscape property being treated
- the amount of active ingredient being applied
- other materials being applied along with the pesticide, such as other pesticides or fertilizers
- cost
- safety
- ease of application

Following is a summary of product formulations and their characteristics.

Solutions. The pesticide dissolves completely and uniformly into a carrier of water or oil. It does not precipitate out, so once dissolved the pesticide mix does not need to be agitated.

Emulsifiable Concentrates. Some pesticides are not water-soluble, but must be applied in a water carrier. A typical emulsifiable concentrate contains the pesticide, a suitable solvent, an emulsifier, and often a wetting agent, sticker, or antifoaming agent. The elements do not settle out, once mixed, so they only need periodic agitation to assure that the mixture stays uniform. The concentration of active ingredient is usually high in emulsifiable concentrates, making them dangerous to handle.

Wettable Powders. These pesticides are of limited solubility in water. They are combined with a filler material such as clay or talc as well as a wetting agent to permit their dispersal in a water carrier. They require continuous agitation to ensure uniform coverage. They are relatively low in cost and easily stored and handled but they are hazardous if inhaled or absorbed through the skin.

Granules and Pellets. Herbicides are often applied in this form. The pesticide is in the form of coarse, solid particles for easy application. The carrier may be sand, clay, ground corn cobs, fertilizer granules, or similar material. No dilution is required, since the percentage of active ingredient is lower than in other formulations, usually between 4 and 10 percent.

Fumigants. The pesticides are in the form of poisonous gases, and they are often highly toxic. They have limited use in landscaping except when applied to covered piles of soil for use in containerized plantings or similar specialized areas.

Flowable Suspensions. These are wettable powders, not soluble, that are suspended in a water-based carrier material. It is usually necessary to agitate the product during application to ensure a uniform coverage.

Dusts. These are fine powders containing the active ingredient in low concentrations and mixed with a high percentage of inert carrier material. Applied dry and directly to the plants, dusts leave a visible residue.

Water-Dispersible Granules. These are granular formulations that are applied in water. They have a low solubility in water, so continuous agitation is required during application to ensure even distribution.

Gels. A gelatinous form of the pesticide, gels are commonly used as the formulation inside containers that are to be returned for recycling. This formulation is often used in water-soluble, premeasured packets.

Additional information about pesticide packaging, application, and use can be found in Chapter 30, Safety in the Landscape Industry.

Integrated Pest Management

A discussion of plant pest control cannot end with the final topic being pesticides because that would imply they are the final or ultimate choice of control measures. In recent years, it has been found to be more logical, more economical, and better for the environment to integrate a variety of techniques in the control of pests. Instead of waiting for a pest to appear and then treating it with a pesticide or anticipating that it will appear and applying a pesticide in advance, it is now considered wiser to monitor the host plants to determine if and when a pest appears and what can best minimize its impact. The technique is termed *integrated pest management* or *IPM*. It requires that the user/owner of the landscape be willing to accept some level of pest presence and not insist upon zero tolerance. The point at which the injury to the landscape's plantings or the number of pests present becomes unacceptable is termed the *action threshold*. The action threshold signals the need for control measures to be applied. It falls to the landscape management professional to know what represents the action threshold, to gain agreement and acceptance of that threshold by the client, and to know what control measures are appropriate.

Understanding how to manipulate the environment so that it favors the growth of the host plants more than that of the pests is vital. Knowing what potential pests are of concern in a particular landscape, their life cycles and appearance at different stages of their life cycle, and the times when they constitute the greatest threat to the host plants is also essential. Landscape monitors must also be able to recognize desirable or harmless insects or other agents. When the action threshold is near, the landscape monitor must advise the landscape manager of the proper control measures, be they cultural, biological, or chemical. Figure 26-2 illustrates the IPM Triangle, a graphic depiction of the relationship between host plant health, pest presence, and the environment within the landscape. The IPM Triangle also illustrates the level and diversity of educational

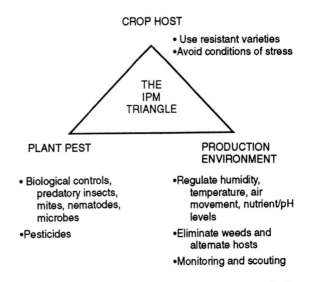

Figure 26-2 Integrated pest management balances the relationship between host, pest, and the environment.

preparation needed by anyone practicing landscape management today.

While the general public's acceptance and understanding of IPM are growing, there is still a need for customer education on the topic. Customers are reluctant to accept and pay for something they do not understand. A customer will understand receiving a bill for a landscape technician to spray the shrubs but may not understand being billed for someone to scout the property for pests and decide that no control measures are warranted. Still other customers may have to be convinced that a low level population of *some* insects on their garden plants is no threat to the landscape and can be kept at low levels with methods other than the blanket application of chemicals.

As public awareness and industry acceptance of IPM programs increase, it will become the expectation, not the exception, throughout the landscape industry. Once every employee can become a knowledgeable spokesperson for environmentally sound pest control, then the landscape profession and the nation will have taken a significant step toward safer and more natural techniques of pest control.

Achievement Review

A. Fill in the blanks in the following table to measure your understanding of which plant irritants are infectious and which are infestious.

	Infectious	Infestious
Biological Agents of Injury		
Insects		
Mites		
Fungi		
Bacteria		
Viruses		
Nematodes		
Rodents, rabbits, deer		
Slugs and snails		
Weeds		
Parasitic plants		
Other Causes of Injury		
Snow, ice, wind, sun scald, hail		
Mowers and other mechanical tools		
Vandalism		
Nutrient deficiency		
Fertilizer burn		
Chemical injury		
Drought		
Poor drainage		
Incorrect light exposure		

B. Define the following terms.
1. parasite
2. metamorphosis
3. pathogen
4. symptom
5. symptom complex
6. chlorosis
7. necrosis
8. weed
9. eradicant
10. protectant

C. Indicate which principle of pest control is being applied when the following measures are taken.

1. A landscape contractor installs plants certified by the grower to be pest-free.
2. Plants are not watered after 2 P.M. to allow them to dry off before evening.
3. A landscape management company hauls away all debris from a planting bed before winter sets in.

4. A diseased street tree is cut down and hauled away before the disease spreads to other trees nearby.
5. The state association of landscapers funds research at their college of agriculture to speed development of a new resistant variety.
6. Healthy nursery plants are sprayed in the spring before the onset of the rainy season.
7. Tulip bulbs that were stored in a bag where diseased bulbs were found are dusted with a pesticide before being planted.
8. An IPM monitor finds a sign of insect pest presence on a single plant and quickly and completely destroys it, leaving the rest of the planting pest-free.
9. The soil to be used in hanging baskets of flowers is fumigated before planting.
10. The local utility company has offered free mulch, produced when it put the cut limbs from its pruning operations through a wood chipper. The landscape company declines the offer.

D. Indicate if the following statements are true or false.

1. Weeds are parasites.
2. Weeds may serve as hosts for insects or pathogens.
3. Weeds are visible problems only. They do not persist unseen in the soil.
4. All weeds are perennial.
5. Dormant weed seeds can persist in the soil for several years.
6. The action threshold signaling the need for control is a common standard that can be applied to all plants and landscape sites.
7. IPM monitoring requires the immediate recognition of when the action threshold for a plant is near.
8. Manipulation of the growing environment can eliminate the need for an IPM program.
9. Wettable powder formulations of pesticides are more easily dissolved in a water carrier than pesticides in solution form.
10. If applied early in the symptom complex, pesticides can cure an infected plant and erase all symptoms of the pest's presence.

Suggested Activities

1. Select a common insect or disease problem for your region of the country. Assume that you do not know a proper control for the problem. Check the following three sources for control recommendations: the local Cooperative Extension Service, a nearby locally owned garden center staff person, the products available at a local chain store such as Wal-Mart or Tru-Value Hardware. Also compare the recommended application rates, time of application, and frequency of application. How similar or dissimilar is the information? Which would you use as a professional and why?

2. Go in search of the signs of pests. Select several nearby plants that are displaying symptoms of injury. Then try to determine if the cause is an insect, a pathogen, another parasite, or something else in the environment. A magnifying glass may help. Look for actual insects, or egg masses or webs, spores or ooze. Perhaps the distribution of the symptoms will suggest that the injury is caused by a nutrient deficiency or an external environmental irritant.

3. Evaluate the impact of weeds on host plant growth. Seed two greenhouse flats with annual flowers. Use weed-free soil in one and untreated soil in the other. Monitor growth of the seedlings over time. If the weeds are permitted to grow unchecked in the one flat, what is the impact on the growth and development of the flower seedlings compared to those in the flat where weeds offer no competition?

CHAPTER 27

Care of the Lawn

Objectives:

Upon completion of this chapter, you should be able to

- describe the operations needed to repair a lawn in the spring season
- explain the meaning of fertilizer analysis statements
- determine the amount of fertilizer needed to cover a specific area of lawn
- compare revolving, oscillating, and automatic sprinklers
- compare reel, rotary, and flail mowers
- explain the ways lawns can be damaged

Of all aspects of the landscape that require maintenance during the year, lawns consume the most time. A lawn necessitates both seasonal care and weekly care. Like any other feature of the landscape, it is easier to maintain if it is installed properly. Thus, the landscaper who is hired to install and maintain a lawn may have an easier job than the landscaper hired to maintain a lawn that was poorly installed by someone else.

Spring Lawn Care

Cleanup

Spring operations begin the season of maintenance. In areas of the country where winters are long and hard, lawns may be covered with compacted leaves, litter, or semi-decomposed thatch. The receding winter may leave behind grass damaged by salt injury, disease, or freezing and thawing.

Small areas can be cleaned of debris with a strong rake. Larger areas of several acres or more require the use of such equipment as power sweepers and thatch removers to accomplish the same type of cleanup.

Rolling of the Lawn

In many central and northern states, the ground freezes and thaws many times during the winter. Such action can cause **heaving** of the turfgrass. Heaving pulls the grass roots away from the soil, leaving them exposed to the drying wind. Heaving also creates a lumpy lawn.

Where heaving occurs, it is advisable to give the lawn a light rolling with a lawn roller in the spring. Rolling presses the heaved turf back in contact with the soil. The roller is applied in a single direction across the lawn, followed by a second rolling at an angle perpendicular to the first.

Two precautions should be observed before rolling a lawn. One is that clay soil should never be rolled, since air can be easily driven from a clay soil and the surface quickly compacted. The other precaution is that no soil should be rolled while wet. The roller can be safely used only after the soil has dried and regained its firmness.

The First Cutting

The first cutting of the lawn each spring removes more grass than the cuttings that follow later in the summer. The initial cutting at 1 1/4 to 1 1/2 inches is done to promote horizontal spreading of the grass. This, in turn, hastens the thickening of the lawn. An additional benefit of the short first cutting is that fertilizer, grass seed, and weed killer that are applied to the lawn reach the soil's surface more easily. Cuttings done later in the year are usually not as short.

Patching the Lawn

If patches of turf have been killed by diseases, insects, dogs, or other causes, it may be necessary to reseed them or add new sod, plugs, or sprigs. Widespread thinness of the grass does not indicate a need for patching. It indicates a lack of fertilization, improper mowing, disease, or insects.

Patching is warranted when bare spots are at least 1 foot in diameter. Seed, sod, plugs, or sprigs should be selected to match the grasses of the established lawn. Plugs can be set directly into the soil using a bulb planter or golf green cup cutter to cut the plug, and then to remove the soil where it is to be planted. With seed, sod, and sprigs, it is best to break the soil surface first with a toothed rake. A mixture of a pound of seed in a bushel of topsoil is handy for patching where seed is to be used,

Figure 27-1. Mulch and moisture must then be applied, as stated earlier.

The timetable for patching is the same as for planting and is related to the type of grass involved.

Aeration

Aeration of a lawn is the addition of air to the soil. The presence of air in the soil is essential to good plant growth. If the lawn is installed properly, the incorporation of sand and organic material into the soil promotes proper aeration. However, where traffic is heavy or the clay content is high, the soil may become compacted. The groundskeeper can relieve the compaction by use of a power aerator, Figure 27-2. There are several types of aerators. All cut into the soil to a depth of about 3 inches and remove plugs of soil or slice it into thin strips.

A topdressing of organic material is then applied to the lawn and a rotary power mower run over it. This forces the organic material into the holes or slits left by the aerator. The plugs of soil left on top

Figure 27-1 Patch seeding of thin areas in established lawns is done by breaking the soil surface and applying a small handful of seed.

Figure 27-2 The aerator is used to remove plugs of soil from compacted lawns, allowing air to enter the soil.

of the lawn may be removed by raking. If the soil plugs are not too compacted, they can be broken apart and left as topdressing. Equipment is made that can aerate and convert the soil plugs to topdressing in a single operation.

Vertical mowing is a technique that can break up the soil plugs left by an aerator or even remove excessive thatch if necessary. It requires a power rake or a mower whose blades strike the turf vertically. It is done when the lawn is growing most rapidly and conditions for continued growth are favorable. For cool-season grasses, late summer or early autumn is the best time. For warm-season grasses, late spring to early summer is best.

The blades of the vertical mower are adjusted to different heights depending upon the objectives of the operator. A high setting is used to break up soil plugs. A lower setting gives deeper penetration into the thatch layer. This makes it easier to remove and relieves compaction of the soil. Deep vertical mowing is only practiced on deep-rooted turfs. Shallow-rooted turfs often grow mainly in the thatch layer. They can be harmed more than helped by vertical mowing.

Clippings

Following a mowing, the cut clippings can be either collected or left on the lawn. While it may seem to be a matter of personal preference, in fact there is a reason to do it a certain way at certain times. It all has to do with the length of the clippings. The clippings contain nutrients that can be used by the grass plants once the clippings decompose and return those nutrients to the soil. Short clippings of healthy turf that result from regular, frequent mowing during the summer decompose rapidly and should not be collected. However, when the clippings are long, they can be problematic. Clumps of dead clippings on the lawn's surface are unattractive. Further, they will not decompose as quickly and may impair the growth of the grass beneath them by reducing the amount of light, retaining excessive moisture that may promote disease, or just smothering the grass. In such situations, the clippings should be collected as the lawn is mowed.

Lawn Fertilization

Much like grass seed, lawn fertilizer is sold in an assortment of sizes and formulations and priced accordingly. Stores selling fertilizers range from garden centers to supermarkets. The professional groundskeeper needs to have a basic knowledge of fertilizer products prior to their purchase. Otherwise, it is difficult to choose among the many brands available.

Nutrient Analysis and Ratio

The fertilizer bag identifies its contents. It displays three numbers that indicate its **analysis**, that is, the proportion in which each of three standard ingredients is present. These numbers, such as 10-6-4 or 5-10-10, indicate the percentage of total nitrogen, available phosphoric acid, and water-soluble potash present in the fertilizer, Figure 27-3. The numbers are always given in the same order and always represent the same nutrients.

With simple arithmetic, fertilizers can be compared on the basis of their **nutrient ratios**. For example, a 5-10-10 analysis has a ratio of 1-2-2. (Each of the numbers has been reduced by dividing by a common factor, in this case 5.) A fertilizer analysis of 10-20-20 also has a ratio of 1-2-2. As the example in the following column illustrates, a 5-10-10 fertilizer supplies the three major nutrients in the same proportion as a 10-20-20 fertilizer, but twice as much of the actual product must be applied to obtain the same amount of nutrients as is contained in the 10-20-20 fertilizer.

50 pounds of 5-10-10 fertilizer contain:	50 pounds of 10-20-20 fertilizer contain:
2 1/2 pounds of N (nitrogen)	5 pounds of N
5 pounds of P_2O_5 (phosphoric acid)	10 pounds of P_2O_5
5 pounds of K_2O (potash)	10 pounds of K_2O

The ratio of the fertilizers is the same, but the amount of nutrients available in a bag of each differs. The 5-10-10 mixtures should be less expensive than the higher analysis material.

Thus, one measure of the quality of a fertilizer is its analysis. The higher the analysis, the greater is the cost. Whether or not a high analysis fertilizer is needed depends upon the individual plant. Generally, residential lawns do not need a high analysis fertilizer.

Trace Elements

In addition to the three major nutrients, fertilizers may contain additional **trace elements.** These nutrients are essential for plant growth but needed in very small amounts. Most soils are not deficient

Figure 27-3 How to interpret fertilizer analysis figures. The nutrients are always shown in the same order.

in trace elements, so their presence in a fertilizer product may be more coincidental than intentional. In those regions of the country where a particular trace element is either deficient or unavailable to the plant because it is bound too tightly within the soil, then its inclusion in the fertilizer is important.

Forms of Nitrogen Content

Another factor influencing the quality and cost of fertilizers is the form of nitrogen they contain. Some fertilizers contain nitrogen in an organic form. Examples include peat moss, peanut hulls, dried blood, tobacco stems, sewage sludge, and cottonseed meal. The nitrogen content of these materials ranges from 1 1/2 to 12 percent, depending upon the particular material. While sewage sludge is used to some extent on golf course turf, organic fertilizers are not widely used for fertilization of grasses because they are too low in nitrogen. Often, the nitrogen present is not in a form that can be used by plants. The best use of organic fertilizers is as soil conditioners that greatly improve the water retention and aeration of the soil.

Chemical forms are the most commonly used fertilizers. They contain a higher percentage of nitrogen. The nitrogen may be quickly available or slowly available; this determines the timing of the nitrogen's release into the soil and uptake by the grass or other plants. It also influences the cost of the fertilizer.

Quickly available fertilizers usually contain water-soluble forms of nitrogen. This means that the nitrogen can be leached (washed) through the soil before the plants take it in through their root systems. Slowly available fertilizers (also called **slow-release**) make their nitrogen available to the plant more gradually and over a longer period of time. The slow-release effect is possible because the nitrogen used is in a form that is insoluble in water. This gives the plants more time to absorb the nitrogen and prevents fertilizer burn of the plant. Slow-release fertilizers are therefore more expensive than the quickly available forms. Slow-release fertilizers are usually labeled as such. This helps the landscaper to know what is being purchased and what to expect as a response from the plants.

Fillers

A final factor affecting the price and quality of a fertilizer is the amount of filler material it contains. This is directly related to the analysis of the product. **Filler material** is used to dilute and mix the fertilizer. Certain fillers also improve the physical condition of mixtures. However, filler material adds weight and bulk to the fertilizer, thereby requiring more storage space.

The following listing compares high analysis fertilizers (those with a high percentage of major nutrients) and low analysis fertilizers (those with a low percentage of major nutrients) on various plants.

In summary, fertilizer cost is determined by three major factors: analysis, form of nitrogen, and amount of bulk filler material. The higher the analysis and the greater the percentage of slow-release nitrogen, the more expensive is the fertilizer.

High Analysis Fertilizer	Low Analysis Fertilizer
Contains more nutrients and less filler	Contains fewer nutrients and more filler
Cost per pound of actual nutrients is less	Cost per pound of actual nutrients is greater
Weighs less; less labor is required in handling	Is bulky and heavy; more labor is required in handling
Requires less storage space	Requires more storage space
Requires less material to provide a given amount of nutrients per square foot	Requires more material to provide a given amount of nutrients per square foot
Requires less time to apply a given amount of nutrients	Requires more time to apply a given amount of nutrients

When to Fertilize Lawns

Lawns should be fertilized before they need the nutrients for their best growth. Cool-season grasses derive little benefit from fertilizer applied at the beginning of the hot summer months; only the weeds benefit from nutrients applied during the late spring. Cool-season grasses should be fertilized in the early spring and early fall. This supplies proper nutrition prior to the seasons of greatest growth. Landscapers should never practice late fall fertilization—it encourages soft, lush growth, which is damaged severely during the winter.

Warm-season grasses should receive their heaviest fertilization in late spring. Their season of greatest growth is the summer.

Amount of Fertilizer

The amount of fertilizer to use is usually stated in terms of the number of pounds of nitrogen to apply per 1,000 square feet. The number of pounds of nitrogen in a fertilizer is determined by multiplying the weight of the fertilizer by the percentage of nitrogen it contains.

Examples

Problem:

How many pounds of actual nitrogen are contained in a 100-pound bag of 20-10-5 fertilizer?

Solution:

100 pounds × 20% N = pounds of N

100 × 0.20 = 20 pounds of N

Problem:

How many pounds of 20-10-5 fertilizer should be purchased to apply 4 pounds of actual nitrogen to 1,000 square feet of lawn?

Solution:

Divide the percentage of N into the pounds of N desired. The result is the number of pounds of fertilizer required.

4 pounds of N desired

÷ 20% = pounds of fertilizer required

4 ÷ 0.20 = 20 pounds of 20-10-5 fertilizer required

"A Comparison Chart for Turfgrasses" in Chapter 21 lists general fertilizer recommendations for various grasses.

When applying fertilizer to lawns, the recommended poundage should be divided into two or three applications. For example, the 4 pounds of nitrogen per 1,000 square feet for bluegrasses and fescues might be applied at the rate of 2 pounds in the early spring and 2 pounds in the early fall. Another possibility is to apply 1 pound in early spring, 1 pound in midsummer, and 2 pounds in early fall. A spreader must be used to assure even distribution of the fertilizer. It is applied in two directions with the rows slightly overlapped, Figure 27-4.

Watering the Lawn

Turfgrasses are among the first plants to show the effects of lack of water, since they are naturally shallow rooted compared to trees or shrubs. The groundskeeper should encourage deep root growth by watering so that moisture penetrates to a depth

of 8 to 12 inches into the soil. Failure to apply enough water so that it filters deeply into the soil promotes shallow root growth, Figure 27-5. Such shallow root systems can be severely injured during hot, dry summer weather.

Infrequent, deep watering is much preferable to daily, shallow watering. The quantity of water applied during an irrigation will depend upon the time

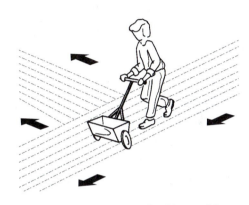

Figure 27-4 Distributing fertilizer with a spreader

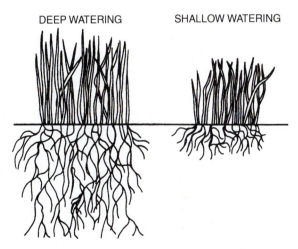

Figure 27-5 Deep watering promotes deep, healthy root growth. Shallow watering promotes shallow rooting and leaves the grass susceptible to injury by drought.

of day and the type of soil. Clay soils allow slower water infiltration than coarser textured sandy soils, but clay soils retain water longer. Therefore, less water may need to be applied to clay soils, or the rate of application may need to be slower, or both. The amount of water given off by a portable sprinkler can be calibrated once and a notation made for future reference. To calibrate a portable sprinkler, set several wide-topped, flat-bottomed cans with straight sides (such as coffee cans) in a straight line out from the sprinkler. When most of them contain 1 to 1 1/2 inches of water, shut off the sprinkler. The amount of time required should be noted for future use. Figures 27-6 and 27-7 illustrate two types of portable sprinklers. In addition, permanently installed irrigation systems are available (at considerable cost) for large turf plantings, Figure 27-8.

The best time of day to water lawns is between early morning and late afternoon. Watering in the early evening or later should be avoided because of the danger of disease; turf diseases thrive in lawns that remain wet into the evening. Watering prior to evening gives the lawn time to dry before the sun sets.

If watering is done at the proper time and to the proper depth, it is necessary only once or twice each week.

Mowing the Lawn

Three types of mowers are available for the maintenance of lawns: the reel mower, the rotary mower, and the flail mower. The **flail mower** is used for turfgrasses that are only cut a few times each year. Reel and rotary mowers are used to maintain home, recreational, and commercial lawns. On a **reel mower**, the blades rotate in the same direction as the wheels and cut the grass by pushing it against a nonrotating bedknife at the rear base of the mower, Figures 27-9 and 27-10. The blades of a **rotary mower** move like a ceiling fan, parallel to the surface of the lawn, cutting the grass off as they revolve, Figure 27-11. Reel mowers are most often used for grasses that do best with a shorter cut, such as bentgrass. Rotary mowers do not cut as evenly or sharply as reel mowers. However, they are satisfactory for lawn grasses that accept a higher cut, such as ryegrass, bluegrass, and fescue. The riding mower, so popular with homeowners, is a rotary mower. Many campuses, parks, and golf course fairways are mown with a large bank of reel mowers, called a gang mower. It is pulled behind a tractor that has been fitted with tires that will not rut the lawn.

In every situation, the blades must be sharp to give a satisfactory cut. Dull or chipped mower

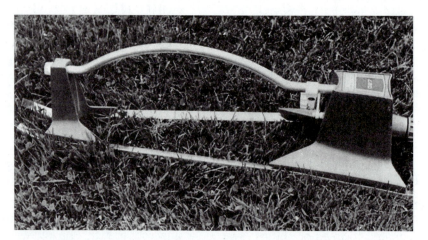

Figure 27-6 One type of lawn sprinkler. An arch of water is cast from side to side. This sprinkler requires periodic relocation.

Figure 27-7 This revolving sprinkler covers a limited area and must be moved manually to each new location. Wind gusts can affect the evenness of the coverage.

Figure 27-9 A powered reel mower

Figure 27-10 Reel mowers provide the best quality cut.

Figure 27-8 An automatic lawn irrigation system in operation (Courtesy of WeatherMatic Irrigation)

blades can result in torn, ragged grass blades that will die and give the lawn an unhealthy color of grey or brown. The sharp blades of any lawn mower should be respected. When powered, they can cut through nearly any shoe. Workers should never mow unless wearing steel-toed work shoes. Hands should never be near the blades while the mower is running. No inconvenience caused by shutting off a mower can equal the instant injury that a power mower can cause to a worker's hand or foot.

"A Comparison Chart for Turfgrasses" in Chapter 21 illustrates the wide range of tolerable mowing heights that exists between and within species. Within the range, the mowing height selected often depends upon how much care the

Figure 27-11 A rotary mower

lawn can be given and what surface quality is expected. Shorter heights require more frequent mowing and watering, and, often, greater pest control efforts. In turn, shorter heights give greater density and finer texture to a lawn. A taller lawn surface will have a slightly coarser texture and take longer to thicken. However, it will not require cutting as often. Longer cutting heights will also withstand hot and dry periods better, since the extra blade length will cast cooling shadows over the soil's surface. Often, fewer weeds are an added benefit of a taller lawn surface. Since not all species are mowed to the same height, mixed-species lawns should be made up of grasses that have similar cutting requirements.

The frequency of mowing is variable. Because the rate of growth of a lawn can vary with the temperature and with the moisture provided, the frequency of the mowing cannot always be precisely specified. Ideally, a lawn should be mown when it is needed, not because a contract specifies cutting on a certain day.

A long-standing rule of thumb is that mowing should remove about one-third of the length of the grass blade. Thus, if the turf is being kept at 1 1/2 inches, it should be mowed when it reaches a height of 2 1/4 inches. If the grass gets too long before cutting, the dead clippings can mar the appearance of the lawn. Then as described earlier, the only alternative is to collect the clippings either in a grass-catcher or with a lawn sweeper or rake. If the grass is cut properly, the clippings will not be excessive. They will decompose rapidly and will not require collection.

The pattern of mowing should be varied regularly to prevent wheel lines from developing in the lawn. Varying the pattern also encourages horizontal growth of the shoots. An easy variation is to mow at a 90-degree angle to the last mowing. If done on the same day with a reel mower, an attractive checkerboard pattern develops. The pattern is not so apparent if done with a rotary mower, but the practice is just as healthy for the lawn.

Damage to Lawns

Like all plants, turfgrasses are susceptible to assorted injuries. Damage can be visual or physical and frequently is both. Causes of lawn injury include:

- Weeds
- Pests
- Drought
- Thatch build-up
- Vandalism

Weeds and Pests

Weeds and pests were described and discussed in detail in Chapter 26. The injuries that they bring to lawns are varied. Weeds usually impact the aesthetic appearance of a lawn. A broad-leaved weed, such as a dandelion, growing within a lawn of fine textured turf grass will not be welcomed. Even coarse textured grasses, such as crabgrass, growing in a lawn of fine textured turf will be unsightly. Since a well-maintained, healthy lawn can discourage the establishment of many weeds, a lawn with a large weed population often indicates additional problems, such as poor soil conditions.

Good landscaping practices are the best defenses against weeds. However, the need for an herbicide is probable in most lawns where appearance of the turf is given a high level of importance. Turf herbicides are available in both liquid and powdered formulations. Some products mix herbicides with fertilizers creating a "weed and feed" product that saves time and labor in its application.

If an herbicide is applied in liquid form, the sprayer should be cleaned afterward and set aside to be used exclusively for that purpose in the future. Small amounts of herbicide remaining in a sprayer can kill valuable ornamental plants if the sprayer is later used to apply another liquid material.

Insect and disease injuries to turf increase in frequency and severity proportionate to the declining health of the lawn. Healthy lawns have fewer insect and disease problems than weak lawns. Because of their comparatively shallow root systems, turf grasses are affected more immediately by environmental conditions and are more immediately responsive to cultural modifications than other landscape plants. Insect damage and diseases can seem to develop overnight if there is a sudden change in the lawn's growing regimen, such as excessive rainfall or a late spring snowfall. The following table, "Common Turf Problems," contains a partial list of the pests and problems that can injure an ordinary lawn.

The best defense against most pest injury is the selection of resistant varieties and the creation of a growth environment that favors the grass more than the pest. For example, watering at night promotes the growth of many fungi that cause turf diseases. Irrigation of lawns should be done earlier in the day, allowing the grass to dry before nightfall. Another example: allowing a thick thatch layer to develop provides a good habitat for certain harmful insects. (Thatch is discussed in the next section.) Large populations of insects in the soil can attract rodents, such as moles, that burrow through the soil to feed on them. So in addition to the injuries caused by the insects, there is additional damage done by the animals that dig after them. Solving the insect problem can indirectly solve the other.

Thatch Build-Up

When grass clippings and other plant tissue on the surface of the lawn do not decompose as rapidly as the growth of new tissue occurs, a matted build-up, termed **thatch**, begins to accumulate. If the layer of thatch is thin, 1/2 inch or less, it can be of benefit to the lawn, serving as a mulch for the grass. As such, it keeps the soil cooler and aids in the retention of moisture. When the thatch becomes too thick, problems begin to occur. Then it begins to harbor insects and pathogenic inoculum. It can also create a cushion over the soil's surface that will permit the wheels of a mower to sink enough to change the mowing height of the lawn.

The best approach to controlling thatch build-up is to prevent it at the outset. The turf should not be allowed to get too high between mowings. If it does occur, the clippings should be collected rather than allowed to remain on the lawn. *Dethatching* is the term used to describe the intentional removal of the thatch layer. It can be done in different ways depending upon the size of the lawn and the type of landscape site. Small lawns can be dethatched using a steel rake to pull the matted layer away from the soil where it can be collected. That technique is impractical on larger lawns. As mentioned earlier, a vertical mower can be used to slice through the thatch and thereby promote better aeration, which in turn can accelerate the decomposition of the thatch. As the segments of thatch are lifted to the lawn's surface, they should be raked or vacuumed away to prevent their settling back to the soil surface.

Common Turf Problems

Turf Insects	Turf Diseases	Other Problems
• Ants	• Anthracnose	• Dogs
• Army worms	• Brown patch	• Gophers
• Bill bugs	• Copper spot	• Ground squirrels
• Chinch bugs	• Dollar spot	• Mice
• Cut worms	• Fairy ring	• Moles
• Grubs	• Fusarium blight	• Human vandalism
• Leaf hoppers	• Leaf spots	• Vehicles and
• Mites	• Net blotch	equipment
• Mole crickets	• Nematodes	
• Periodical cicadas	• Powdery mildew	
• Scale	• Pythium blight	
• Sod webworm	• Red thread	
• Weevils	• Rots	
• Wireworms	• Rusts	
	• Slime molds	
	• Smuts	
	• Snow molds	

Drought

Periods of severe water shortage can harm a lawn. The grass will turn brown and enter a stage of **dormancy** (nongrowth). If the drought continues, the entire lawn can be killed. Irrigation is the major defense against drought. In areas of predictable drought (arid regions), tolerant varieties of grasses should be selected at the time of installation. A greater lawn height is also helpful in withstanding periods of drought.

Vandalism

Vandalism is impossible to control if the vandals are determined. Lawns rutted by vehicles are unattractive and are common abuses of residential and recreational landscapes. Locked gates and the strategic placement of trees can sometimes help by making vehicular access difficult for the would-be vandal. Education to increase public awareness of the value of the landscape and the responsibilities of good citizens is the only real solution of vandalism.

Achievement Review

A. Define the following terms.

1. heaving
2. aeration
3. fertilizer analysis
4. low analysis fertilizer
5. slow-release fertilizer
6. herbicide

B. What does 10-6-4 on a bag of fertilizer mean?

C. Would 10-6-4 fertilizer be considered a high analysis or low analysis product? Why?

D. Of the three fertilizers listed below, which two have the same ratio of nutrients?

10-20-10
5-10-5
5-10-15

E. Why might the prices of two 50-pound bags of fertilizer differ greatly?

F. At what time of the year are warm-season grasses fertilized? Cool-season grasses?

G. How many pounds of actual nitrogen are contained in a 50-pound bag of 12-4-8 fertilizer?

H. How many pounds of 10-10-10 fertilizer should be purchased to fertilize a 2,000 square-foot lawn of bluegrass and fescue if it is applied at the recommended rate?

I. How much water is needed to deeply soak an average lawn?

J. What is the best time of day to water lawns?

K. Describe how to repair a lawn by seeding, sodding, sprigging, and plugging.

L. Match the type of sprinkler with its characteristic:

a. revolving
b. oscillating
c. automatic

1. casts water in an arched pattern
2. permanently installed system for large landscapes
3. casts water in a circular pattern

M. Match the type of mower with its characteristic:

a. flail
b. reel
c. rotary

1. blades move like a ceiling fan, parallel to the lawn's surface
2. used for grasses that are only cut a few times each year
3. the blades rotate in the same direction as the wheels

N. List five different ways that lawns can be damaged.

O. What should be done with the clippings of a lawn that is mowed weekly and why?

Suggested Activities

1. Study "A Comparison Chart for Turfgrasses" in Chapter 21 and match grasses growing in your local area that could be blended and grown together successfully. Compare grasses on the basis of texture, frequency of fertilization, mowing height, and preferred soil type.

2. Calibrate one or more types of portable sprinklers following the directions given in this chapter. Determine the length of time each requires to apply 1 inch of water and measure the area of coverage.

3. Visit a lawn equipment dealership. Ask the owner to show the various models of mowers, spreaders, sprayers, rollers, and rakes that are stocked for lawn maintenance.

4. Have an equipment field day, perhaps in association with other nearby schools. Invite equipment dealers to bring selected pieces of power equipment to the school for demonstration and/or student use.

5. Measure thatch layers from different types of turf area, such as a well-maintained residential lawn, a public park, a cemetery, an athletic field, a golf green. With a sharp knife, cut down into each turf area and remove a small plug of sufficient size to permit the thatch to be seen and its depth measured. Compare the depth of the thatch with the frequency of mowing and the type of grass.

CHAPTER 28

Winterization of the Landscape

Objectives:

Upon completion of this chapter, you should be able to

- list those elements of the landscape that require winter protection
- describe eight possible types of winter injury
- explain nine ways to protect against winter injury

Winter injury is any damage done to elements of the landscape during the cold weather season of the year. The injury may be due to natural causes or to human error. It may be predictable or totally unexpected. At times winter injury can be avoided, while at other times it can only be accepted and dealt with.

Winter injury attacks most components of the outdoor room. Plants, paving, steps, furnishings, and plumbing are all susceptible to damage from one or more causes.

Types of Winter Injury

While the types of winter damage are almost unlimited, there are several that commonly occur. The landscaper should be especially aware of these. Injuries are caused by one of two agents:

nature or human beings. There are many different examples within these two general categories.

Natural Injuries

The severity of winter weather can cause extensive damage to plant materials in the landscape.

Windburn results when evergreens are exposed to strong prevailing winds throughout the winter months. The wind dries out the leaf tissue, and the dehydrated material dies. Windburn causes a brown to black discoloration of the leaves on the windward side of the plant. Very often, leaves further into the plant or on the side opposite the wind show no damage. Broad-leaved evergreens are the most susceptible to windburn because they have the greatest leaf surface area exposed to drying winds. To protect themselves, many broad-leaved evergreens roll their leaves in the winter to reduce the amount of exposed surface area, Figure 28-1.

Figure 28-1 Rolled and discolored leaves show the effects of windburn on this rhododendron.

Needled evergreens can also burn. If burn has occurred, brown-tipped branches are apparent in the spring when new growth is beginning. As with broad-leaved forms, windburn on conifers is likely to be confined to the outermost branches on the most exposed side of the plant.

Temperature extremes can also cause injury to plants. Plants that are at the limit of their hardiness (termed *marginally hardy*) may be killed by an extended period of severely cold weather. Others may be stunted when all of the previous season's young growth freezes.

After some cold winters, certain plants may show no sign of injury except that their spring flower displays are absent. This results if the plant produces its flowers and leaves in separate buds. The weather may not be cold enough to affect the leaf buds, but freezes the more tender flower buds. This is especially common with forsythia and certain spireas in the northern states.

Unusually warm weather during late winter can also cause plant damage. Fruit trees may be encouraged to bloom prematurely, only to have the flowers killed by a late frost. As a result, the harvest of fruit can be greatly reduced or even eliminated. Spring flowering bulbs can also be disfigured if forced into bloom by warm weather that is followed by freezing winds and snow.

Sunscald is a special type of temperature-related injury. It occurs when extended periods of warm winter sunshine thaw the above-ground portion of a plant. The period of warmth is too brief to thaw the root system, however, so it remains frozen in the ground, unable to take up water. Above-ground, the thawed plant parts require water, which the roots are unable to provide. Consequently, the tissue dries out and a scald condition results.

Sunscald is especially troublesome on evergreens planted on the south side of a building.

It also occurs on newly transplanted young trees in a similar location. The young, thin bark scalds easily and the natural moisture content of the tissue is low because of the reduced root system.

Heaving affects turfgrass, hardy bulbs, and other perennials when the ground freezes and thaws repeatedly because of winter temperature fluctuations. The heaving exposes the plants' roots to the drying winter wind, which can kill the plants.

Ice and snow damage can occur repeatedly during the winter. The sheer weight of snow and/or ice on plant limbs and twigs can cause breakage and result in permanent destruction of the plant's natural shape, Figures 28-2 and 28-3. Evergreens are most easily damaged because they hold heavy snow more readily than deciduous plants. Snow or ice falling off a pitched roof can split foundation plants in seconds.

Plants that freeze before the snow settles on them are even more likely to be injured. Freezing reduces plant flexibility and causes weighted twigs to snap rather than bend under added weight.

Unfortunately, the older and larger a plant is, the greater is the damage resulting from heavy snowfalls and ice storms. There are numerous recorded accounts of the street trees of entire cities being destroyed by a severe winter storm.

Animal damage to plants results from small animals feeding on the tender twigs and bark of plants, especially shrubs and young trees. Bulbs are also susceptible. Entire floral displays can be destroyed by the winter feeding of small rodents. Shrubs can be distorted and stunted by removal of all young growth. In cases where the plant becomes girdled (with the bark around the main stem completely removed), the plant is unable to take up nutrients and eventually dies.

Human-Induced Injuries

Certain types of injury are created by people during wintertime landscape maintenance. Some types of injury are due to carelessness on the part of groundskeepers. Other types are the predictable result of poor landscape design. A large number are injuries created because the landscape elements are hidden beneath piles of snow.

Salt injury harms trees, shrubs, bulbs, lawns, and paving. Often the damage does not appear until long after the winter season passes. Thus, the cause of the injury may go undiagnosed.

The salt used to rid walks, streets, and steps of slippery ice becomes dissolved in the water it creates. The saline solution flows off walks and into nearby lawns or planting beds. Paving sometimes crumbles under heavy salting. Poured concrete is especially sensitive to this treatment, Figure 28-4.

Figure 28-2 Evergreen trees can be broken and even suffer permanent damage from a heavy snow.

Figure 28-3 The weight of ice on the branches of deciduous plants can break and misshape them.

Figure 28-4 Excess salt had a damaging effect on these concrete steps. This type of damage is unnecessary.

Salt is toxic to nearly all plant life. The resultant injury to lawns appears as strips of sterile, barren ground paralleling walks, Figure 28-5. Injury can also be seen on the lower branches of evergreens.

Snowplow damage can occur to plant and construction materials for several reasons. A careless plow operator may push snow onto a planting or a bench. This often results when those unfamiliar with the landscape are hired to do the snowplowing. Other injuries from plowing are the result of design errors. Plants, outdoor furniture, and light fixtures should not be placed near walks, parking

Figure 28-5 Dead grass edging the walk in the summer is a symptom of excessive winter salting.

areas, or streets where they will interfere with winter snow removal.

Damage to lawns can result when the plow misses the walk and actually plows the grass, scraping and gouging the lawn, Figure 28-6. The grass may not survive if this occurs repeatedly.

Rutting of lawns is the result of heavy vehicles driving and parking on softened ground. When the surface layer of the soil thaws but the subsoil remains frozen, surface water is unable to soak in. Users of the landscape accustomed to finding the ground firm may be unaware of damage caused by vehicles tem-

Figure 28-6 A scraped lawn is the result of a snowplow that has missed the unmarked sidewalk.

porarily parked on soft lawns. The soil becomes badly compacted, resulting in unsightly ruts.

Reducing Winter Injury

Some types of winter damage can be eliminated by properly winterizing the landscape in the preceding autumn. Other types can be reduced by better initial designing of the grounds. Still other winter injuries can only be minimized, never totally eliminated.

Windburn can be eliminated in the design stage of the landscape if the planner selects deciduous plant materials rather than evergreen materials. This problem can be avoided in the garden by the use of deciduous shrubs on especially windy corners. If evergreens are important to the design or already exist in the garden, windburn can be reduced by erecting burlap shields around shrubs, Figure 28-7. The use of an antitranspirant may also reduce water loss from plants and thereby reduce the effects of windburn. The antitranspirant must be applied in the autumn and again in late winter. While antitranspirants are fairly expensive, they are more practical for the protection of large evergreens than burlap shields.

Temperature extremes can be only partially guarded against. Where wind chill (lowering of temperature because of the force of wind) is a factor, plants should be located in a protected area. Wrapping the plant in burlap also helps. This has proven to be an effective technique for protecting tender flower buds on otherwise hardy plants.

Certain plants, such as roses, can be cut back in the fall and their crowns mulched heavily to assure insulation against the effects of winter. Likewise, any plant that can be damaged by freezing and thawing of the soil should be heavily mulched after the ground has frozen to insulate against premature thawing.

If the landscaper is trying to prolong the lives of annual flowers in the autumn and a frost is forecast, the foliage can be sprinkled with water prior to nightfall. This helps to avoid damage caused by a light frost. The water gives off enough warmth to keep the plant tissue from freezing.

Figure 28-7 Burlap shields protect broad-leaved evergreens from winter windburn.

Sunscald of young transplants lessens as the plants grow older and form thicker bark. It can be avoided on trees during the first winter of growth by wrapping the trunks of the trees with paper or burlap stripping. For other types of sunscald, such as that affecting broad-leaved evergreens, the same remedies practiced for windburn and temperature extremes are effective. Wrapping the plants in burlap or the use of antitranspirants gives protection.

Some sunscald can be avoided by the designer. Vulnerable species of plants should not be placed on the south side of a building, nor should they be placed against a reflective white wall that will magnify the sun's effect on the above-ground plant parts.

Heaving of the turf is impossible to prevent completely. The best defense against it is encouraging deep rooting through proper maintenance and good landscaping practices during the growing season.

Bulbs, groundcovers, and other perennials can be protected from heaving through application of a mulch after the ground has frozen. The mulch acts to insulate the soil against surface thawing.

Ice and snow damage to foundation plants can be avoided if the designer is careful not to place plants beneath the overhanging roof line of a building. If the groundskeeper must deal with plants already existing in a danger area, the use of hinged, wooden A-frames over the plants can help to protect them, Figure 28-8. As large pieces of frozen snow and ice tumble off the roof, the frame breaks them apart before they can damage the plants.

To aid plants that have been split or bent because of heavy snow accumulation, the groundskeeper must work quickly and cautiously. A broom can be used to shake the snow off the weighted branches. However, snow removal must be done gently and immediately after the snow stops. If the branches are frozen or the snow has become hard and icy, removal efforts will cause more damage than benefit.

If breakage of plants occurs during the winter, the groundskeeper should prune the damaged parts as soon as possible. This prevents further damage to the plants during the rest of the winter.

Certain plants, such as upright arborvitae and upright yews, become more susceptible to heavy snow and ice injury as they mature. Often the damage cannot be repaired. Large and valuable plants in a landscape can be winterized in the autumn by tying them loosely with strips of burlap or twine. (Do not use wire.) When prepared in this manner, the branches cannot be forced apart by the heavy snows of winter, and splitting is avoided

Animal damage can be prevented by either eliminating the animals or protecting the plants from their feeding. While rats, mice, moles, and voles are generally regarded as offensive, plantings are damaged as much or more by deer, rabbits, chipmunks, and other gentler kinds of animals. Certain rodenticides (substances that poison rodents) may be employed against some of the undesirable animals that threaten the landscape. In situations where the animals are welcome but their winter feeding damage is not, a protective enclosure of fine mesh wire fencing around the plants helps to discourage animals from feeding there, Figure 28-9.

Salt injury to plants and paving need not be as bad as was illustrated earlier if caution is exercised by the groundskeeper. Salt mixed with coarse sand

Figure 28-8 Hinged wooden A-frames protect foundation plantings from damage caused by sliding ice and snow.

Figure 28-9 An enclosure of fine mesh wire placed around young plants before winter can offer protection from the damage caused by rabbits, mice, and other small animals.

does a better job than either material used separately. The sand provides traction on icy walks, and a small amount of salt can melt a large amount of ice. Excessive salt has no value; it only kills plants and destroys paving. In very cold temperatures, salt does not melt ice; therefore, in these cases it serves no purpose. Salt can turn compacted snow, which is comparatively safe for traffic, into inches of slush, which is messy and even more slippery. The problem of salt injury can be solved by reducing the amount of salt spread on walks and streets during the winter season. The addition of sand distributes the salt more evenly and provides grit for better traction.

Snowplow damage is to be expected if a designer places plants too close to walks and roadways. Therefore, one obvious solution to the problem begins with the designer. When planning landscapes for areas in which winter is normally accompanied by a great deal of snow, the designer should avoid placing shrubs or other items near intersections or other places where snow is likely to be pushed.

Another type of plow damage is the result of the plow driver's inability to see objects beneath the snow. If possible, all objects such as outdoor furniture and lights should be removed from plow areas prior to the winter season. If it is not possible to move them, low objects should be marked with tall, colored poles that can be seen above the snow.

Whenever possible, snow blowers should be used instead of plows. These machines are much less likely to cause damage.

Rutting of lawns usually results from the practice of permitting individuals to park cars on lawns. The best solution to the problem is to avoid doing so. Otherwise, sawhorses or other barriers offer a temporary solution.

Achievement Review

A. Indicate whether the following types of winter injury are caused by natural conditions (N) or human error (H).

1. crumbled paving resulting from too much salt
2. dry, dead twigs on the windward side of a pine tree
3. dried, blistered bark on the trunk of a recently transplanted tree
4. deep ruts in the lawn in front of a house
5. dead branches in a shrub following an unusually cold winter
6. failure of a shrub to flower in the spring

7. perennials lying on the surface of the soil in the early spring with roots exposed to the drying air
8. dead grass next to a walk heavily salted during the winter
9. an upright evergreen split in the center by snow sliding off a roof
10. a young tree girdled at the base

B. Of the types of winter damage in the following list, which ones could be reduced or prevented by proper winterization of the landscape during the late autumn?

1. crumbled paving

2. sunscald on new transplants
3. breakage of outdoor furniture by snow plows
4. rutting of the lawn by automobiles
5. foundation plants broken by falling snow
6. sunscald on broad-leaved evergreens
7. tree limbs broken off by an ice storm
8. bulbs heaved to the surface of the soil
9. windburn on evergreens
10. flooded basement caused by melting snow

Suggested Activities

1. Look for signs of winter injury in nearby landscapes. Find windy corners where evergreens are planted and check for windburned tips. Visit a shopping center, campus, or park where salt is used on walks and parking lots. Look at the paving and nearby plantings for signs of damage. Note the placement of plants in relation to walk intersections and other places where snow may be piled.

2. Conduct ice melting tests. Freeze four pie plates of water. Apply four different mixtures of salt and/or sand to the surfaces and see which melts first. In the first pan, use all salt; in the second, half salt and half sand; in the third, one-quarter salt and three-quarters sand; and in the fourth, all sand. What conclusions can be drawn from the trials?

   NOTE: Returning the treated pans of ice to the freezer (approximately 20 degrees F) will assure that no natural melting occurs. Check the pans every 15 minutes for observations.

3. Demonstrate the damaging effects of salt upon plant life. Grow some experimental plants in advance. Root each plant in a separate container. Apply only water to some of the plants for a week. To others, apply varying dilutions of a salt water solution. To a third group, apply water to the soil, but mist the foliage with salt water several times each day. Record observations of each treatment daily.

4. Demonstrate the reduction of water loss from plant tissue caused by antitranspirants. Purchase a small bottle of antitranspirant from a garden center, or write to a manufacturer and request a complimentary sample. Dilute with water as specified on the label. Using a small pump sprayer, apply the antitranspirant to one of two comparable samples of cut evergreens in vases of water. Be certain to apply it evenly over all needle surfaces. The untreated sample should dry out sooner than the treated one, indicating greater water loss.

5. Repeat the demonstration in #4 using a sun lamp or a fan to simulate the drying effects of sun and wind and the protection offered by antitranspirants.

CHAPTER 29

Pricing Landscape Maintenance

Objectives:

Upon completion of this chapter, you should be able to

- describe the values of cost analysis to a maintenance firm
- list the features of a landscape maintenance cost analysis
- describe unit pricing
- prepare a maintenance cost estimate

The Need for Cost Analysis in Landscape Maintenance

An accurate analysis of the costs of specific tasks done by a landscape maintenance firm has several values:

- It assures that all costs to the firm are recognized.

- It permits a fair price to be charged to the customer.

- It allows a comparison of the profitability of different tasks.

- It can compare the efficiency of different crews performing the same task.

The first two benefits were discussed in Chapter 14. They are equally important in landscape maintenance. The last two warrant explanation also. A new landscape maintenance firm often believes that all jobs are good jobs. "No job is too big or too small" is the way their advertisements often read. Actually, as a firm grows, not all jobs can be accepted. There may be insufficient time or personnel to respond to every client request. The tasks that return the greatest profit for the labor invested will need to be emphasized over those that return less profit. A carefully prepared cost analysis can illustrate these different profit potentials.

The performance efficiency of several crews or laborers doing the same task can also be compared with a maintenance cost analysis. Assuming the cost of materials and the site conditions to be the same, the labor time required is the only comparable variable. Carefully kept work records documenting the crew size and hours of labor required to complete a task permit the performance comparison.

Features of the Maintenance Cost Analysis

A cost analysis of a landscape maintenance job includes

- a listing of all tasks to be performed
- the total square footage area involved for each service
- the number of times each service is performed during the year
- the time required to complete each task once
- the time required to complete each task annually
- the cost of all materials required for each task
- the cost of all labor required for each task

Unit Pricing

The most precise method of tracking job costs and determining the correct price to charge a client is **unit pricing**. It is also the method that requires the most time to set up because of the amount and precision of the data that is required. It is not a system that permits *guess-timating*. The system reduces all area dimensions and material quantities to common and measurable units, such as *a thousand square feet* or *an acre*. It measures the quantities and costs of materials used for particular projects in terms of those same units (for example, the application of one pound of nitrogen per thousand square feet of lawn area at a cost of $3.00 per thousand square feet). The initial conversion of material costs purchased in bulk quantities to unit costs that permit their use in the company's pricing system may take some time, but it is not difficult math and can be done.

Determining the time requirements for separate tasks and getting them to match the units used for calculating material costs is more difficult. The time required to perform a task is directly affected by the number of people on the crew, the type, size, and speed of the equipment being used, site conditions such as terrain, obstacles, and weather, and even the efficiency and/or work ethic of the workers. There are numerous time charts published by industry trade groups and the media that offer *average* unit times required to do typical landscape tasks, such as lawn mowing, shrub installation, mulching, or pruning. They are at best guidelines and should not be used as anything more than that. As a landscape company prepares to set up its unit pricing system, it must do some careful calculating of the time its crew people require to perform specific tasks using specific equipment and operating under specific conditions. These calculations can then be converted to the required time units, such as *minutes per thousand square feet*. It is important to realize that when one of the variables changes, e.g., the terrain goes from flat to rough or a larger mower is used, the time required per unit will change as well.

Calculations for Cost Analysis

To apply the technique properly requires practice. Study the following examples and their explanations.

Example 1

Problem:

To calculate the cost of mowing 10,000 square feet of lawn with an 18-inch power mower 30 times each year.

Necessary Information:

It takes 5 minutes to mow 1,000 square feet. The laborer receives $6.50 per hour. There are no material costs.

Solution:

Maintenance Operation: Lawn mowing with 18-inch power mower

Square footage area involved	10,000 sq. ft.
Number of times performed annually	30
Minutes per 1,000 sq. ft.	5
Total annual time in minutes (1)	1,500
Material cost per 1,000 sq. ft.	none
Total material cost	none
Wage rate per hour	$6.50
Total labor cost (2)	$162.50
Total cost of maintenance operation per year	$162.50

Note:

(1) To obtain the total annual time in minutes:

   a. divide the square footage of area involved by 1,000 [10,000 sq. ft. ÷ 1,000 = 10]

   b. multiply by minutes per 1,000 sq. ft. [10 × 5 = 50 minutes]

   c. multiply by number of times performed annually [50 minutes × 30 = 1,500 minutes]

   d. enter answer under total annual time in minutes

(2) To obtain the total labor cost:

   a. divide the total annual time in minutes by 60 minutes [1,500 minutes ÷ 60 minutes = 25 hours]

   b. multiply by the wage rate per hour [25 hours × $6.50 = $162.50]

   c. enter answer under total labor cost

Example 2

Problem:

To calculate the cost of mulching 2,000 square feet of planting beds with wood chips, 4 inches deep.

Necessary Information:

The task is done once each year. It requires 30 minutes to mulch 1,000 square feet, 4 inches deep. The laborer receives $6.50 per hour. The wood chips cost $165.00 per 1,000 square feet of coverage.

Solution:

Maintenance Operation: Mulching plantings with wood chips, 4 inches deep

Square footage area involved	2,000 sq. ft.
Number of times performed annually	1
Minutes per 1,000 sq. ft.	30
Total annual time in minutes (1)	60
Material cost per 1,000 sq. ft.	$165.00
Total material cost (2)	$330.00
Wage rate per hour	$6.50
Total labor cost (3)	$6.50
Total cost of maintenance operation per year (4)	$336.50

Note:

(1) To obtain the total annual time in minutes:

   a. divide the square footage of area involved by 1,000 [2,000 sq. ft. ÷ 1,000 = 2]

   b. multiply by minutes per 1,000 sq. ft. [2 × 30 = 60 minutes]

   c. multiply by number of times performed annually [60 minutes × 1 = 60 minutes]

   d. enter answer under total annual time in minutes

(2) To obtain the total material cost:

   a. divide the square footage of area involved by 1,000 [2,000 sq. ft. ÷ 1,000 = 2]

   b. multiply by material cost per 1,000 sq. ft. [2 × $165.00 = $330.00]

   c. multiply by number of times performed annually [$330.00 × 1 = $330.00]

   d. enter answer under total material cost

(3) To obtain the total labor cost:

   a. divide the total annual time in minutes by 60 minutes [60 minutes ÷ 60 minutes = 1 hour]

   b. multiply by the wage rate per hour [1 hour × $6.50 = $6.50]

   c. enter answer under total labor cost

(4) To obtain the total cost of maintenance operation per year:

   a. add total material cost and total labor cost [$330.00 + $6.50 = $336.50]

   b. enter answer under total cost of maintenance operation per year

The Completed Cost Estimate

A full cost estimate for maintenance is simply an enlargement of the previous examples. For convenience, all of the maintenance operations that deal with the same area of the landscape are grouped together in the estimate. Study the following example and note the calculation of all figures.

Data

I. A landscape requires the following maintenance tasks and equipment:

   a. 24,500 square feet of lawn cut 30 times each year with a power riding mower

   b. 500 square feet of lawn cut 30 times each year with an 18-inch power hand mower

   c. all lawn areas fertilized twice each year

   d. 5,000 square feet of shrub plantings fertilized once each year

   e. shrubs pruned once each year

   f. 400 square feet of flower beds requiring soil conditioning once each spring

   g. flowers planted once each year

   h. flowers hand weeded 10 times each year

   i. flower beds cleaned and prepared for winter once each autumn

II. Calculated time requirements for the maintenance tasks:

   a. a power riding mower cutting 1,000 square feet of lawn in 1 minute

   b. an 18-inch power mower cutting 1,000 square feet of lawn in 5 minutes

   c. lawn fertilization requiring 3 minutes per 1,000 square feet for spreading

   d. shrub fertilization requiring 5 minutes per 1,000 square feet

   e. pruning time for shrubs averaging 60 minutes per 1,000 square feet

   f. soil conditioning for flower beds requiring approximately 200 minutes per 1,000 square feet

   g. flower planting requiring 600 minutes per 1,000 square feet

   h. weeding of flowers requiring 60 minutes per 1,000 square feet

   i. cleanup of flower beds in the autumn requiring 400 minutes per 1,000 square feet

III. All laborers receive wages of $6.50 per hour.

IV. Material costs:

   a. lawn fertilizer at $5.00 per 1,000 square feet

   b. shrub fertilizer at $6.00 per 1,000 square feet

   c. conditioning materials for flower beds at $3.00 per 1,000 square feet

   d. flowers for planting averaging $125.00 per 1,000 square feet

V. Overhead and profit. To convert the cost analysis to a cost estimate, additional charges for overhead and profit must be added. As described in Chapter 14, these charges may be figured as a percentage of the total cost of the maintenance operation. In this example, overhead costs of 10 percent ($54.85) and profit allowance of 25 percent ($137.11) could be added to the $548.45 cost to the firm. The price quotation to the client would then be $740.41.

NOTE: Many firms apply different overhead and profit percentages to different parts of a project, rather than using single percentages, as in this example.

The Completed Cost Estimate

Maintenance Operation*	Sq. Footage Area Involved*	Number of Times Performed Annually*	Minutes Per 1,000 Sq. Ft.*	Total Annual Time in Minutes**	Material Cost Per 1,000 Sq. Ft.*	Total Material Cost**	Wage Rate Per Hour*	Total Labor Cost**	Total Cost of Maintenance Operation Per Year**
Lawn									
Mowing – rider	24,500 sq. ft.	30	1	735	None		$6.50	$79.63	$ 79.63
Mowing – 18" power	500 sq. ft.	30	5	75	None		$6.50	$ 8.13	$ 8.13
Fertilization	25,000 sq. ft.	2	3	150	$ 5.00	$250.00	$6.50	$16.25	$266.25
Shrubs									
Fertilization	5,000 sq. ft.	1	5	25	$ 6.00	$ 30.00	$6.50	$ 2.73	$ 32.73
Pruning	5,000 sq. ft.	1	60	300	None		$6.50	$32.50	$ 32.50
Flowers									
Soil conditioning	400 sq. ft.	1	200	80	$ 3.00	$ 1.20	$6.50	$ 8.65	$ 9.85
Planting	400 sq. ft.	1	600	240	$125.00	$ 50.00	$6.50	$26.00	$ 76.00
Hand weeding	400 sq. ft.	10	60	240	None		$6.50	$26.00	$ 26.00
Autumn cleanup	400 sq. ft.	1	400	160	None		$6.50	$17.36	$ 17.36
									$548.45

Notes: * All entries in this column came directly from the data given.
      ** All entries in this column were calculated using methods in the earlier examples. Students should practice the calculations to ensure their understanding of the methods.

Practice Exercise

Complete a cost estimate based upon the following data. Add additional overhead costs of 10 percent and a profit allowance of 25 percent.

I. A landscape requires the following maintenance tasks and equipment:
   a. 12,000 square feet of lawn cut 25 times a year with a power riding mower
   b. 600 square feet of lawn cut 25 times each year with a 25-inch power hand mower
   c. all lawn areas fertilized twice each year
   d. all lawn areas raked once each spring with a 24-inch power rake
   e. 700 square feet of shrub plantings cultivated with hoes twice each year
   f. shrubs pruned once each year
   g. shrubs fertilized once each year
   h. 250 square feet of flower beds requiring soil conditioning once each spring
   i. flowers in 250 square feet planted once each year
   j. flowers in 250 square feet weeded by hand 10 times each year
   k. flower beds in 250 square feet cleaned and prepared for winter once each autumn

II. Time requirements for the maintenance tasks include:
   a. power riding mower cutting 1,000 square feet of lawn in 1 minute
   b. the 25-inch power hand mower cutting 1,000 square feet of lawn in 3 minutes
   c. lawn fertilization requiring 3 minutes per 1,000 square feet for spreading
   d. lawn raking with a 24-inch power rake requiring 10 minutes per 1,000 square feet
   e. hand hoe cultivation of the shrubs requiring 60 minutes per 1,000 square feet
   f. pruning of shrubs averaging 60 minutes per 1,000 square feet
   g. shrub fertilization requiring 5 minutes per 1,000 square feet
   h. soil conditioning for flower beds requiring approximately 200 minutes per 1,000 square feet
   i. flower planting requiring 600 minutes per 1,000 square feet
   j. weeding of flowers requiring 60 minutes per 1,000 square feet
   k. cleanup of flower beds in the fall requiring 400 minutes per 1,000 square feet

III. All laborers receive wages of $7.00 per hour.

IV. Material costs include:
   a. lawn fertilizer at $5.00 per 1,000 square feet
   b. shrub fertilizer at $6.00 per 1,000 square feet
   c. conditioning materials for flower beds at $3.00 per 1,000 square feet
   d. flowers for planting at $125.00 per 1,000 square feet

Achievement Review

A. List four ways in which a cost estimate benefits a landscape maintenance firm.

B. What seven items of data are required before beginning a maintenance cost estimate?

C. Figure the total annual time in minutes for a task that is done five times a year, involves 6,000 square feet of area, and requires 7 minutes per 1,000 square feet to accomplish.

D. Figure the total material cost for mulch that is purchased to cover 1,500 square feet of area and costs $125.00 per 1,000 square feet. The mulch is applied once each year.

E. Calculate the total labor cost for a job that requires a total of 420 minutes. The worker assigned to the job is paid $6.00 per hour.

CHAPTER 30

Safety in the Landscape Industry

Objectives:

Upon completion of this chapter, you should be able to

- identify the people most at risk
- identify the potential hazards and dangers associated with landscape practices
- describe the governmental and industry efforts to promote safe landscape practices
- describe those safe practices that must be implemented by individual employees and their company supervisors

It's a Risky Business

Every line of work has its hazards, and the landscape industry is no exception. Because they work with sharp tools and heavy equipment, field personnel are the first ones thought of when discussing potential ways to get hurt as landscapers, but the reality is that everyone who works for a landscape company, regardless of the mix of work being done, can be injured if he or she does not work correctly. Working correctly means doing each task carefully and in accordance with all recommended or required safety procedures. Safe practice can also mean protecting people who do not even realize that they are at risk, such as the company's clients. There are no shortcuts in safe practice, and there are no exceptions to those who can be injured.

The Potential Dangers

Consider the following dangers that are associated with the landscape industry. Some are so obvious that nearly everyone could predict them. Others may not have occurred to you, and they may not occur to an unskilled and/or untrained worker asked to perform for the company. Most of us are prone to thinking and acting in accordance with what we know and what is familiar to us. Someone who understands chemistry is more likely to think about safe pesticide application than someone who has never even taken a high school chemistry

Task or Tool	Potential Danger
Hand tools, such as spades, saws, pruners, axes	Cuts, bruises, splinters, blisters, pinching
Power hand tools such as chain saws, string trimmers, electric hedge trimmers, mowers	Deep cuts, amputation, propelled rocks, cutting unseen electric wires, eye injuries, hearing damage
Large power tools such as rototillers, backhoes, trenchers, forklifts	Operators unfamiliar with correct use, poor maintenance, eye injuries, hearing damage
Motorized vehicles such as trucks, tractors, bulldozers, cranes	Operators unfamiliar with correct use, poor maintenance, incorrect loading, worn parts
Chemical application as with pesticides, fertilizers, herbicides	Poisonous to people and animals, drift or runoff onto nontarget plants or into waterways, skin absorption, eye injuries, children or pets playing in application zones too soon, incorrect application rates
Climbing	Falling
Combustible materials	Burns, injuries to eyes and body as well as property
Heavy lifting	Muscle and bone injuries
Painting and staining	Eye injury, fume inhalation, and possible skin absorption
Working without gloves or other types of hand protection	Blisters, splinters, thorns, cuts, poison ivy and other rashes, absorption of chemicals through the skin
Working with dusty materials such as cement, stone dust, or mulches	Inhalation of hazardous dust and eye damage
Digging in unfamiliar areas	Cutting into utility lines resulting in property damage and possible injury to the worker
Working in sunlight	Skin damage resulting from overexposure
Snowplowing	Accidents resulting from fatigue, unseen objects, and/or poor visibility
Working in hazardous public areas such as golf course fairways, or along railroad tracks or highways	Unintentional injury from activities common to the public area, such as golf balls, traffic, trains

course. A man or woman who is skilled mechanically will probably identify the dangers associated with the misuse of equipment more readily than someone who has little or no experience with power tools and equipment.

The Probable Causes

It is easier for some people to understand the real and potential *causes* of landscape injuries and dangers than it is for them to understand and admit to some of the *reasons* why those injuries occur. The belief that accidents happen to other people, not us, is a widely held belief until a personal accident brings it home. People in general, and young people in particular, have a false sense of invincibility that can make something as logical and simple as wearing a seat belt seem unnecessary. Were it not for government enforcement of speed limits, child restrainer seats, and similar safety laws, there would be far less voluntary application of these logical safety procedures. The lack of participation would not be based on lawless disregard, but rather on the belief by a majority of individuals that they can personally avoid any injury. That same attitude of personal invincibility can be brought to the workplace, and that is why government at all levels, the landscape industry in general, and the companies that comprise the industry nationwide are now giving increasing levels of attention to safety in the workplace. Those levels range from broad regulations imposed by government and enforced by government agents to safety pep talks given by a foreman to his or her crew before they start work each morning. Knowing what to do when someone is injured is important, but knowing how to prevent the injury in the first place is just as important. The landscape industry now realizes that workers are not as naturally inclined toward accident prevention as they should be, so it is up to the leaders within each company to initiate the action and to enforce those regulations established by governmental authorities.

Safety Regulations Established by Government

Government officials try to set safety guidelines that protect both workers and the general public. Theirs is not an easy job because different people can interpret the same issue differently. There are individuals and groups that are focused on the protection of the natural environment and often interpret any form of chemical application as a threat to the well-being of the environment. They may attempt to persuade government officials to regulate against any use of chemical pesticides by individuals or companies. Members of industries such as the landscape industry will often attempt to counter or modify proposed application regulations by requesting government officials to temper their legislation so that chemicals needed to control insects, diseases, or weeds, and chemicals for other beneficial uses can be applied to accomplish a project successfully. Because extremists are common on all sides of regulatory issues, officials of the federal and state governments try to remain objective and do what is best for the citizenry at large. Members of the landscape industry must stay abreast of what is happening at all legislative levels, not only because doing so makes them better-informed citizens, but so that their point of view is considered when new regulations are being formulated.

The safety regulations of government take several forms.

- **vehicle safety regulations**
  licensing
  inspection
  safety markings

- **chemical regulations**
  certification of applicators
  approval and restriction of certain products
  specification of allowable rates of application
  Material Safety Data Sheets
  approval or restriction of certain products in certain geographic areas

notification requirements before application
container disposal restrictions

- **worker safety regulations**
  licensing for vehicle operation
  certification for pesticide application
  Workers' Compensation insurance
  OSHA (Occupational Safety and Health
  Administration)

Vehicle Safety

As vehicle size and complexity of operation increase, the knowledge required of a driver extends beyond what is needed to operate an automobile. Most states require special licenses for drivers of large trucks or other heavy equipment. The need for a commercial driver's license or other certification of the driver's training in the safe operation of oversized vehicles is common throughout the nation. It is a violation of law for a company to permit an employee to operate a vehicle for which he or she is not properly licensed. Also, most insurance companies will not insure a landscape company that permits vehicles to be operated by an employee who has been convicted of an alcohol-related charge, such as driving while intoxicated. Many states require an immediate test for alcohol or drugs in the system of the driver who has been involved in a vehicle accident. There is no faster way for an employee to destroy his or her career in the landscape industry than through drug or alcohol abuse. An employee who cannot operate the company's vehicles is of limited value to the company.

Many state governments require that all vehicles operated on the roadways of the state be inspected annually at stations approved by the state. The intent of these inspections is to assure that there are no broken, missing, or altered safety features that would make the vehicle a hazard to others. Typical features checked during a safety inspection are brakes, headlights, taillights, brake lights, wiper blades, muffler, and horn. Many states also include a check of the vehicle's pollution control devices during these annual inspections. Not all states cur-

rently require annual safety inspections, but the number is increasing.

Every state also now requires that slow-moving and oversized vehicles used on public roadways be marked with eye-catching, reflective signs that caution oncoming motorists to slow down when approaching. Large vehicles are also required to have audible beeper systems that are activated whenever the vehicle is put into reverse gear, Figure 30-1.

Chemical Safety

The outreach of government is not so lengthy that a representative can be on-site every time a chemical is applied by a landscaper or other applicator of chemical products. What government officials can

Figure 30-1 Slow-moving vehicles, such as this tractor, bear a triangle with fluorescent colors that are visible both day and night. Added safety features include the roll bar.

do is regulate the production and distribution of the products by the manufacturers. They can also regulate *where* a chemical product may or may not be used, *on what* it may be used, and *by whom* it may be used. Government regulations also try to assure that everyone who can be affected when a chemical is used, including the worker, nearby workers, and the general public, are aware of the product and protected against any hazardous effects.

All chemical products created for uses that permit their entrance into the environment and directly or indirectly into the food chain and atmosphere must be approved by the U.S. Environmental Protection Agency (EPA) before they can be introduced for public or commercial use. All manufacturers and distributors of pesticides are required by federal law to provide explicit information about their products' ingredients, formulations, toxicity, and proper rates of application, and about the specific pests controlled and proper means of safe handling. This information is provided on the pesticide label, Figure 30-2. The signal words that indicate the product's level of toxicity should be committed to memory by everyone who must work with these important but dangerous tools of the landscape business, Table 30-1.

Equally important is the Material Safety Data Sheet (MSDS) that is prepared for each chemical product and included with it at delivery, Figure 30-3. The MSDS provides a complete compilation of all the information a user should know about the product. Information includes the manufacturer, the product's chemical name, physical properties, and chemical reactivity; fire and explosion data; health hazard data including symptoms of exposure and carcinogenicity; special protection and handling information; and an emergency telephone number to call at any time for information about treatment for exposure or spillage. MSDS sheets should be kept in a place where they can be easily referenced if needed. Copies should be posted where everyone working with the pesticides can read them.

Even when EPA-approved, some pesticides are further regulated within a state by state or county officials. Usually on the advice of the state's college of agriculture, a pesticide's use may be

Figure 30-2 A typical pesticide label (Courtesy of Chevron Chemical Company)

restricted or entirely banned because of concern for its impact on the underground or surface water supply or its dangers to wildlife or nearby population centers.

Some chemical pesticides are available in reduced strength formulations for purchase and use by homeowners and amateur gardeners but are available in full-strength concentrations only to trained industry professionals. As far back as the 1970s, the EPA required states to develop and administer programs that require persons who apply restricted pesticides in their work to have their knowledge measured and certified by their state on a regular and recurring basis. Pesticide certification programs now require anyone wanting to purchase commercial strength restricted pesticides to first pass a written test that measures

Table 30-1 Levels of Pesticide Toxicity and Safety Equipment Needed

Pesticide Signal Word	Level of Toxicity	Label Symbol	Special Equipment Required
Danger—Poison	High	Skull and crossbones	Rubber boots, gloves, rubber pants, hat, and raincoat; face shield and gas mask with its own air supply
Warning	Moderate	None	Same as for high toxicity
Caution	Low	None	Rubber boots and gloves; respiratory equipment recommended if used indoors

his or her basic knowledge of pesticides and proper application. Once certified, the pesticide applicator must be recertified on a regular basis. While the states vary in their methods of recertification, it is often done by requiring ongoing training of the certified applicators through their attendance and completion of workshops and short courses held throughout the year at different venues around their state. The purchase or application of a restricted pesticide by anyone who is not a current certified pesticide applicator is a violation of the law and can subject the person to fines and/or imprisonment.

Local officials will usually regulate how a company is permitted to dispose of its empty chemical containers. Often a community does not want the empty containers put into its landfill or recycling center, so it will publish directives for the safe and proper disposal of containers having toxic residues in accordance with their local standards.

Workers' Compensation

Should employees become injured or ill due to work-related causes, they can receive financial assistance through their state's Workers' Compensation program. Every state in the nation offers this insurance in one form or another. It is a no-fault type of insurance and is available to all employees who seek compensation for injuries that they receive while performing their jobs. There are two types of benefits available through Workers' Compensation: one is the payment of medical expenses that are connected with an injury or disability incurred on the job; the other is wage replacement to help the injured worker recover some of his or her lost income. What percentage of the worker's lost income will be replaced is determined by the degree of injury and varies among the states.

The cost of Workers' Compensation insurance is paid by the employer. The rate charged a company is based upon how many claims have been filed by employees of that company. A company that has had many employees file claims of injury will have to pay more for its insurance than one that has had few or none. A large number of injury claims will greatly increase the costs of operation for a company and reduce its profits. That in turn reduces what it can do with its profits and limits its growth and prevents it from improving the compensation of all its employees. If employees can see no other benefits to safety in the workplace, they will usually understand that which affects the size of their paycheck.

TELEPHONE: 518-234-5315

 CHEMICAL & FERTILIZER CORPORATION

BOX 123
COBLESKILL, NEW YORK 12043 U.S.A.

MATERIAL SAFETY DATA SHEET

Conforms to U.S. Department of Labor Bureau of Labor Standards

SECTION I

MANUFACTURER'S NAME AJAX Chemical & Fertilizer Corp.	EMERGENCY TELEPHONE NO. 518-234-5315
ADDRESS *(Number, Street, City, State, and ZIP Code)* Box 123 Jayridge Road, Cobleskill, NY 12043	
CHEMICAL NAME AND SYNONYMS Oxy-Doxyl	TRADE NAME AND SYNONYMS Oxy-D 23
CHEMICAL FAMILY Insecticide-Nematicide	FORMULA Chemical mixture

SECTION II HAZARDOUS INGREDIENTS

	%	TLV
A. ACTIVE INGREDIENTS Methyl N'N'-dimethyl-N-		
(Methylcarbamoyl)-Oxy) -2-Thioxamimidate CAS NO. 23135-22-0	10	NA
B. SOLVENTS	%	TLV
C. OTHER Inert	PURPOSE	% 90

SECTION III PHYSICAL DATA

BOILING POINT (°F.)	NA	SPECIFIC GRAVITY (H₂O =1)	NA
VAPOR PRESSURE (mm HG.) Negligible		PERCENT VOLATILE BY VOLUME (%)	NA
VAPOR DENSITY (AIR =1)	NA	EVAPORATION RATE (=1)	NA
SOLUBILITY IN WATER 28G/100 ML (Oxamyl)		Bulk Density	27.5 lbs. cu ft.
APPEARANCE AND ODOR Blue-green granular			

SECTION IV FIRE AND EXPLOSION HAZARD DATA

FLASH POINT (Method used) NA	FLAMMABLE LIMITS NA	Lel	Uel

EXTINGUISHING MEDIA On small fires use dry chemical, carbon dioxide, foam or water spray.

SPECIAL FIRE FIGHTING PROCEDURES
If area is heavily exposed to fire and if conditions permit this extinguishing with water spray. If conditions permit, cool containers with water if exposed to fire.

UNUSUAL FIRE AND EXPLOSION HAZARDS
Wear self contained breathing apparatus. Protective inhalation equipment should be worn in the vicinity of the fire until the ashes are cold. All unprotected people should be removed from the area and upwind.   NA NOT Applicable or not available

Figure 30-3 Example of a Material Safety Data Sheet (MSDS)

The Occupational Safety and Health Administration (OSHA)

The U.S. Congress created the Occupational Safety and Health Administration (OSHA) in 1970 to make the American workplace a safer place for our citizens. Among its mandates, OSHA was directed to:

- encourage the reduction of workplace hazards by both employers and employees and further encourage their implementation of new and existing health and safety programs

- provide for research into new ways of dealing with job-related safety and health problems

- establish responsibilities and rights for both employers and employees that will make them separately responsible for better safety and health, while still making them dependent on each other for their success

- maintain monitoring systems for both reporting and keeping records of job-related injuries and illness

- develop mandatory job-related health and safety standards for the nation

- oversee the development of health and safety standards and programs within the 50 states

The authority of OSHA is nationwide and reaches into nearly every business. The only exceptions are self-employed individuals, farms where only members of the immediate family are employed, or work sites that are protected by other federal agencies or laws, such as mines.

The federal act that established OSHA requires all employers to "furnish a place of employment that is free from recognized hazards that are causing or are likely to cause death or serious physical harm to their employees." OSHA is empowered by the federal government to set the standards of safety for specific industries, including the landscape industry. Usually a new standard or modification of an existing safety standard is established by OSHA following public hearings and other forums that enable them to receive input from the members of the industry being affected and by the public at large. All employers are required to abide by all of the OSHA regulations that affect their business. The safety standards of OSHA generally fall into two categories of regulation:

1. *hazard communication standards* that require employers to be proactive in training their employees in the safe use of hazardous chemicals; such training includes how to read product labels, access and understand the information in a material safety data sheet, apply particular chemicals safely, recognize when a hazardous chemical has been released into the work environment, and know what actions to take to protect against a chemical during application as well as during a spill;

2. *work environment issues* that address the physical working conditions of the workplace. These standards deal with such issues as the availability of facilities including a cafeteria, vending machines, and places for employees to take breaks, the cleanliness of these facilities, and even the restriction of smoking by employees while on the job.

To enforce their standards, OSHA inspectors have the authority to conduct inspections of companies at the workplace. Inspections may be unannounced and often result from complaints submitted by employees or outside observers who believe they have witnessed unsafe activity or hazardous situations within a company. Also, businesses that by the nature of their work subject employees or customers to an above average level of risk can expect more frequent inspections and regulations by OSHA. If OSHA finds a company in violation of its standards, the inspectors are empowered to impose citations and penalties on the company.

Safety Awareness Is Promoted by the Landscape Industry

Even as government officials continue to establish standards of safety and send inspectors to check for adherence to those standards, the landscape industry has been proactive in developing its own programs to promote safe practices. On the national level, the four major industry trade organizations (American Nursery and Landscape Association, Associated Landscape Contractors of America, Professional Lawn Care Association of America, and Professional Grounds Maintenance Society) have developed programs for their members that include workplace safety posters, safety videos in English and Spanish for training workers, safety handbooks, safety award programs, and regular health and safety updates during national conferences, Figure 30-4. Association newsletters and trade magazines include articles on worker health and safety, and they recognize specific companies that have outstanding safety records. By giving positive recognition to companies that place the safety of their workers at the top of their list of professional obligations, these associations create role models and benchmarks for others in the industry to emulate. When a company receives an award for safety that can be taken back to the workplace and displayed prominently for all employees to see, it accomplishes at least two things. First, it allows the employees to own a piece of the recognition and

share in the pride. Second, it contributes to an overall positive morale among the workers. They are reassured that each morning they are coming to work for a company that cares about their safety and asks them to share in the responsibility of keeping their workplace safe. In addition, nearly every state and many regions within the states have their own trade organizations that sponsor member health and safety programs. To those industry efforts can be added the programs, videos, and workshops on safety that are prepared and funded by many manufacturers of products and equipment used by landscape companies. While these efforts are largely focused on companies' products, since it is in their best interest to promote the hazard-free use of what they are selling, companies nevertheless contribute to the overall safety of the national landscape industry.

Safe Practice at the Work Site

The *buck* for the personal health and safety of landscape employees *stops* at the actual workplace. Safety cannot only be read about, it must be practiced constantly and consistently. The leaders of a company must make every effort to train their employees in the safe practices of their work. They must actually observe, evaluate, and correct employee performance until they are certain that every employee understands how to perform his or her tasks correctly and safely and does so every time. If a new employee is being trained by a senior employee, it becomes the responsibility of the senior employee to include all safety and health information as training progresses. No newly hired employee, even if experienced, should be permitted to begin work until he or she has been observed, evaluated, and found knowledgeable of all appropriate safety practices.

Many companies have established their own safety programs and employee incentives. For example, the crew having the best safety record at the conclusion of a season may be rewarded with a special dinner or paycheck bonus. An employee

Figure 30-4 Examples of safety manuals prepared for the landscape industry. These in the photo were prepared by the Associated Landscape Contractors of America.

who makes a safety suggestion that is accepted and implemented by the company receives some sort of special recognition. Simple though the incentives may be, they keep the employees mindful of the need to continuously work safely.

Individual workers have the right and the obligation to be safe and to work safely at their job site. Shoes should be heavy soled and made of thick leather. Canvas shoes, sandals, or other leisure types of footwear should never be worn in the landscape workplace. For some tasks, the work shoes should have steel toe reinforcement. When working with mowers, string trimmers, or chemical sprays, or at any task where the worker's legs and/or arms could be contacted by projectiles or chemicals, the worker should not wear shorts or short sleeves. Long sleeves and full-length trousers also give protection against the sun, insect bites, and poison ivy. If hot weather makes long sleeves uncomfortable, the employee should apply both sunscreen and insect repellent at regular intervals throughout the day.

Employees who work in or around trees or other sites where there is a danger of things falling from above should wear hard hats for protection. The use of power saws, lawn mowers, string trimmers, and other noisy tools necessitates ear plugs. Safety glasses are also needed whenever there is particulate matter in the air, as when string trimming or mixing concrete, Figure 30-5.

When pesticides or other chemicals are being prepared and applied, workers must be even more careful. Body coverings become increasingly more protective as the toxicity level of the chemicals increase. A worker should check the signal words on the label to determine the toxicity class of the chemical. *Danger* identifies a highly toxic Class I chemical, *Warning* indicates a moderately toxic Class II chemical, and *Caution* identifies a slightly toxic Class III or IV chemical. The MSDS should also be consulted for additional information about the potency of the chemical. The objective of the worker's information gathering is to assure that the proper type of clothing, footwear, eye protection, respiratory gear, and gloves are selected and worn before beginning work with the chemical. For some dry chemicals having a low toxicity, long trousers

and booties to cover the worker's shoes may be sufficient protection. Leather shoes, as recommended above, should *not* be worn during chemical application, since the leather may absorb and retain the chemical. Rubberized footwear with protective and disposable covers are most suitable. For sprays and more toxic chemicals, a full protective suit, gloves, respirator, and face protection may be needed. Not all protective suits are alike, so it is important that the one chosen for use actually protects the worker from the chemical. Before a product is applied, the material of the protective suit and gloves should be checked to see if it prevents the chemical from soaking through or brings about an unexpected reaction such as fabric color change.

The safety glasses used when mowing or string trimming are inadequate for protecting the chemical applicator's face. Goggles that fit snug against the wearer's face are the best protection for the eyes. If the sprays are being directed overhead, as when an arborist treats trees, the applicator should also wear a protective hood made of material identical to that pre-tested for the suit.

Not all respirators are the same, so the employee should consult the pesticide label for the approval number that indicates the type of respirator to be worn when the chemical is applied. Some respirators only filter particulate matter. Others give protection against toxic gases and vapors. Each type has an approval number assigned to it by the National Institute for Occupational Safety and Health, and that is what the applicator should check for on the label of the pesticide or other chemical before beginning.

To assure that customers, children, pets, and others unaware of the danger posed by the application of chemicals on a project site do not inadvertently come in contact with the material, many communities require that prior notification of the chemical's application be made by the company. Depending upon the state and the community, a company may be required to post flags or other signs at prescribed intervals around the treated site up to 72 hours before application occurs. They may also be required to notify nearby neighbors so that they can keep their pets and children away from

Figure 30-5A Crewmen loading a tractor for transport to a work site. Having two workers assures the safe maneuvering and balanced placement of the tractor on the flatbed. Both crewmen are wearing long trousers, hard hats, heavy gloves, and leather work boots with sturdy toes and soles.

Figure 30-5B The tractor is securely chained to the flatbed before transport begins.

the treated site until the chemical no longer poses a danger.

Even Safety Has Its Problems

It would be expected that the landscape company that follows all of the laws, regulations, and sug-gestions for safe and healthy practice would be problem-free, at least where its job site perfor-mance is concerned. That is usually the case, but there have been and continue to be some prob-lems, especially in the application of pesticides. Some employees try to avoid wearing proper pro-tective equipment because it makes them hot and uncomfortable. Some companies try to reuse protective gear that should be discarded after one use or do not store or clean their reusable spray gear properly. Some clients are alarmed when they see company employees dressed like astronauts spraying chemicals on their home grounds. In extreme cases, some landscape workers find themselves confronted by environmental activists who try to prevent the application of chemicals they believe are harmful to life. Each of these

problems requires the leadership of a landscape company to be steadfast in his or her commitment to safety and healthful practice by all employees of the company whenever and wherever they are working. It is also advisable for the leaders of the company to be vocal proponents of public education in their community to help others understand and appreciate all that the landscape industry is doing to assure a safe and ethical impact on the environment.

Achievement Review

A. Listed below are several types of landscape workers. Describe specific types of injury that could occur if precautions were not taken.

  1. arborist
  2. lawn maintenance specialist
  3. landscape contractor
  4. small-engine repairman
  5. heavy equipment operator

B. List three specific requirements of federal and state governments that deal with the safe operation of motorized vehicles in the workplace.

C. List at least five ways that federal and state governments regulate the safe use of chemicals in the workplace.

D. Explain the differences among pesticides signaled by these three words on the product label: Danger, Warning, and Caution.

E. Why is it harmful to a company when a large number of its employees file Workers' Compensation insurance claims?

F. What can cause OSHA to select a company for an unannounced safety inspection?

G. *Essay*: Write a short statement that could be used in an employee handbook to impress on newly hired young employees the important role they will play in the safe and healthy operation of the company. The statement should specifically answer the question "Why should they care about safety?"

Suggested Activities

1. Invite a representative of OSHA to visit the class and explain in greater detail how the agency operates.

2. Have students select a piece of landscape equipment that interests them, and prepare a three-minute safety talk suitable for presentation to a crew of unskilled workers. Let the class role-play as the crew members and ask appropriate questions during and following the talk.

3. Schedule a presentation by a professional safety officer. If there is not one employed by a local landscape company, there may be one in the school district or at a nearby police or fire department or hospital.

The Final Word

Customer Service

This text began with consideration of the meaning of a word. The word was *landscaper*. Many of the skills and crafts necessary to the practice of landscaping have been discussed in the chapters of the text. It is now appropriate to close the text with another word and an explanation of its meaning. The final word is *service*, and without an acceptance of its meaning and importance as well as knowledge of how and to whom it should be applied, no one is fully prepared to be a successful and professional landscaper.

Landscaping is a service industry. It is labor intensive and involves the satisfaction of customer needs. A company is only as successful as its clients allow it to be. The key to understanding what service means and requires is acceptance of some new descriptions of old ideas.

Few businesses today have not encountered the concept of total quality management in the popular press as well as in the literature of their industry. The landscape industry is no exception. Total quality management, or TQM as it is frequently abbreviated, is a system of business operation that is driven by the objective of meeting and exceeding client expectations through ongoing improvements in the processes that comprise the business. It replaces the old idea of quality control, as established by the company, with the more factual realization that performance by a company is only qualitative if the client perceives it as such. It matters little if the company praises its own performance. Only praise by the customer who pays the bill truly measures whether the project was done satisfactorily or not. Thus modern landscape companies must do everything possible to ascertain the true desires of their clients at the outset of a project if they are to provide the quality of service that will meet and exceed their clients' expectations. This is no easy task, since each client is different and each expresses or fails to express his or her desires for the project differently. For example, two different clients may each want a patio constructed. Each may have offered nearly similar specifications to the landscaper during the discussion phase of the project; e.g., same size, materials, furnishings, and budget. What was not specified to the company was that one of the clients had in his mind a memory of a patio his family had when he was a child, and he is going to measure the new one against his sentimentalized remembrance. The other client cares more about what her friends and neighbors will think of the patio than anything else about it. She is most concerned that it be visually stunning to impress her guests. Meeting and exceeding the expectations of the two clients will require personal attention to each one by staff members who are able and willing to provide it. If the service meets the quality expectations of the clients, each is likely to return to the company as a future customer. If they are not pleased with the service provided, it is unlikely that they will provide repeat business and highly probable that they will tell their friends and family of their dissatisfaction.

The provision of such personalized service to clients requires all employees of the landscape company to realize that they are in customer service no matter what their job. Even if they seldom or never have direct contact with the bill-paying clients, they must be shown and convinced that the way they perform their jobs affects the ultimate impression of the company's performance by the client. The on-site crew must not only perform their tasks competently, they must look and behave professionally. Dirty, tattered clothing or uniforms that don't match convey a different image than a well-groomed crew dressed in matching uniforms, working with freshly painted equipment, and driving clean trucks. One crew may fit the bad image of landscapers that white-collar professionals often have, while the other crew exceeds the client's expectations, and pleasantly so. Even if the finished project could be done equally well by both crews, the crew perceived as providing the best service would probably be the one that presented the best

image. Similar perceptions of professionalism or lack thereof are conveyed by the way crew members behave at the work site. Smoking, loud talk, profanity, wasting time, blaring radios, and spitting can all contribute to client dissatisfaction and make the installation of their new landscape an ordeal rather than a pleasure.

Another instance where clients are sensitive to their treatment by a company occurs when the clients approach a crewperson or supervisor with a question or request and find that employee powerless to respond. While it is difficult for some owners and managers of companies to empower their employees enough to address client concerns when and where they arise, it is an important step toward making employees feel involved with the company, not merely used by it. The level of client satisfaction is raised when clients' questions are regarded as sufficiently important to warrant immediate responses.

In addition to the crewpersons, others in the company affect the client's perception of service and quality. The number of rings before the telephone is answered and the tone of the person's voice all create an impression. Languishing on hold, listening to squeaky recorded music, is certain to frustrate any caller. Voice mail systems and answering machines, unless used carefully and selectively, can also alienate customers. A landscape is a living thing, and clients want to talk about it with living people, not recorded voices.

Treatment by salespersons, the speed and accuracy of billing, and the follow-up concern for client satisfaction are among the many benchmarks by which clients evaluate the services of a landscape company. Since people remember bad experiences, such as rudeness or an unreturned phone call, more than good experiences, one inattentive employee can undermine the superb performances of a dozen other employees. The commitment to total quality management must be company wide. All employees must be taught to understand and value it, then expected to practice it. If they cannot or will not, despite repeated training, then they must be let go. Accepting the philosophy of TQM is as difficult for some executives as it is for others in the company.

TQM negates their role as overseers and bosses and instead requires them to be leaders and coaches. TQM is most easily accepted by team players, and most difficult for strong-willed, dominant personalities to accept.

Some would argue that providing a quality of service described as acceptable by the client rather than the company will add greatly to the cost of building landscapes. That has proven to be untrue. It takes about the same amount of time and materials to build a landscape improperly as it does to do it correctly the first time. The defective landscape, which must be fixed after the customer complaint, ends up costing the company more in additional labor expenses and diminished reputation. It also costs a company more to recruit and train new employees to replace departed, disgruntled employees than it does to offer sufficient incentives to retain trained, contented employees.

The original proponents of total quality management also challenged traditional business thinking by redefining the word *client*. The term seems simple: the client is the recipient of goods and services. The new interpretation includes the concept of both external clients and internal clients in a business. The external clients are the ones who order and pay for landscape projects. They are the ones for whom every worker must be convinced to do his or her best every minute of every working day. They make the final judgment about the true quality of the project, and they are the ones whose expectations are to be surpassed. The internal client is the newer expansion of the term. It refers to the company personnel, each of whom is part of one or more systems that deliver some piece of the service and/or product to the external client. If a system or process for the provision of a service or product can be likened to a chain, then each employee can be seen as being linked to those other employees who are responsible for functions or materials that either precede or follow the contribution of that particular employee. In such a situation, each employee within a process or system can be seen as a client of the employee whose step in the process precedes his or hers. Once the client-employee has contributed to the process, it is handed off to the next employee in the chain. Now the client-employee

becomes the supplier to the next employee in the sequence, his client. Seen this way, it is easier to accept the proposition that every employee within the landscape company is at times the supplier and at times the client of other suppliers.

The not-so-easy part is gaining an agreement from all employees to accept no defective products or performances from their internal suppliers. If an employee does not pass along work to her internal client in the best possible state, the end product or service will not be as good as it can be. It will wind up either costing the company money to remedy it, or disappointing the external client rather than exceeding expectations. Defective performance must be identified at its source and the internal supplier given the opportunity to correct the defect before returning it to the internal client. To gain acceptance of this important quality factor by employees, the leadership of the company must remove the stigma traditionally attached to errors. Employees must be permitted to make mistakes and correct them without fear of being criticized or reprimanded. This is where the role of the leaders as coaches and cheerleaders for their employees, rather than vigilantes seeking criminals, is critical.

Each process or system within the company must be broken down into its component steps, then studied, made better, the improved process implemented, and the entire process repeated again, and again, and again. TQM is never ending, and it does not happen quickly. Neither can it be accomplished with motivational posters and shiny plaques. It is a management style that calls upon all employees to participate. It uses their expertise to improve its products and services. It rewards their creativity rather than discouraging independent thinking. In the end, it fulfills the expectations of both internal and external clients, and even surpasses those expectations. In a service industry such as the landscape industry, catering to a clientele whose own companies are teaching them about TQM, the expectation level of clients grows higher each year. There are a number of good companies in every major market in the country. The ones that will be most successful will be the ones that keep in mind and excel at the final word . . . *service*.

Appendix A

Examples of Landscape Designs

The illustrations that follow are examples of landscape designs done by professionals currently working in the landscape industry. The designs include residential areas, small commercial sites, and recreational areas. These plans are the work of designers with college preparation ranging from two to five years. The designers are employed as landscape contractors, landscape architects, or recreational planners.

Students will gain several insights by studying the plans. First, each plan can be seen as a graphic explanation of the designer's ideas. Some are mainly concepts, with few precise details, aimed at selling the designer's proposals to a client. Others are detailed instructions of what, where, and how elements of the landscape are to be developed.

Second, notice how the graphics vary. Some of the plans are highly mechanical in appearance, due to the use of lettering templates and waxed press-on letters. Others are less rigid in appearance, due to free-hand lettering and a looser graphic style. All of the designs are of professional quality. The graphic technique used depends upon how much time the project warrants at a particular stage in its development, and how much competition there is among designers to gain a potential client.

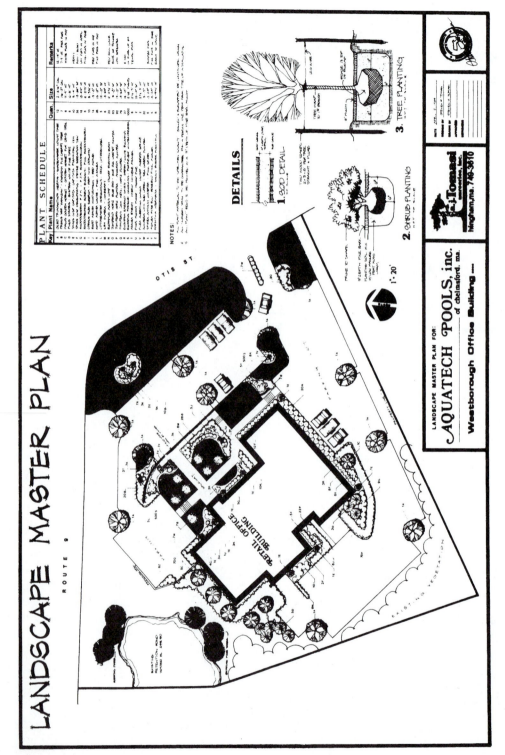

(Courtesy of A. J. Tomasi Nurseries, Inc., Pembroke, MA)

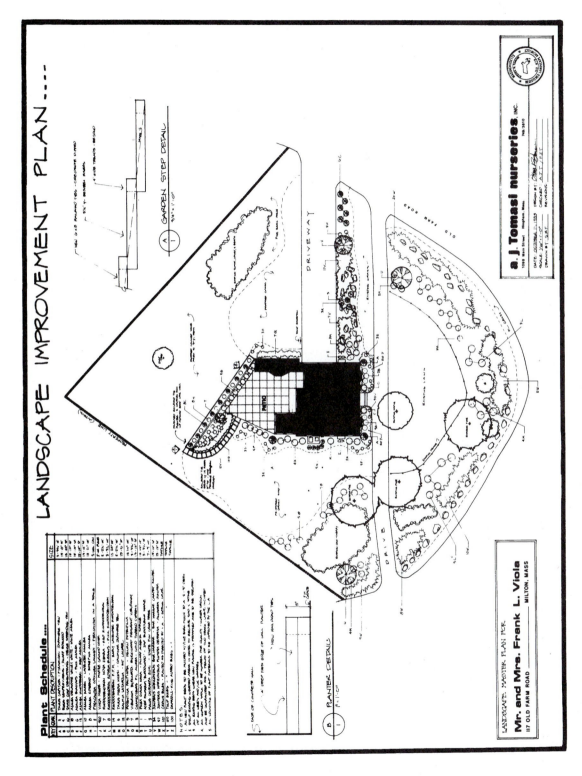

(Courtesy of A. J. Tomasi Nurseries, Inc., Pembroke, MA)

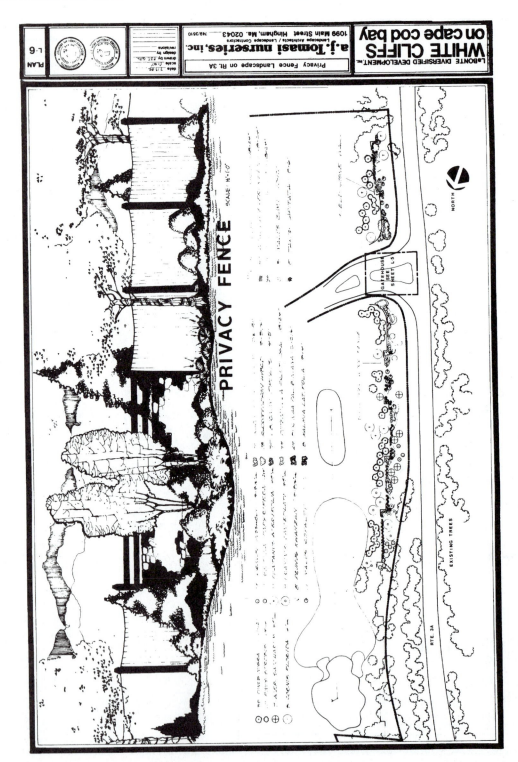

(Courtesy of A. J. Tomasi Nurseries, Inc., Pembroke, MA)

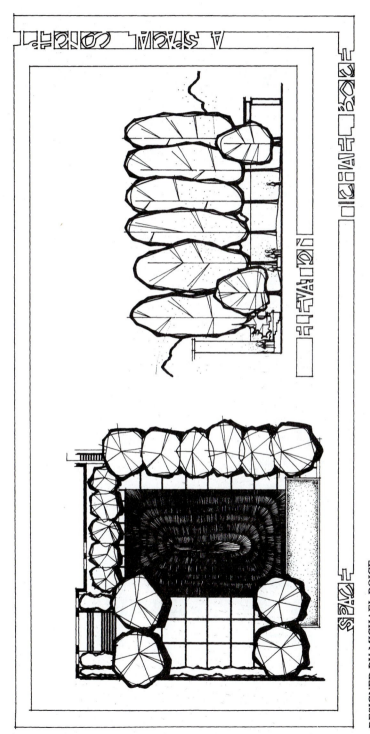

A SMALL CORNER

RESIDENTIAL

ELEVATION

PLAN

DESIGNED BY MICHAEL BOICE
ALBANY, NY

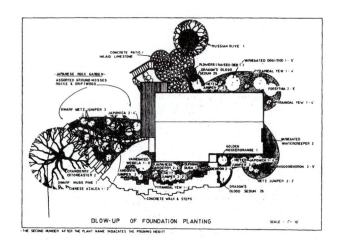

BLOW-UP OF FOUNDATION PLANTING

SCALE · 1" = 10'

- THE SECOND NUMBER AFTER THE PLANT NAME INDACATES THE PRUNING HEIGHT

DESIGNED BY
MICHAEL E. BOICE

MAKE THE BOICE CHOICE !

SCALE · 1' = 10'   BLOW-UP

SCALE · 1' = 20'

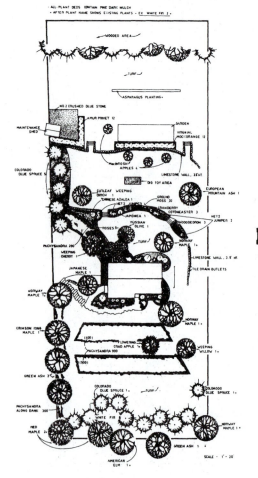

- ALL PLANT BEDS CONTAIN PINE BARK MULCH
- AFTER PLANT NAME SHOWS EXISTING PLANTS - EX: WHITE FIR 2

SCALE · 1' = 20'

PLANT LIST

BOTANICAL	-SHRUBS-	COMMON	NO. USED
AZALEA MOLLIS		CHINESE AZALEA	2
BERBERIS THUNBERGI ATROPURPUREA		RED JAPANESE BARBERRY	1
CORNUS ELEGANTISSIMA VARIEGATA		VARIEGATED DOGWOOD	1
COTONEASTER APICULATA		CRANBERRY COTONEASTER	3
EUONYMUS ALATUS		BURNING BUSH	1
FORSYTHIA INTERMEDIA LYNWOOD GOLD		LYNWOOD GOLD FORSYTHIA	2
JUNIPERUS CHINENSIS 'HETZI'		HETZ JUNIPER	15
JUNIPERUS CHINENSIS 'HETZI' COMPACTA		DWARF HETZ JUNIPER	3
JUNIPERUS HORIZONTALIS PLUMOSA		ANDORRA JUNIPER	3
JUNIPERUS SQUAMATA MEYERI		MEYER JUNIPER	3
PHILADELPHUS CORONARIUS AUREUS		GOLDEN MOCKORANGE	1
PHILADELPHUS VIRGINALIS		VIRGINAL MOCKORANGE	12
PIERIS JAPONICA		JAPONICA (ANDROMEDA)	4
PINUS MUGHUS		DWARF MUGO PINE	1
RHODODENDRON SPP		RHODODENDRON	3
TAXUS CUSPIDATA		JAPANESE YEW (PYRAMIDAL)	2
WEIGELA FLORIDA VARIEGATA		VARIEGATED WEIGELA	1
	-TREES-		
ABIES CONCOLOR		WHITE FIR	8
ACER PALMATUM		JAPANESE MAPLE	1
BETULA PENDULA LACINIATA		CUT LEAF WEEPING BIRCH	1
ELAEAGNUS ANGUSTIFOLIA		RUSSIAN OLIVE	1
MALUS SPP		MacINTOSH APPLES	4
PICEA PUNGENS		COLORADO BLUE SPRUCE	5
PRUNUS SUBHIRTELLA 'PENDULA'		WEEPING CHERRY	1
SORBUS AUCUPARIA		EUROPEAN MOUNTAIN ASH	1
	-VINES & GROUND COVER-		
EUONYMUS FORTUNEI ARGENTEO-MARGINATUS		VARIEGATED WINTERCREEPER	3
PACHYSANDRA TERMINALIS		PACHYSANDRA	1400
SEDUM DRAGONS BLOOD		'DRAGON'S BLOOD SEDUM	50
		GROUND MOSS	100

PLANTING   PLAN   FOR

MR. & MRS. C. MASICK

COBLESKILL,   N.Y.

DESIGNED BY MICHAEL BOICE
ALBANY, NY

SITE PLAN

DESIGNED BY MARK MAGNONE
SCHENECTADY, NY

(Courtesy of Mark Magone, Landscape architect)

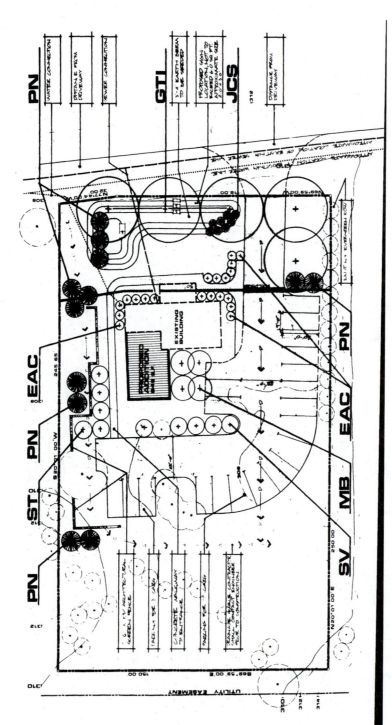

DESIGNED BY MARK MAGNONE
SCHENECTADY, NY
(Courtesy of Mark Magone, Landscape architect)

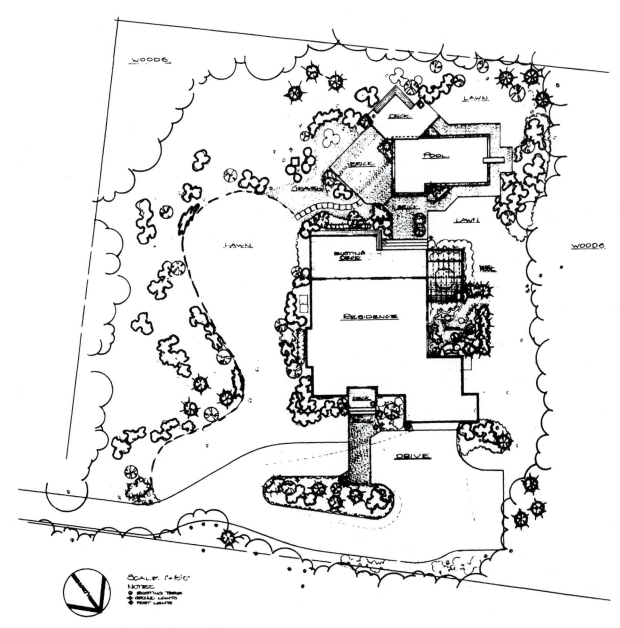

DESIGNED BY GOLDBERG AND RODLER
HUNTINGTON, NY

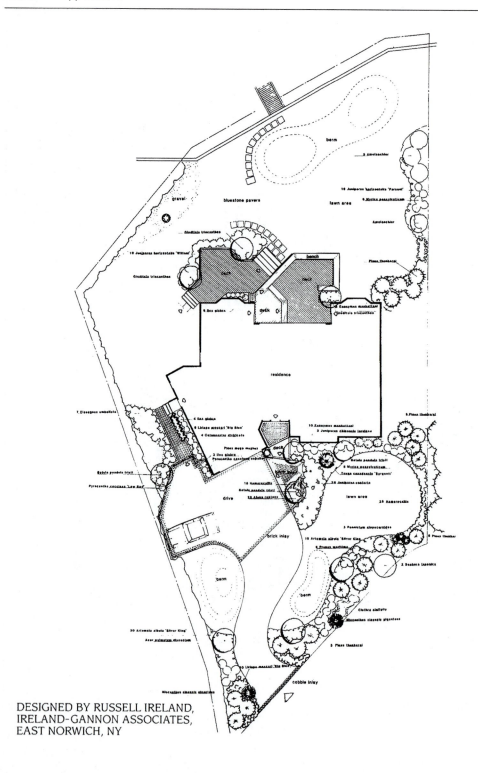

DESIGNED BY RUSSELL IRELAND,
IRELAND-GANNON ASSOCIATES,
EAST NORWICH, NY

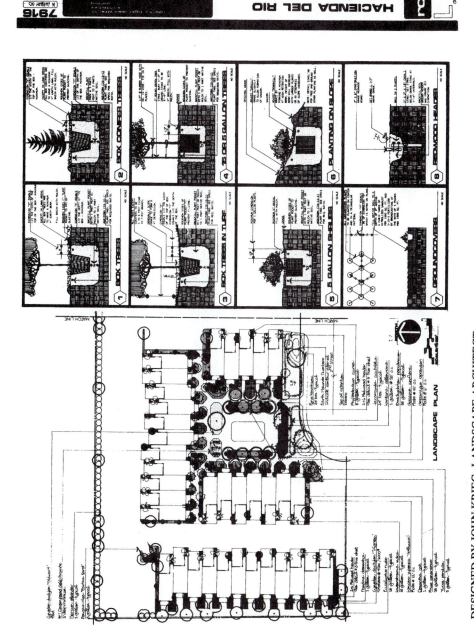

DESIGNED BY JOHN KRIEG, LANDSCAPE ARCHITECT
ROBT. C. BIGLER ASSOC., ARCHITECTS
PHOENIX, ARIZONA

(Courtesy of John C. Krieg, Landscape architect)

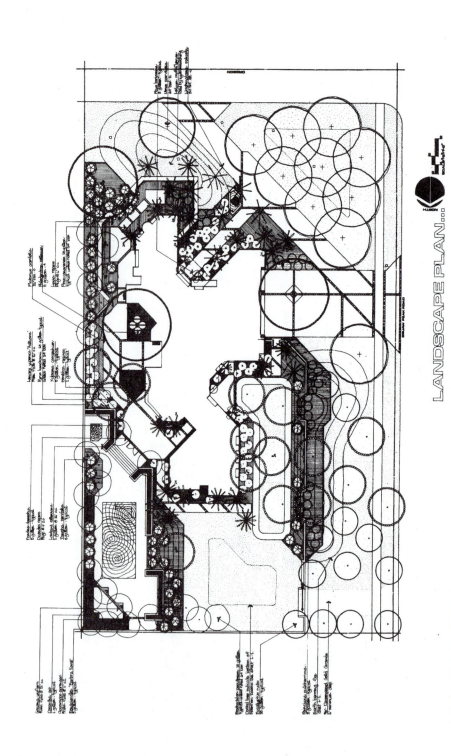

LANDSCAPE PLAN

DESIGNED BY JOHN KRIEG, LANDSCAPE ARCHITECT
ROBT. C. BIGLER ASSOC., ARCHITECTS
PHOENIX, ARIZONA

(Courtesy of John C. Krieg, Landscape architect)

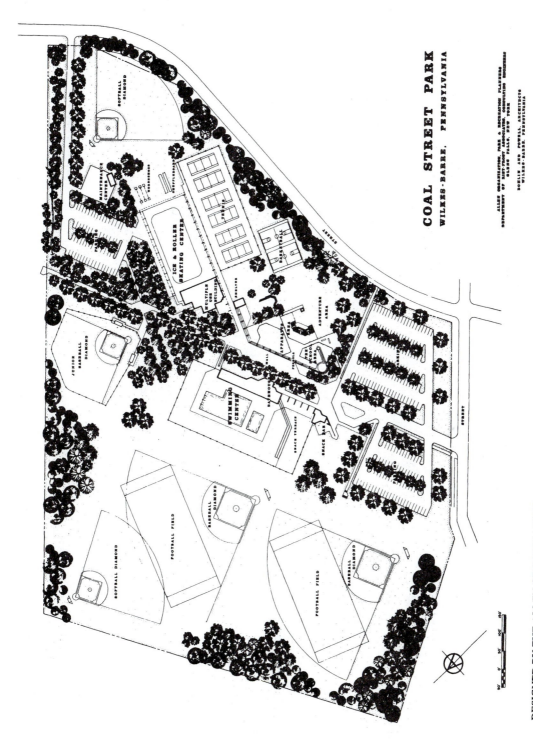

COAL STREET PARK
WILKES-BARRE, PENNSYLVANIA

ALLEN ORGANIZATION, PARK & RECREATION PLANNERS
DEPARTMENT OF HET-FACOT ASSOCIATES, CORPORATION ENGINEERS
GLENS FALLS, NEW YORK

SOBLIN AND POWELL, ARCHITECTS
WILKES-BARRE, PENNSYLVANIA

DESIGNED BY THE ALLEN ORGANIZATION
GLENS FALLS, NY

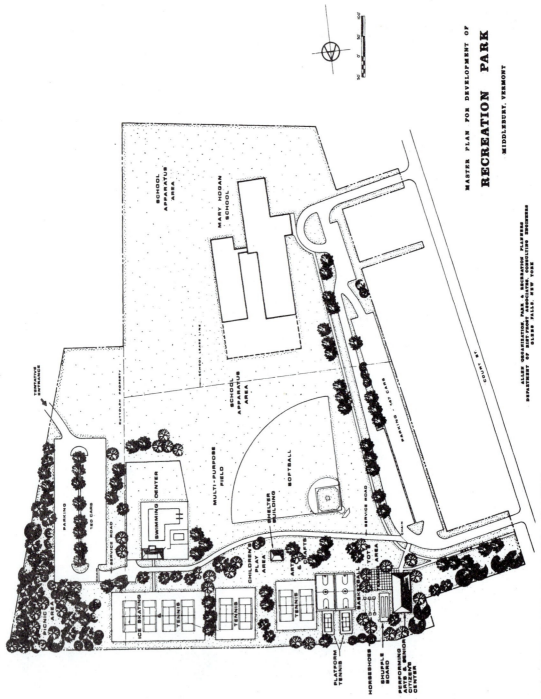

MASTER PLAN FOR DEVELOPMENT OF

RECREATION PARK

MIDDLEBURY, VERMONT

ALLEN ORGANIZATION, PARK & RECREATION PLANNERS
DEPARTMENT OF RIST FRONT ASSOCIATES, CONSULTING ENGINEERS
GLENS FALLS, NEW YORK

DESIGNED BY THE ALLEN ORGANIZATION
GLENS FALLS, NY

Appendix B

Tools of the Trade

Specialized Hand Tools and Their Functions

Tool and Name	Function	Tool and Name	Function
Grass Shears	Used to trim grass along walks, roadways, the edge of planting beds, and around trees, posts, etc.	Pruning Saw	Removes any tree or shrub part that cannot be easily cut with the lopping shears. Usually parts are 1 inch or more in diameter.
Pruning Shears	Used to trim tree and shrub twigs up to 1/2 inch in diameter.	Crosscut Saw	Removes large limbs and small trees. The saw has additional general uses.
Lopping Shears	Used to trim tree and shrub twigs from 1 inch to 1 1/2 inches in diameter.	Grass Hook	For reducing the height of overgrown grass areas. It requires the user to bend over.
Hedge Shears	Prunes shrubs grown closely spaced as hedges. These shears are only used on young, tender new growth.		

Specialized Hand Tools and Their Functions *(continued)*

Tool and Name	Function	Tool and Name	Function
Grass Whip	For reducing grass height without bending over. (Once reduced in height, a lawn mower can be used on the grass.)	Spading Shovel	A combination tool with uses similar to both spades and shovel. It can be used for digging as well as scooping.
Spading Fork	Used for turning over the soil when it is not too hard or compacted. Also used for lifting bulbs in the fall.	Scoop	Good for moving loose materials such as crushed stone, peat moss, soil, etc. Scoops have high sides. They are not used for digging.
Spades	Obvious general uses in digging. Spades have flatter shapes than shovels. They penetrate the soil more easily but have less scooping capability.	Manure Fork	The best tool for moving coarse, lightweight materials such as straw, wood chips, etc.
Shovel	Used for cleaning loose soil from planting holes and other scooping uses. A shovel has sides that a spade does not have.		

Specialized Hand Tools and Their Functions (continued)

Tool and Name	Function	Tool and Name	Function
Single-Bit and Double-Bit Axes	Obvious chopping uses. Especially useful in tree removal and for cutting up fallen timber.	Lawn Comb	An excellent rake for collection of leaves and coarse debris from lawn surface.
Weed Cutter	Removes annual weeds by cutting them off at ground level. Not very effective against biennial and perennial weeds.	Shrub Comb	Used for raking debris from small areas between shrubs.
Toothed Rakes	Used for heavy-duty raking that requires a strong tool. Commonly used in preparation of lawn seed beds and cultivation of planted beds.	Bulb Planter	Used to install flowering bulbs.
Broom Rake	Very useful in places where a lightweight, springy rake is needed. Very good for collecting debris and clippings from lawn surface.	Push Hoe	Similar to a scuffle hoe. It is good for rooting out weeds.

Specialized Hand Tools and Their Functions (continued)

Tool and Name	Function	Tool and Name	Function
Hand Trowel	Used for transplanting bedding plants into flower beds, borders, and boxes.	Post Hole Digger	Prepares holes for the support posts of fences.
Transplanting Hoe	Uses are similar to those of a hand trowel. It has less adaptability for other types of digging.	Garden Hoe	Widely used for breaking up the soil prior to planting. It is also good for cultivating planted beds and for weed removal.
Scuffle Hoe	Useful in weeding and cultivating in planted beds. It cuts off weeds and loosens surface soil.	Hand Cultivator	Loosens the surface soil in flower beds and around shrubs.

Specialized Hand Tools and Their Functions *(continued)*

Tool and Name	Function	Tool and Name	Function
Pick	Used for breaking up hard rocky soil. It has two pointed ends for gouging into the soil.	Cutter Mattock	Stronger than a grading hoe. Its uses are similar. It has two flat ends.
Sprayer	Needed to apply pesticides, antitranspirants, and other chemicals in liquid form. Sprayers are available in a wide range of sizes.	Spreader	Used for the application of fertilizer, seed, and other dry turf products.
Grading Hoe	Loosens hard or compacted soil during preparation for planting. Has a sharpened flat end.		

Power Tools and Their Functions

Tool and Name	Function	Tool and Name	Function
Walk Behind Mower	Professional lawn mowing equipment.	Line and Blade Trimmer	To trim grass and plant material from areas unable to be mown with the walk behind mower.
Backpack Blower	Removes leaves and trash from walks and lawn areas.	Lawn Edger	To trim the grass along the edge of the sidewalk.

Equipment Maintenance

All equipment requires preventive maintenance to ensure that it is in proper operating condition. Preventive maintenance lengthens the useful life span of machinery and results in reduced equipment repair and replacement costs. The manager, mechanic, or equipment operator must keep records of when maintenance services are performed on a machine and how many hours the machine is in use. The equipment manual contains a preventive maintenance schedule based on the number of hours of equipment operation. The person responsible for equipment maintenance should rigidly adhere to this schedule.

The list of maintenance services necessary after 50, 100, 300, and 600 hours of equipment use is fairly typical and will help the reader to understand the importance of equipment maintenance records. However, the manager should consult the manual for each piece of equipment under his or her supervision to determine exact maintenance schedule specifications. The oil level must always be checked before a machine is used.

Generalized Preventive Maintenance Schedule

At 50 hours

Check tire pressure and water level in battery.

Clean air filter (wash or replace if necessary).

Remove corrosion from battery terminal connections.

Check oil level in transmission and hydraulic system.

Grease lubrication points.

At 100 hours

Clean engine.

Drain oil and refill engine crankcase.

Lubricate clutch and throttle linkage.

Tighten loose screws and nuts.

At 300 hours (or every six months)

Inspect and clean spark plugs.

Check ignition point gap.

Change hydraulic oil filters.

Check fan and drive belts.

At 600 hours (or yearly)

Change air cleaner and spark plugs.

Touch up with paint.

Hand Tool Maintenance

Remove rust from metal tools and apply rust-inhibiting materials to metal surfaces; repair split or broken wood handles; reshape the heads of driving or driven tools; sharpen blades or cutting tools. Pruners' and shears' sharpening frequency depends on the usage and the hardness of the steel. Saw sharpening should be done by a professional service and involves the tooth sharpness and tooth set. An indication of the need to have a saw sharpened is being able to run your fingers up and down the saw teeth without getting cut. If the saw binds and becomes stuck in the groove, the teeth may need to be set.

Storing Hand Tools and Small Power Equipment

- Store equipment and fuel in dry, ventilated area. Storage spaces within a garage area, a storage building, or a good shed are all workable options.

- Remove batteries and fully charge them before storage.

- Clean all dirt, grass, and debris from engines.

- Disconnect spark plug wire(s) or remove spark plugs based on manufacturer's recommendations.

- Inspect power equipment – belts, blades, etc. – and make any needed repairs.

- Drain all gasoline from tanks into a container for disposal.

- Do not blend gasoline, gasohol, or alcohol with diesel fuel, as this is a potential for an explosive hazard.

- Do not store equipment in the same location as fertilizers or swimming pool chemicals. These chemicals are extremely corrosive to metal parts on equipment.

- Start and run engines until they quit to be sure the carburetor is dry to prevent diaphragms from sticking together.

- Before storing hand tools, be sure to clean them and apply a light coat of oil to all metal surfaces to prevent rust from forming.

- For hand tools, check to see if any repairs are needed.

Appendix C

Professional and Trade Organizations

Agronomy, Soil

AMERICAN SOCIETY OF AGRONOMY
677 S. Segoe Road
Madison, WI 53711
<http://www.agronomy.org>

CROP SCIENCE SOCIETY OF AMERICA
677 S. Segoe Road
Madison, WI 53711
<http://www.crops.org>

SOIL AND WATER CONSERVATION SOCIETY OF
AMERICA
7515 NE Ankeny Road
Ankeny, IA 50021

SOIL SCIENCE SOCIETY OF AMERICA
677 S. Segoe Road
Madison, WI 53711
<http://www.soils.org>

WEED SCIENCE SOCIETY OF AMERICA
PO Box 7050
Lawrence, KS 66044
<http://www.wssa.net>

Botany

AMERICAN ASSOCIATION OF BOTANICAL GAR-
DENS AND ARBORETA
351 Longwood Road
Kennett Square, PA 19348
<http://www.aabga.org>

AMERICAN BOTANICAL COUNCIL
PO Box 144345
Austin, TX 78714
<http://www.herbalgram.org>

BOTANICAL SOCIETY OF AMERICA
1735 Neil Avenue
Columbus, OH 43210
<http://www.botany.org>

Bulbs

INTERNATIONAL BULB SOCIETY
PO Box 330
Sanger, CA 93657
<http://www.bulbsociety.com>

NETHERLANDS FLOWER BULB INFORMATION
CENTER
30 Midwood St. Street
Brooklyn, NY 11225
<http://www.bulb.com>

Business Management

GARDEN INDUSTRY OF AMERICA, INC.
2501 Wayzata Boulevard
Minneapolis, MN 55440

LAWN AND GARDEN MARKETING AND DISTRIBU-
TION ASSOCIATION
1900 Arch Street
Philadelphia, PA 19103
<http://www.lgmda.org>

NATIONAL BUSINESS ASSOCIATION
5151 Beltline Road
Dallas, TX 75254
<http://www.nationalbusiness.org>

UNITED STATES SMALL BUSINESS ASSOCIATION
409 3rd Street
Washington, DC 20416
<http://www.sba.gov>

Fertilizers, Chemicals

FERTILIZER INSTITUTE
Union Center Plaza
820 First Street NE, Suite 430
Washington, DC 20002
<http://www.tfi.org>

NATIONAL FERTILIZER SOLUTIONS ASSOCIATION
11701 Borman Drive
St. Louis, MO 63146

POTASH/PHOSPHATE INSTITUTE
655 Engineering Drive Suite 110
Norcross, GA 30092
<http://www.ppi-far.org>

Flowers, Plants

ALL-AMERICA ROSE SELECTIONS, INC.
221 N. LaSalle Street
Chicago, IL 60601
<http://www.rose.org>

AMERICAN CAMELLIA SOCIETY
PO Box 1217
Fort Valley, GA 31030
<http://www.camellias-acs.com>

AMERICAN DAHLIA SOCIETY
345 Merritt Avenue
Bergenfield, NJ 07621
<http://www.dahlia.org>

AMERICAN HIBISCUS SOCIETY
PO Box 12073W
St. Petersburg, FL 33733
<http://www.americanhibiscus.org>

AMERICAN IRIS SOCIETY
PO Box 14750
Richmond, VA 23221
<http://www.irises.org>

AMERICAN ORCHID SOCIETY
16700 AOS Lane
Delray Beach, FL 33446
<http://www.orchidweb.org>

AMERICAN PRIMROSE SOCIETY
PO Box 210913
Auke Bay, AK 95821
<http://www.americanprimrosesoc.org>

AMERICAN RHODODENDRON SOCIETY
11 Pinecrest Drive
Fortuna, CA 95540
<http://www.rhododendron.org>

AMERICAN ROSE SOCIETY
PO Box 30,000
Shreveport, LA 71130
<http://www.ars.org>

BEDDING PLANTS INTERNATIONAL
525 SW 5th Street, Suite A
Des Moines, IA 50309
<http://www.bpint.org>

BROMELIAD SOCIETY INTERNATIONAL
PO Box 12981
Gainesville, FL 32604
<http://www.bsi.org/mainhtm>

HOLLY SOCIETY OF AMERICA
309 Buck Street
Millville, NJ 08332
<http://www.hollysocam.org>

NORTH AMERICAN LILY SOCIETY, INC.
PO Box 272
Owatonna, MN 55060
<http://www.lilies.org>

Government Agencies

USDA ANIMAL AND PLANT HEALTH INSPECTION
SERVICE
12th and Independence Avenue SW
Washington, DC 20250
<http://www.aphis.usda.gov>

USDA NATURAL RESOURCES CONSERVATION
SERVICE
PO Box 2890
Washington, DC 20013
<http://www.ncrs.usda.gov>

Horticulture

AMERICAN HORTICULTURAL SOCIETY
7931 E. Boulevard Drive
Alexandria, VA 22308
<http://www.ahs.org>

AMERICAN SOCIETY FOR HORTICULTURE
SCIENCE
113 S. West Street # 200
Alexandria, VA 22314
<http://www.ashs.org>

ASSOCIATION FOR WOMEN IN HORTICULTURE
PO Box 75093
Seattle, WA 98125
<http://www.awhort.org>

HORTICULTURE RESEARCH INSTITUTE, INC
1250 I Street, NW, Suite 500
Washington, DC 20005

Landscaping

AMERICAN SOCIETY OF LANDSCAPE
ARCHITECTS
636 I Street, NW
Washington, DC 20001
<http://www.asla.org>

ASSOCIATED LANDSCAPE CONTRACTORS OF
AMERICA
150 Eldon Street
Suite 270
Herndon, VA 20170
<http://www.alca.org>

ASSOCIATION OF PROFESSIONAL LANDSCAPE
DESIGNERS
1924 North Second Street
Harrisburg, PA 17102
<http://www.apld.com>

COUNCIL OF TREE AND LANDSCAPE APPRAISERS
1000 Vermont Avenue, NW
Suite 300
Washington, DC 2005

IRRIGATION ASSOCIATION
6540 Arlington Boulevard
Falls Church, VA 22042

NATIONAL LANDSCAPE ASSOCIATION
1000 Vermont Avenue, NW
Suite 300
Washington, DC 20036
<http://www.anla.org>

PROFESSIONAL GROUNDS MANAGEMENT
SOCIETY
720 Light Street
Baltimore, MD 21230
<http://www.pgms.org>

Nurseries

AMERICAN NURSERY AND LANDSCAPE
ASSOCIATION
1000 Vermont Avenue, NW
Suite 300
Washington, DC 20005
<http://www.anla.org>

ORNAMENTAL GROWERS ASSOCIATION
PO Box 67
Batavia, IL 60510

PERENNIAL PLANT ASSOCIATION
3383 Schirtzinger Road
Hilliard, OH 43026

WHOLESALE NURSERY GROWERS OF AMERICA
1000 Vermont Avenue, NW
Suite 300
Washington, DC 20005
<http://www.anla.org>

Outdoor Living

CALIFORNIA REDWOOD ASSOCIATION
405 Enfrente Drive
Suite 200
Novato, CA 94949
<http://www.calredwood.org>

GARDEN INDUSTRY OF AMERICA, INC.
2501 Wayzata Boulevard
Minneapolis, MN 55440

GOLF COURSE SUPERINTENDENTS ASSOCIATION
OF AMERICA
1421 Research Park Drive
Lawrence, KS 66049
<http://www.gcsaa.org>

KEEP AMERICA BEAUTIFUL, INC.
1010 Washington Boulevard
Stamford, CT 06901
<http://www.kab.org>

NATIONAL SPA AND POOL INSTITUTE
2111 Eisenhower
Alexandria, VA 22314
<http://www.nspi.org>

Pest Control

ASSOCIATION OF NATURAL BIOCONTROL
PRODUCERS
10202 Cowan Heights Drive
Santa Ana, CA 92705
<http://www.anbp.org>

ENTOMOLOGICAL SOCIETY OF AMERICA
9301 Annapolis Road
Suite 300
Lanham, MD 20706
<http://www.entsoc.org>

NATIONAL PEST MANAGEMENT ASSOCIATION
8100 Oak Street
Dunn Loring, VA 22027
<http://www.pestworld.org>

WEED SCIENCE SOCIETY OF AMERICA
PO Box 1897
810 East 10th Street
Lawrence, KS 66044
<http://www.wssa.net>

Power Equipment, Parts

OUTDOOR POWER EQUIPMENT INSTITUTE, INC.
341 S. Patrick Street
Old Town Alexandria, VA 22317
<http://www.mow.org>

POWER TOOL INSTITUTE
1300 Sumner Avenue
Cleveland, OH 44115
<http://www.powertoolinstitute.com>

SERVICE DEALERS ASSOCIATION
PO Box 4315
Dallas, TX 75208
<http://www.servicedealers.com>

Seeds

AMERICAN SEED TRADE ASSOCIATION
225 Reinekers Lane
Alexandria, VA 22314
<http://www.amseed.com>

AMERICAN SEEDMENS' ASSOCIATION
298 E. McCormick Avenue
State College, PA 16801
<http://www.atlanticseedmen.com>

Soil Conditioners

CANADIAN SPHAGNUM PEAT MOSS INFORMA-
TION BUREAU
7 Oasis Court
St. Alberta
Alberta, Canada T8N 6X2
<http://www.peatmoss.com>

MULCH AND SOIL COUNCIL
10210 Leatherleaf Court
Manassas, VA 20111
<http://www.nbspa.org>

PERLITE INSTITUTE, INC.
710 East Ogden Avenue
Suite 600
Naperville, IL 60563
<http://www.perlite.org>

Trees

AMERICAN FORESTRY ASSOCIATION
910 17th Street, NW
Suite 600
Washington, DC 20006
<http://www.americanforests.org>

AMERICAN SOCIETY OF CONSULTING
ARBORISTS
15245 Shady Grove Road
Suite 130
Rockville, MD 20850
<http://www.asca-consultants.org>

ARID ZONE TREES
9750 E. Germann Road
Mesa, AZ 85208
<http://www.aridzonetrees.com>

INTERNATIONAL SOCIETY OF ARBORICULTURE
PO Box 3129
Champaign, IL 61825
<http://www.isa-arbor.com>

NATIONAL ARBORISTS ASSOCIATION
3 Perimeter Road, Unit 1
Manchester, NH 03103
<http://www.natlarb.com>

NATIONAL CHRISTMAS TREE ASSOCIATION
1000 Executive Parkway
Suite 220
St. Louis, MO 63141
<http://www.christree.org>

SOCIETY OF AMERICAN FORESTERS
5400 Grosvenor Lane
Bethesda, MD 20814
<http://www.safnet.org>

SOCIETY OF MUNICIPAL ARBORISTS
7000 Olive Boulevard
St. Louis, MO 63130
<http://www.urban-forestry.com>

U.S. NATIONAL ARBORETUM
3501 New York Avenue, NE
Washington, DC 20002
<http://www.usna.usda.gov>

Turf

CALIFORNIA SOD PRODUCERS ASSOCIATION
926 J Street
Suite 815
Sacramento, CA 95814

THE LAWN INSTITUTE
1501 Johnson Ferry Road, NE
Suite 206
Marietta, GA 30062
<http://www.lawninstitute.com>

PROFESSIONAL LAWN CARE ASSOCIATION OF
AMERICA
1000 Johnson Ferry Road, NE
Suite C-135
Marietta, GA 30068
<http://www.plcaa.org>

TURF PRODUCERS INTERNATIONAL
1855-A Hicks Road
Rolling Meadows, IL 60008

TURFGRASS INFORMATION CENTER
3-W Main Library
East Lansing, MI 48824
<http://www.lib.msu.edu/tgif>

Wildflowers

NATIONAL WILDFLOWER RESEARCH CENTER
4801 LaCrosse Avenue
Austin, TX 78739
<http://www.wildflower.org>

Glossary

Accent plant A plant that is more distinctive than many plants, but does not attract the eye as much as a specimen plant.

Acclimatization The preparation of plants for a reduced light setting.

Adobe A heavy soil common to the southwestern United States.

Aeration The addition of air into the soil; it is accomplished during soil conditioning with materials such as sand or peat moss. It can be encouraged in established lawns by the use of machines called aerators.

Aesthetic Attractive to the human senses.

Alkaline Characterized by a high pH.

Ames lettering guide A plastic device used with a T-square to produce guidelines for hand lettering.

Analysis The proportion in which each of three standard ingredients is present: nitrogen, available phosphoric acid, and water-soluble potash.

Angle The relationship between two joined straight lines.

Annual A plant that completes its life cycle in one growing season.

Antitranspirant (*also* antidesiccant) A liquid sprayed on plants to reduce water loss, transplant shock, windburn, and sunscald.

Arid A term used in the description of landscapes where there is little usable water.

Axonometric view A drawing that permits multiple sides of an object to be seen in a single, measurable view.

Bacteria Single-celled, parasitic pathogens.

Balance The even distribution of materials on opposite sides of a central axis. There are three kinds of balance: symmetric, asymmetric, and proximal/distal.

Balled and burlapped A form of plant preparation in which a large part of the root system is retained in a soil ball. The ball is wrapped in burlap to facilitate handling during sale and transplanting.

Bare root A form of plant preparation in which all soil is removed from the root system. The plant is lightweight and easier to handle during sale and transplanting.

Base map A graphic depiction of the site features that were collected, measured, and inventoried during the site analysis stage of the landscape process.

Bedding plant An herbaceous plant preseeded and growing in a peat pot or packet container.

Bid A statement of what work, materials, and standards of quality will be provided by a landscape firm in return for the price specified. Once agreed to by a client and the landscape firm, it is legally binding upon both.

Bid bond A security deposit submitted by a contractor at the time of bidding to guarantee that the bid will be honored if selected.

B.L.A. Bachelor of Landscape Architecture. The first professional degree for those who study landscape architecture.

Branching habit A combination of the number of branches, the average size of branches, the flexibility of branches, and the vertical or horizontal direction of their growth.

Broadcast When material such as seed or fertilizer is distributed evenly over a site.

Bulb A flowering perennial that survives the winter as a dormant fleshy storage structure.

Calibration The adjustment of a piece of equipment so that it distributes a given material at the rate desired.

Caliche A highly alkaline soil common to the southwestern United States.

Canopy The collective term for the foliage of a tree.

Capital Money used to finance a business.

Change order An amendment to the original contract to which all parties agree. Once signed, it becomes part of the original contract.

Chemical forms Material that is synthesized and not naturally occurring.

Chlorosis The yellowing of plant tissue for reasons other than lack of light.

Client The person or organization that owns and provides the financing for a project.

Closed specifications Specifications that do not allow for the substitution of specific products with similar products.

Compaction A condition of soil in which all air has been driven out of the pore spaces. Water is unable to move into and through the soil.

Compass A graphic design tool used for the construction of circles.

Complete fertilizer A fertilizer containing nitrogen, phosphorus, and potassium, the three nutrients used in the largest quantities by plants.

Conditioning Preparation of soil to make it suitable for planting.

Containerized A form of plant preparation for sale and transplanting. When purchased, the plant is growing with its root system intact within a plastic, metal, or tarpaper container.

Contingency Allowance added to insure that unpredictable costs will be covered.

Contour interval The vertical distance between contour lines.

Contour lines Broken lines found on a topographic map. They represent vertical elevation.

Contract An agreement between two or more parties that is legally binding.

Contractor Party in contract with the client or the client's representatives.

Cool-season grass A type of grass that grows best in temperate regions and during the cooler spring and fall months.

Corporation A form of business operation that makes the business a legal entity, separate from its owners.

Cost Refers to the recovery of expenditures.

Cost estimate An itemized listing of the expenses in an operation. It can be applied to a single task or a total project.

Crop seed Cash crop seeds, often included in turf seed mixes. They are undesirable species for lawns.

Crotch The point on a tree at which two branches or a branch and the trunk meet.

Crown The point at which aboveground plant parts and the root system meet.

Cultipacker A large lawn seeder pulled by a tractor.

Cut A grading practice that removes earth from a slope.

Dead-heading The removal of dead flower blossoms from herbaceous bedding plants.

Deciduous A type of plant that loses its leaves each autumn.

Degree The unit of measurement for angles.

Density The number of leaf shoots that a single grass plant will produce.

Design-build firm A landscape business that provides both design and construction services.

Design specifications All details regarding materials, methods, and performance standards are included in the specifications.

Diameter The distance across a circle as measured through the exact center.

Diazo machine A duplicating machine that makes positive copies from vellum tracings onto heavy paper.

Diazo process Duplication process used to reproduce landscape drawings in large numbers. Makes a direct positive copy using specially treated paper, ultraviolet light, and an ammonia vapor developer.

Dormancy (adj.: dormant) A period of rest that perennial plants experience during the winter season. They continue to live, but have little or no growth.

Downstream With regard to irrigation, describes the direction of water flow away from the source.

Drafting powder An absorbent granular material used to prevent smudging and keep drawings free of skin oil.

Drainage The act of water passing through the root area of soil. Soil is well drained if water disappears in 10 minutes or less from a shrub or tree planting.

Drainage tile A plastic or clay tube buried beneath the soil that collects excess water from the soil and carries it away.

Drop spreader A device for the application of granular materials such as grass seed and fertilizer. The material is dispensed through holes in the bottom of a hamper as the spreader is pushed across the lawn.

Edging The cutting of a sharp line of separation between a planting bed and the adjacent lawn. Also, a product, usually metal or plastic strip, that is installed as a physical separator between planting beds and lawns or between lawns and paved areas.

Effluent water Treated sewage.

Elevation view A scaled drawing with two dimensions, one horizontal and one vertical.

Enclosure materials Materials that form the walls of the outdoor room.

Enrichment A contribution made to the outdoor room by a landscape item that is not an element of a wall, ceiling, or floor.

EPA The U.S. Environmental Protection Agency.

Eradicant A pesticide that is applied to a plant when a pathogen or insect has arrived and must be killed immediately.

Erosion The wearing away of the soil caused by water or wind.

Espalier A form of pruning in which plants are trained against a fence or wall. The effect is vinelike and two dimensional.

Estimate An approximation of the price that a customer will be charged for a landscape project.

Estimator The individual usually assigned to do the landscape take-off.

Etiolation A condition resulting in a plant due to a lack of light. The plant turns yellow, leaf size is reduced, and stems become weak and spindly.

Evergreen A type of plant that retains its foliage during the winter. There are needled forms (such as pine, spruce, hemlock, and fir) and broad-leaved forms (such as rhododendron, pieris, euonymus, and holly).

Exotic plant A plant that has been introduced to an area by human beings, not nature.

Fauna Animal life.

Fertilization The addition of nutrients to the soil through application of natural or synthesized products called *fertilizers*.

Fertilizer analysis The percentage of various nutrients in a fertilizer product. A minimum of three numbers on the fertilizer package indicates the percentage of total nitrogen (N), available phosphoric acid (P_2O_5), and water-soluble potash (K_2O), in that order.

Fill A grading practice that adds earth to a slope.

Filler material Used to dilute and mix fertilizer, adding weight and bulk.

Flail mower A mower used for turfgrasses that are only cut a few times each year.

Flora Plant life.

Flower bed A free-standing planting made entirely of flowers with no background of shrub foliage.

Flower border A flower planting used in front of a planting of shrubs. The shrubs provide green background for the blossoms.

Foam board Two sheets of thin cardboard bonded to a central sheet of styrofoam. It is commonly used for mounting designs for presentations.

Focal point A point of visual attraction. A focal point can be created by color, movement, shape, size, or other characteristics.

Focalization of interest Principle of design that selects and positions visually strong items in the landscape composition.

Foliage texture The effect created by the combination of leaf size, sunlight, and shadow patterns on a plant.

Foliar burn A reaction to chemicals in fertilizer.

Footcandle The amount of light produced by a candle at a distance of one foot.

Foundation planting The planting next to a building that helps it blend more comfortably into the surrounding landscape.

Fungus, (plural fungi) Multicelled, parasitic pathogens.

General contractor The firm directly responsible to the client for the construction of a project.

Girdling The complete removal of a strip of bark around the main stem of a plant. After girdling, the ability of nutrients to pass from roots to leaves is lost, causing the eventual death of the plant.

Grading Changing the form of the land.

Graft A man-made bond between two different plants, one selected for its aboveground qualities (scion) and the other for its belowground qualities (stock).

Grass seed blend A combination of two or more cultivated varieties of a single species.

Grass seed mixture A combination of two or more different species of grass.

Groundcover A low-growing, spreading plant, usually 18 inches or less in height.

Groundskeeper A professional engaged full-time in landscape maintenance.

Hardiness The ability of a plant to survive through the winter season.

Hardscape Design materials that are not living plant materials. The term usually is applied to the constructed materials of a landscape.

Heading back A pruning technique that shortens a shrub branch without totally removing it.

Heaving An action that causes shallowly rooted plants, such as grasses, groundcovers, and bulbs, to be forced to the surface of the soil. The action results from repeated freezing and thawing of the soil surface.

Herbaceous A type of plant that is nonwoody. It has no bark.

Herbicide A chemical used to kill weeds.

Holdfasts Special appendages of certain vines that allow them to climb.

Humus Created by decaying organic matter, it aids the soil in moisture retention.

Hydroseeder A spraying device that applies seed, water, fertilizer, and mulch at the same time.

Hydrozoning Grouping plants on the basis of their water needs.

Incurve The center of a corner planting bed and a natural focal point.

Inert material Filler material that has no purpose other than to carry and dilute active ingredients in a mixture.

Inoculum The infectious form of a pathogen.

Inorganic A material that does not contain carbon.

Insect Small members of the Animal Kingdom that are parasitic on plants. All have six legs, and most have two sets of wings.

Intangible A quality denoting something that cannot be touched.

Irrigation Supplying water to plants through artificial means.

Jump-cut   A pruning technique for the removal of large limbs from trees without stripping bark from the trunk. It involves a series of three cuts.

Landscape architect   A licensed professional who practices landscape planning, usually on a scale larger than residential properties.

Landscape contractor   A professional who carries out the installation of landscapes.

Landscape designer   A professional who devotes all or part of a work day to the design of landscapes.

Landscape installation   The actual construction of the landscape.

Landscape maintenance   The care and upkeep of the landscape after installation.

Landscape management   The extended care of existing landscapes, usually under terms of a contract.

Landscape nursery worker   A professional who is concerned with the sale and installation of landscape plants and related materials.

Landscaping   A profession involving the design, installation, and maintenance of the outdoor human living environment.

Lateral bud   Any bud below the terminal bud on a twig.

Latin binomial   The unique two-part scientific name that identifies each plant and animal.

Leaching   The dissolving of materials (such as nutrients) in the water which is present in soil, causing the material to quickly pass the point at which plant roots can benefit from them.

Lead branch   The most important branch of a tree; cannot be removed without destroying the distinctive shape of the tree.

Lettering Machine   A device that types an assortment of letter styles and sizes onto a transparent tape. The tape can then be applied to a landscape design.

Lettering template   A stencil used to form letters. It is an alternative to free-hand lettering.

Light compensation point   The point of exact balance between photosynthesis and respiration by plants.

Light quality   The color of light emitted by a particular source.

Lime   A powdered material used to correct excess acidity in soil.

Line-width variation   A variation of hand-lettering style where the designer uses an instrument with a chiseled tip, so lines that are thin and wide can be made with the same instrument.

Loam   Soil that contains approximately equal amounts of clay, silt, and sand (a desirable condition).

Lux   The amount of illumination received on a surface one meter from a standard light source (an international measurement).

Macro-range   Viewer sees the landscape from the most distant vantage point.

Marker paper   Smooth, hard surfaced paper used for marker renderings.

Market   The geographic area from which a business attracts most of its customers.

Masses   Solid vertical areas of plantings, buildings, walls, or land forms.

Massing plants   Plants that serve to fill large amounts of space both on the ground and in the air.

Masterformat™   A standardized system of organization for project manuals.

Metamorphosis   A term comprising the changes an insect undergoes as it grows.

Microclimate   Regions that provide atypical growing conditions.

Microcomputer   A small computer that adapts well to the business needs of landscape firms.

M.L.A.   Master of Landscape Architecture. The second professional degree for those who study landscape architecture.

Morphology   Physical structure of plants.

MSDS   Material Safety Data Sheet.

Mulch   A material placed on top of soil to aid in water retention, prevent soil temperature fluctuations, or discourage weed growth.

Mylar film   Polyester drafting film used for drawing when durability of the original work is important.

Native plant A plant that evolved naturally within a certain locale.

Naturalized plant A plant that was introduced to an area as an exotic plant, but which has adapted so well that it may appear to be native.

Necrosis A symptom of plant injury in which the tissue becomes dried and dies. Necrosis may be localized and limited or extensive.

Nematode A microscopic, worm-like pathogen.

Newsprint paper Inexpensive paper used for preliminary sketches and designs.

Noncollusion affidavit A document affirming that bidders have not discussed a project prior to bidding in order to establish a minimum bid among themselves.

Noxious weeds Persistent weeds defined by law in most states. They are perennial and difficult to control. The presence of these weeds in a grass seed mix indicates that the seed is of low quality.

Nutrient ratio A comparison of the proportion of each nutrient in a fertilizer to the other nutrients in the same fertilizer. Example: A fertilizer with a 5-10-5 analysis has a 1-2-1 ratio of ingredients.

Open specifications Specifications that permit substitutions after approval by the designer or owner.

Organic Consisting of modified plant or animal materials.

Outcurve The sides of a corner planting.

Overhead Operational costs of doing business; not individual job costs.

Parasite An injurious agent that is biological and infectious or infestious. It is incapable of manufacturing its own food and derives its sustenance from other organisms.

Partnership A form of business operation engaged in by two or more persons.

Pastels Chalky crayons made of powdered pigment and used in color rendering.

Patching A method of restoring turf to a sparse lawn. Warranted when bare spots are at least 1 foot in diameter; seed, sod, plugs, or sprigs should be selected to match the grasses of the established lawn.

Pathogen The infectious agent in plant diseases.

Perennial A plant that lives more than two growing seasons. It usually is dormant during the winter.

Performance bond A security provided by the contractor to provide the owner with sufficient money to have a project completed should the contractor default during the performance of the contract.

Performance specifications Establish the standard of performance or service that must be met by the feature being described.

Perspective view A drawing that permits multiple sides of an object to be seen in a single view. Dimensions are not measurable.

Pesticide A chemical used for the control of insects, plant diseases, or weeds.

pH A measure of the acidity or alkalinity of soil. A pH of 7.0 is considered neutral. Ratings below 7.0 are acidic, above 7.0 alkaline (basic).

Photocopy process Duplication process also used to reproduce landscape drawings in large numbers; operates much like an office copier, using black toner to reproduce the image.

Photosynthesis Production of food by the plant.

Physiology Biological functioning of plants.

Pinching The removal of the terminal shoot from branches of plants to promote fuller, denser growth.

Plan view A measurable drawing seen as though the viewer's line of sight is perpendicular to the surface.

Plant list An alphabetical listing of the botanical names of plants used in a landscape plan, their common names, and the total number used.

Plug A small square, rectangle, or circle of sod, cut about two inches thick.

Plugging A method of lawn installation that uses cores of live, growing grass.

Price The outlay of funds in payment for the provision of goods and/or services.

Principles of design Standards by which designs can be created, measured, discussed, and evaluated.

Prismacolors® Colored pencils used in rendering.

Process A sequence of steps, in the form of decisions or activities, that result in the accomplishment of a goal. Also, a sequence of stages during which something is created, modified, or transformed.

Project manual A bound compilation of the documents pertaining to a landscape (or other construction) project.

Propagation The reproduction of plants. It may be sexual or asexual (by vegetative cuttings, layering, etc.).

Proportion Principle of design concerned with the size relationships between all the features of the landscape.

Proprietary specifications Those that require the use of specific products.

Protectant A pesticide applied to a plant before a pathogen or insect arrives to kill the injurious agent when it does arrive.

Protractor A graphic design tool for measuring angles.

Pruning The removal of a portion of a plant for better shape, improved health, size control, or more fruitful growth.

Pubescence The presence of fine hairs on the surface of leaves.

Puddling Compaction of soil to such a degree that water will not soak into it.

Purity The percentage, by weight, of the pure grass seed in a mixture.

Quadrille paper Ruled paper divided 4, 5, 8, or 10 squares per inch and used for first-draft drawings.

Quickly available fertilizer A fast-action fertilizer that has its nitrogen in a water-soluble form for immediate release into the soil.

Radius One-half of the diameter of a circle.

Reel mower A mower used for home, recreational, and commercial lawn maintenance. The blades rotate in the same direction as the wheels and cut the grass by pushing it against a non-rotating bedknife at the rear base of the mower.

Respiration The use of food reserves for growth and maintenance by plants.

Rhizome A horizontal underground stem. New shoots are sent to the surface some distance out from the parent plant. Each new plant develops its own root system and becomes independent of the parent plant.

Rhythm and line Principle of design. Something is repeated at a standard interval, creating a rhythm. Lines establish the shape and form of a landscape.

Riser The elevating portion of a step.

Root systems Range from total tap root systems, with a large single root growing straight down into the soil, to full fibrous systems, with thousands of fine, hairlike roots spreading out in all directions.

Rotary mower A mower used for home, recreational, and commercial lawn maintenance. The blades move like a ceiling fan, parallel to the surface of the lawn, cutting the grass off as they revolve.

Rotary spreader A device for the application of granular material such as grass seed and fertilizer. The material drops from a hamper onto a rotating plate and is propelled outward in a semicircular pattern.

Scaffold branch A lateral branch of a tree.

Scale (engineer's) A measuring tool that divides the inch into units ranging from 10 to 60 parts.

Scion Shoot portion of a graft.

Seed analysis A breakdown of the contents of the seed package on which it appears; must appear on every package of seeds sold.

Serif A decorative stroke attached to a letter to create an ornate appearance.

Shrub A multistemmed plant smaller in size than a tree.

Silhouette The outline of an object viewed as dark against a light background.

Simplicity One of the principles of design. The intent of its application is to control the potential complexity of a landscape's design resulting from the use of too many different colors, angles, materials, textures, shapes, etc.

Site An area of land having potential for development.

Slope A measurement that compares the horizontal length to the vertical rise or fall of land. The measurement can be determined from a topographic map.

Slow-release fertilizer A slow-action fertilizer in which the nutrient contents are in a form not soluble in water. The nutrients are released more slowly into the soil for more efficient intake by plants.

Sodding A method of lawn installation that uses strips of live, growing grass. It produces an immediate effect on the landscape, but is more costly than seeding.

Software Supportive material that allows a computer to carry out specific functions.

Soil texture The composition of a soil as determined by the proportion of sand, silt, and clay that it contains.

Species A category of plant classification distinguishing the plant from all others.

Specifications Written requirements for the installation of a landscape. They are usually prepared and made available to contractors before the bidding for a contract begins.

Specimen plant A plant that is highly distinctive because of such qualities as flower or fruit color, branching pattern, or distinctive foliage. Its use creates a strong focal point in a landscape.

Spore The reproductive form of fungi. A type of inoculum.

Spreader A garden tool used for the even distribution of materials such as grass seed and fertilizer.

Sprig A piece of grass shoot. Sprigs are commonly used to establish warm-season grass plantings.

Sprinkler irrigation Water applied under pressure over the tops of plants.

Stock Root portion of a grafted plant.

Stolon A stem that grows parallel to the ground. New plants develop from it and become independent of the parent plant.

Stolonizing A means of installing warm season lawn grasses. Pieces of grass shoots (sprigs) are distributed evenly over the lawn site and covered lightly with soil, then rolled or disked.

Subcontractor Firm in contract with the prime contract holder to provide selected services for the accomplishment of a project.

Sucker A succulent branch that originates from the root system. The vegetation of suckers is abnormal and undesirable.

Sunscald A temperature-induced form of winter injury. The winter sun thaws the aboveground plant tissue, causing it to lose water. The roots remain frozen and thereby unable to replace the water. The result is drying of the tissue.

Symbols Drawings that represent overhead views of trees, shrubs, or other features of a landscape plan.

Symptom A host plant's response to a pathogenic or insect irritant.

Symptom complex The sum of all the symptoms expressed by a host plant from the time it is initially infected until it either recovers or dies.

Systemic herbicide A chemical weed killer that is absorbed into the weed plant, usually through the roots and sometimes through the leaves. It then moves through the entire plant, killing all parts of it.

Take-off The calculation of quantities from plans and specifications.

Tangible A quality denoting something that is touchable.

Taxonomist A person who specializes in the classification of plants and animals.

Technical specifications Written descriptions of the technical standards that the work must meet; often used for portions of projects that are not easily verified for correctness after construction has been completed.

Tender A condition of plants that implies their lack of tolerance to cold weather.

Tendrils Special appendages of certain vines that allow them to climb.

Terminal bud The end bud on a twig.

Terrain The rise and fall of the land.

Texture A description of the coarseness or fineness of a plant compared to other nearby plants.

Thatch Dead, semidecomposed grass clippings on the surface of soil.

Thinning out A pruning technique that removes a shrub branch at or near the crown of the plant.

Topiary A form of pruning in which plants are severely sheared into unnatural shapes such as animals or chess pieces.

Topography A record of an area's terrain.

Trace elements Nutrients essential to the growth of many plants, but needed in far less amounts than the major elements.

Trajectory Path of irrigation water as it is propelled through the air.

Transfer film Paper-thin, transparent plastic sheeting with a clear adhesive backing used for reproduction of selected graphics in desired locations on layout sheets.

Transplant To relocate a plant.

Tread That portion of a step on which the foot is placed.

Triangle A three-sided graphic design tool. It commonly has either a 30°-60°-90° or a 45°-90°-45° combination of angles.

Trickle irrigation Water supplied directly to the root zone of plants.

T-square A long straightedge that takes its name from its shape. It is a graphic design tool.

Twining One method by which certain vines are able to climb.

Unit cost The price of the smallest available form of an item described in a cost estimate.

Unit pricing The reduction of all landscape area dimensions and material quantities to a common measurement, such as thousand square feet or acre.

Unity Principle of design in which all separate parts contribute to the creation of the total design.

Vellum Thin, transparent paper used for tracing designs.

Vertical mowing A technique that requires a power rake or a mower whose blades strike the turf vertically. It is done to break up the soil plugs left by an aerator or to remove excessive thatch.

Virus The smallest of the plant pathogens, it has an uncertain Kingdom classification.

Voids The horizontal open areas of the landscape.

Warm-season grass A grass that grows best in warmer regions of the country and during the summer months.

Water sprout A succulent branch that grows from the trunk of a tree. The vegetation of water sprouts is abnormal and undesirable.

Weed A plant growing where it is not wanted and with no economic value.

Weep hole A means of preventing water buildup behind a retaining wall.

Windburn Drying out of plant tissue (especially evergreens) by the winter wind.

Winter injury Any damage done to elements of the landscape during the cold weather season of the year.

Working drawing A copy of a landscape design done on heavy paper or plastic film. The working drawing is used repeatedly during actual construction of the landscape and must be very durable.

Wound paint A sealing paint used over plant wounds of 1 inch or more in diameter after pruning.

Xeriscaping® Techniques of landscaping that conserve water.

Zoning The regulations of a community that govern what uses can be made of different areas within the community.

Index